Introduction to Classical Electrodynamics

INTRODUCTION TO CLASSICAL ELECTRODYNAMICS

Y K LIM

World Scientific

Published by

World Scientific Publishing Co. Pte. Ltd.
5 Toh Tuck Link, Singapore 596224
USA office: 27 Warren Street, Suite 401-402, Hackensack, NJ 07601
UK office: 57 Shelton Street, Covent Garden, London WC2H 9HE

Library of Congress Cataloging-in-Publication Data
Lim, Y. K.
Introduction to classical electrodynamics
1. Electrodynamics. I. Title
QC631.L56 1986 537.6 85-29620
ISBN-13 978-9971-978-51-8 -- ISBN-10 9971-978-51-2
ISBN-13 978-9971-978-85-3 (pbk) -- ISBN-10 9971-978-85-7 (pbk)

British Library Cataloguing-in-Publication Data
A catalogue record for this book is available from the British Library.

PREFACE

The present volume has been prepared to meet the general requirements in electrodynamics of the B. Sc. (Honours) degree. It starts with a review of the concepts of the four electromagnetic field vectors **E**, **B**, **D** and **H** and follows with the introduction of Maxwell's field equations from an empirical basis as the fundamental axioms. Different types of the electromagnetic field — electrostatic, magnetostatic, quasi-stationary and rapidly varying fields — are then developed in turn as special cases of time dependence of varying importance. The propagation of electromagnetic waves across a boundary between different media, in a material medium, and as guided waves in waveguides is then considered. Finally, special relativity is introduced as an extension of classical electrodynamics.

No attempt has been made for either rigorous treatment or comprehensive coverage. Rather, emphasis has been placed on logical development of the subject as a coherent theory. It is hoped that the treatment would enable the reader to correlate his knowledge of the diversive electric and magnetic phenomena in the proper perspective and gain a better understanding of electromagnetism.

The suffix notation method of vector analysis has been liberally employed. A summary covering the required techniques is included as

Appendix. The author is forever grateful to his lecturer, the late Prof. E. O. Hercus of the University of Melbourne, for his initiation to this remarkable method.

September, 1985 Y. K. LIM

CONTENTS

Chapter I

FUNDAMENTAL CONCEPTS AND EXPERIMENTAL LAWS

Electrodynamics deals with the fields and radiation of moving charges. In describing the interaction between charges it is convenient, both mathematically and physically, to consider it, not as forces that act at a distance, but as the force exerted by the field set up by one charge on the other. This approach is in fact essential for charges in relative motion as electromagnetic effects are found to propagate with finite velocity. The four field vectors, **E, B, D** *and* **H,** *which are fundamental in Maxwell's electromagnetic theory are introduced and discussed in this chapter in a phenomenological manner. In addition, a short review is made of the experimental laws which lead to Maxwell's equations.*

1.1 Electric Field Intensity E

Electric field is said to exist at a point where a stationary particle experiences a force on account of its charge. The *electric field intensity or electric field strength* **E** is defined as the force per unit charge acting on a small positive charge q' introduced at that point. Let **F** be the electric force acting on the test charge, then by definition

$$\mathbf{E} = \lim_{q' \to 0} \frac{\mathbf{F}}{q'} \quad . \tag{1.1}$$

The limit $q' \to 0$ is required in order that the introduction of the test charge will not significantly influence the source; the field can then be described independently of the presence of a test charge. The finite magnitude of the elementary charge e does not permit the limiting process to be realized even in principle. The definition applies, therefore, to macroscopic phenomena only. For microscopic processes, the field is usually defined in terms of its source, assuming that the macroscopic laws governing the field-source relationship still apply.

The simplest type of electric field is one that is set up by stationary charges, the electrostatic field. We shall confine ourselves in the first instance to free space. Experimentally, the electrostatic forces two stationary charges q and q' in vacuum exert on each other are given by *Coulomb's law*, which states that the force $\mathbf{F}$ experienced by q' is

$$\mathbf{F} = \frac{qq'}{4\pi\varepsilon_0} \frac{\mathbf{r}}{r^3} \quad , \tag{1.2}$$

where ε_0 is a constant characteristic of the vacuum or free space called the *vacuum permittivity*, and $\mathbf{r}$ the position vector of q' with respect to the location of q. If q' is a test charge, the electrostatic field intensity at $\mathbf{r}$ due to the presence of q is by Eq. (1.1)

$$\mathbf{E} = \frac{q}{4\pi\varepsilon_0} \frac{\mathbf{r}}{r^3} = -\frac{q}{4\pi\varepsilon_0} \nabla\left(\frac{1}{r}\right) \quad . \tag{1.3}$$

Since forces add up like vectors, the electrostatic field $\mathbf{E}$ at a point of position vector $\mathbf{r}$ due to a source consisting of n

discrete charges $q_1, q_2, \ldots, q_n$ located at $\mathbf{r}'_1, \mathbf{r}'_2, \ldots, \mathbf{r}'_n$ respectively, all position vectors being with respect to some fixed origin, can be expressed as

$$\mathbf{E} = \sum_{s=1}^{n} \mathbf{E}_s = \frac{1}{4\pi\varepsilon_0} \sum_{s=1}^{n} q_s \frac{\bar{\mathbf{r}}_s}{\bar{r}_s^3}$$

$$= -\frac{1}{4\pi\varepsilon_0} \sum_{s=1}^{n} q_s \bar{\nabla}^{(s)} \left(\frac{1}{\bar{r}_s}\right) \quad , \qquad (1.4)$$

where $\bar{\mathbf{r}}_s = \mathbf{r} - \mathbf{r}'_s$ and $\bar{\nabla}^{(s)}$ denotes the gradient with respect to the components of $\bar{\mathbf{r}}_s$.

In general we shall refer to the point with position vector $\mathbf{r}(x_i)$, where the field is to be evaluated or under consideration, as the *field point* and the points represented by $\mathbf{r}'_s$ where the source elements are situated as the *source points*. If $\mathbf{r}' \equiv (x'_i)$ is one such source point, the radius vector $\bar{\mathbf{r}}$ of the field point with respect to it will have components $\bar{x}_i = x_i - x'_i$. Then for any function $f(\bar{\mathbf{r}})$ we have the following relationship when either $\mathbf{r}'$ or $\mathbf{r}$ is varied[a]:

$$\frac{\partial f}{\partial \bar{x}_i} = \left(\frac{\partial f}{\partial x_i}\right)_{x'} = -\left(\frac{\partial f}{\partial x'_i}\right)_{x} \quad ,$$

or
$$\bar{\nabla} f = \nabla f = -\nabla' f \quad . \qquad (1.5)$$

Returning to the system of charges, if we vary the position $\mathbf{r}$ of the field point while keeping the source points fixed and calculate curl $\mathbf{E}$, we shall obtain

---

[a] The notation $(\ )_{x'}$ means that x'_i, where $i = 1,2,3$, are all kept fixed. In addition, the partial derivative $\partial/\partial x_i$ signifies, as usual, that among the three components of $\mathbf{r}$ only x_i is varied.

$$\nabla \times \mathbf{E} = -\frac{1}{4\pi\varepsilon_0}\sum_{s=1}^{n} q_s \nabla \times \bar{\nabla}^{(s)}\left(\frac{1}{\bar{r}_s}\right)$$

$$= -\frac{1}{4\pi\varepsilon_0}\sum_{s=1}^{n} q_s \bar{\nabla}^{(s)} \times \bar{\nabla}^{(s)}\left(\frac{1}{\bar{r}_s}\right) = 0 \quad .$$

This analysis can be readily extended to the case of a continuous charge distribution, so that for electrostatic fields in general

$$\nabla \times \mathbf{E} = 0 \quad . \tag{1.6}$$

It follows from this equation that $\mathbf{E}$ can be expressed as the gradient of a scalar function $\Phi(x_i)$:

$$\mathbf{E} = -\nabla\Phi \quad , \tag{1.7}$$

since the curl of the gradient of a scalar function vanishes identically.

The work done in carrying unit charge from a point A to a point B against electrostatic force is

$$-\int_A^B \mathbf{E}\cdot d\mathbf{r} = \int_A^B \nabla\Phi \cdot d\mathbf{r} = \int_A^B d\Phi = \Phi_B - \Phi_A \quad , \tag{1.8}$$

as
$$\nabla\Phi \cdot d\mathbf{r} = \frac{\partial\Phi}{\partial x_j}dx_j = d\Phi \quad ,$$

where Φ_A and Φ_B are values of the scalar function Φ at points A and B respectively. The integral is seen to be independent of the path followed. Hence if we integrate $\mathbf{E}$ along an arbitrary closed loop C we obtain

$$\oint_C \mathbf{E}\cdot d\mathbf{r} = 0 \quad . \tag{1.9}$$

This shows that the electrostatic field is conservative. The work done in carrying a charge from one location to another in the field is stored up as the potential energy of the system. Furthermore, if we arbitrarily fix the zero level of the potential energy at A, say, then Φ_B is unique for any given point B. It is called the *potential* at B. It is convenient, though not necessary, to take for A an infinity point so that the potential at a point is

$$\Phi = -\int_\infty \mathbf{E}\cdot d\mathbf{r} \tag{1.10}$$

with the line of integration starting at infinity.

With this definition the potential due to a system of n discrete charges can be conveniently expressed in terms of the source quantities:

$$\begin{aligned}\Phi(\mathbf{r}) &= \frac{1}{4\pi\varepsilon_0}\sum_{s=1}^{n} q_s \int_\infty \bar{\nabla}^{(s)}\left(\frac{1}{\bar{r}_s}\right)\cdot d\mathbf{r} \\ &= \frac{1}{4\pi\varepsilon_0}\sum_{s=1}^{n} q_s \int_\infty \bar{\nabla}^{(s)}\left(\frac{1}{\bar{r}_s}\right)\cdot d\bar{\mathbf{r}}_s \\ &= \frac{1}{4\pi\varepsilon_0}\sum_{s=1}^{n} \frac{q_s}{\bar{r}_s} \end{aligned} \tag{1.11}$$

since for any s, $d\mathbf{r} = d\bar{\mathbf{r}}_s$, $\mathbf{r}'_s$ being kept fixed. For a continuous distribution with volume charge density ρ in V and surface charge density σ on its boundary S, the above becomes

$$\Phi(\mathbf{r}) = \frac{1}{4\pi\varepsilon_0}\int_V \frac{\rho(\mathbf{r}')dV'}{\bar{r}} + \frac{1}{4\pi\varepsilon_0}\int_S \frac{\sigma(\mathbf{r}')dS'}{\bar{r}} \tag{1.12}$$

where $\bar{\mathbf{r}} = \mathbf{r} - \mathbf{r}'$ and the integrals are to be taken over the source coordinates.

Consider the integral of the intensity $\mathbf{E}$ of the electrostatic field of a system of discrete charges over an arbitrary closed surface S,

$$\oint_S \mathbf{E}\cdot d\mathbf{S} = \frac{1}{4\pi\varepsilon_0}\sum_{s=1}^{n} q_s \oint_S \frac{\mathbf{r}_s\cdot d\mathbf{S}}{\bar{r}_s^3}$$

$$= \frac{1}{4\pi\varepsilon_0}\sum_{s=1}^{n} q_s \oint_S d\Omega_s \quad ,$$

where $d\Omega_s$ is the solid angle subtended by the surface element $d\mathbf{S}$ at $\mathbf{r}'_s$, the location of q_s. The closed surface integral $\oint_S d\Omega_s$ equals 4π if q_s is enclosed by S and zero if otherwise. Hence

$$\oint_S \mathbf{E}\cdot d\mathbf{S} = \frac{1}{\varepsilon_0}\sum_s{}' q_s \quad , \tag{1.13}$$

where the summation $\sum_s{}'$ extends only to all charges enclosed by S. If we map a vector field by curved lines, the *lines of force*, whose tangent at a point gives the direction of the vector and whose number crossing a unit area perpendicular to the vector at that point is proportional to the magnitude of the vector, this equation may be interpreted as stating that the net number of electric lines of force leaving a charge is proportional to its magnitude. For a continuous system with charge density ρ, Eq. (1.13) becomes

$$\oint_S \mathbf{E}\cdot d\mathbf{S} = \frac{1}{\varepsilon_0}\int_V \rho dV \quad ,$$

where V is the volume enclosed by S. On the other hand, the divergence theorem gives

$$\oint_S \mathbf{E}\cdot d\mathbf{S} = \int_V \nabla\cdot\mathbf{E}\, dV \quad .$$

Since S is arbitrary, the last two equations give

$$\nabla \cdot \mathbf{E} = \frac{\rho}{\varepsilon_0} \tag{1.14}$$

at any point in an electrostatic field. The expressions for curl $\mathbf{E}$ and div $\mathbf{E}$ summarize the properties of the electrostatic field. Note that both of these follow from Coulomb's law.

There are other electric fields besides the electrostatic field, e.g. the electric field of a moving charge or the electric field generated by a time-varying magnetic field. In general an electric field need not be conservative. Consider an electric field $\mathbf{E}$ which includes a non-conservative part $\mathbf{E}'$, then integrating along an arbitrary closed loop C we obtain

$$\oint_C \mathbf{E} \cdot d\mathbf{r} = \oint_C \mathbf{E}' \cdot d\mathbf{r} \quad . \tag{1.15}$$

The non-vanishing loop integral of $\mathbf{E}$ is called the *electric loop tension* of the closed path, or the *electromotive force* if the path is along a circuit. It represents the net work done per unit charge by the field in moving a charge once around C.

1.2 Electric Displacement D

Matter consists of atoms, which are positively charged nuclei surrounded by negative electron clouds, or molecules, which are made up of two or more atoms arranged in a definite geometrical pattern. In a conductor the outer electrons of the atoms are only loosely bound and more or less free to move. In fact, due to the wave nature of the electron, these free electrons, which are carriers of electric currents, may move freely throughout the periodic lattice formed by the atoms or molecules. Resistance arises only from the imperfectness of the lattice. Thus if a conductor is placed in an electric field the free electrons, which carry negative charges, will move opposite to the field to one end of the conductor, leaving excess positive charges on the other. These end charges themselves will set up a field which tends to cancel the external field. The process of charge migration will continue until the resultant field inside the conductor is reduced to zero. Thus under static conditions there can be no electric field inside a conductor.

In a dielectric the electrons are not free to move but may be displaced on an atomic scale. Upon the application of an external field the opposite charges within a single atom will separate slightly and become an electric dipole. This effect is called *electronic polarization*. When molecules are formed, the charge distribution of the constituent atoms is distorted with the outer electrons drawn towards the stronger binding atoms and a molecule may be regarded as a combination of oppositely charged ions. In an applied field the positive and negative ions are displaced with respect to each other giving rise to *ionic* or *atomic polarization*. Some molecules are formed with an asymmetric charge distribution and consequently a permanent dipole moment. These molecules, which exhibit dipole characteristics even in the absence of an external field, are called polar molecules. In the absence of an external field they are randomly oriented but tend to swing into alignment with the external field when one is applied, giving rise to *orientation* or *dipolar polarization*.

Restricting ourselves to macroscopic phenomena, we may thus consider a dielectric in an electric field as a continuous distribution of elementary dipoles and define the net dipole moment per unit volume as its *polarization* $\mathbf{P}$. Each of the above-mentioned effects may contribute to the total polarization. The elementary dipoles will themselves give rise to an electric field which must also be taken into account.

Consider two equal but opposite charges $-q$ and $+q$ separated by a distance 2ℓ as shown in Fig. 1.1 The potential Φ at a point P, whose position vector with respect to the centre O of the dipole is $\mathbf{r}$, is by Eq. (1.11)

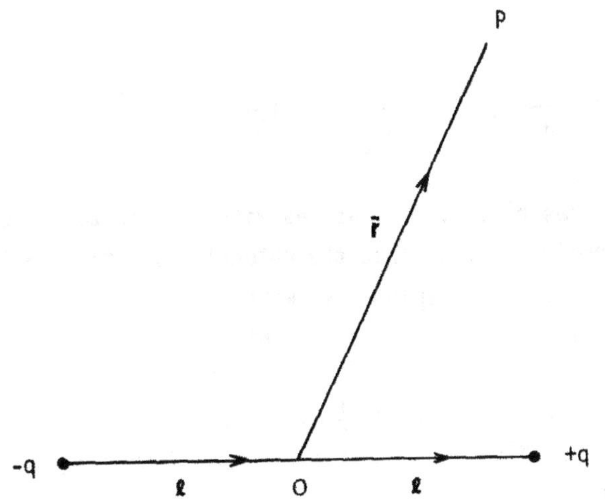

Fig. 1.1 Two charges -q and +q separated by a distance 2ℓ. They form a dipole of moment $\mathbf{p} = 2q\boldsymbol{\ell}$.

$$\Phi = \frac{q}{4\pi\varepsilon_0}\left(\frac{1}{|\bar{r}-\boldsymbol{\ell}|} - \frac{1}{|\bar{r}+\boldsymbol{\ell}|}\right)$$

$$= \frac{q}{4\pi\varepsilon_0\bar{r}}\left\{\left(1-\frac{2\bar{r}\cdot\boldsymbol{\ell}}{\bar{r}^2}+\frac{\ell^2}{\bar{r}^2}\right)^{-\frac{1}{2}} - \left(1+\frac{2\bar{r}\cdot\boldsymbol{\ell}}{\bar{r}^2}+\frac{\ell^2}{\bar{r}^2}\right)^{-\frac{1}{2}}\right\}$$

$$\cong \frac{2q}{4\pi\varepsilon_0}\frac{\bar{r}\cdot\boldsymbol{\ell}}{\bar{r}^3} \quad ,$$

where we have assumed $\ell \ll \bar{r}$ and consequently retained only the first-order terms in $\frac{\ell}{\bar{r}}$ in the expansion. A dipole is defined as a system of two equal and opposite charges -q and +q, such that as the displacement from the negative to the positive charge $2\boldsymbol{\ell} \to 0$, the vector $2q\boldsymbol{\ell}$ tends to a finite limit $\mathbf{p}$, called the dipole moment of the system. The potential of the field of a dipole of moment $\mathbf{p}$ is therefore

$$\Phi = \frac{\mathbf{p} \cdot \bar{\mathbf{r}}}{4\pi\varepsilon_0 \bar{r}^3} = -\frac{1}{4\pi\varepsilon_0} \mathbf{p} \cdot \bar{\nabla} \left(\frac{1}{r}\right) \quad , \tag{1.16}$$

where $\bar{\nabla}$ denotes partial derivatives with respect to the components of $\bar{\mathbf{r}}$. It furthermore follows that the potential due to a continuous dipole distribution in a volume V with dipole moment per unit volume given by polarization $\mathbf{P}(\mathbf{r}')$ is the integral

$$\Phi = -\frac{1}{4\pi\varepsilon_0} \int_V \mathbf{P} \cdot \bar{\nabla} \left(\frac{1}{r}\right) dV' \quad .$$

Using Eq. (1.5) and the divergence theorem we can write this as

$$\begin{aligned}
\Phi &= \frac{1}{4\pi\varepsilon_0} \int_V \mathbf{P} \cdot \nabla' \left(\frac{1}{r}\right) dV' \\
&= \frac{1}{4\pi\varepsilon_0} \int_V \nabla' \cdot \left(\frac{\mathbf{P}}{r}\right) dV' - \frac{1}{4\pi\varepsilon_0} \int_V \frac{\nabla' \cdot \mathbf{P}}{\bar{r}} dV' \\
&= \frac{1}{4\pi\varepsilon_0} \oint_S \frac{\mathbf{P} \cdot d\mathbf{S}'}{\bar{r}} - \frac{1}{4\pi\varepsilon_0} \int_V \frac{\nabla' \cdot \mathbf{P}}{\bar{r}} dV' \quad ,
\end{aligned} \tag{1.17}$$

where S is the boundary of V.

A comparison of Eq. (1.17) and Eq. (1.12) shows that the continuous dipole distribution gives rise to a potential which is the sum of two parts: the potential of an equivalent surface charge distribution on S of density

$$\sigma' = P_n \tag{1.18}$$

which is the component of $\mathbf{P}$ normal to the surface element $d\mathbf{S}'$, and the potential of an equivalent volume charge distribution in V of density, called the *polarization charge density*,

$$\rho' = -\nabla \cdot \mathbf{P} \quad . \tag{1.19}$$

We may introduce into a dielectric medium charges which are not related to any atomic or molecular system. These we shall refer to as "true" charges to distinguish them from the bound charges. Then for a point in a material medium Eq. (1.14) should be written as

$$\nabla \cdot \mathbf{E} = \frac{1}{\varepsilon_0}(\rho + \rho') = \frac{1}{\varepsilon_0}(\rho - \nabla \cdot \mathbf{P}) \quad , \tag{1.20}$$

where ρ denotes the volume density of true charges. Rearranging the terms we obtain

$$\nabla \cdot (\mathbf{E} + \frac{\mathbf{P}}{\varepsilon_0}) = \frac{\rho}{\varepsilon_0} \quad .$$

This equation would have the same form as the corresponding equation for vacuum, Eq. (1.14), if we introduce a new field vector

$$\mathbf{D} = \varepsilon_0 \mathbf{E} + \mathbf{P} \tag{1.21}$$

called the *electric displacement, electric induction,* or *dielectric displacement.* In terms of **D**, Eq. (1.20) becomes

$$\nabla \cdot \mathbf{D} = \rho \quad . \tag{1.22}$$

The divergence theorem then gives

$$\oint_S \mathbf{D} \cdot d\mathbf{S} = \int_V \nabla \cdot \mathbf{D} \, dV = q \tag{1.23}$$

for any arbitrary closed surface S of volume V enclosing a total net charge q. The simplicity of this equation is one of the reasons for introducing displacement **D**. In vacuum, where **P** vanishes, Eq. (1.23) reduces immediately to Eq. (1.13). Both these equations are known as Gauss' flux theorem or Gauss' law of electrostatics.

While electric displacement **D** is defined by Eq. (1.21) in terms of electric intensity **E** and polarization **P**, some definite relationship in general exists between **P** and **E**. For a linear and isotopic medium, **P** is parallel and proportional to **E** so that

$$\mathbf{P} = \varepsilon_0 \chi\, \mathbf{E}$$

where χ is known as the *electric susceptibility* or the *polarizability* of the medium. It follows then for such a medium that

$$\mathbf{D} = \varepsilon_0(\chi + 1)\mathbf{E} = \varepsilon\, \mathbf{E} \quad . \tag{1.24}$$

The constant of proportionality ε is called the *permittivity* or the *dielectric constant* of the medium. Note that ε is always greater than ε_0, the vacuum permittivity, since **P** and **E** are in the same direction.

1.3 Magnetic Induction B

Since the magnetic effect of an electric current was first demonstrated by Oersted in 1920 it has become clear that all magnetic phenomena can be accounted for in terms of interactions between currents. Even magnetized bodies derive their properties from the microscopic circulating currents in the interior and there is no basic necessity of introducing the concept of magnetic charge. We shall first consider interactions of currents in vacuum.

Interactions between current systems may be formulated in terms of current elements. A *current element* is the product of the current I and the length element $d\mathbf{r}$ of the circuit in which it flows. Ampère's experiments have shown that the mutual forces between two current-carrying linear circuits can be expressed empirically as

$$\mathbf{F}_2 = \frac{\mu_0}{4\pi} I_1 I_2 \oint_{C_1} \oint_{C_2} \frac{d\mathbf{r}_2 \times (d\mathbf{r}_1 \times \mathbf{r}_{21})}{r_{21}^3} \tag{1.25}$$

where $\mathbf{F}_2$ is the force acting on circuit C_2 by circuit C_1, $\mathbf{r}_{21}$ the radius vector of length element $d\mathbf{r}_2$ with respect to element $d\mathbf{r}_1$ as shown in Fig. 1.2, I_1 and I_2 the currents in the respective

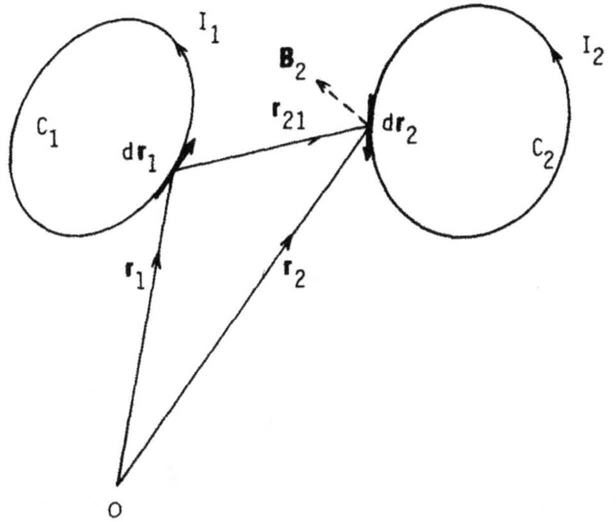

Fig. 1.2 Two linear circuits C_1 and C_2 carrying currents I_1 and I_2 respectively.

circuits, and μ_o a constant characteristic of the vacuum called the *vacuum permeability*. $\mathbf{F}_1$, the force acting on circuit C_1 by C_2, can be similarly expressed.

Equation (1.25) is asymmetrical with respect to indices 1 and 2 and appears to violate Newton's third law. This however is not true as the asymmetry disappears when the integration is partially carried out. Writing Eq. (1.25) as

$$\mathbf{F}_2 = \frac{\mu_o}{4\pi} I_1 I_2 \oint_{C_1} \oint_{C_2} \frac{1}{r_{21}^3} (d\mathbf{r}_2 \cdot \mathbf{r}_{21} d\mathbf{r}_1 - d\mathbf{r}_1 \cdot d\mathbf{r}_2 \mathbf{r}_{21}) \quad ,$$

we note that the first term on the right-hand side contains the integral

$$\oint_{C_2} \frac{\mathbf{r}_{21}}{r_{21}^3} \cdot d\mathbf{r}_2 = -\oint_{C_2} \nabla^{(21)}\left(\frac{1}{r_{21}}\right) \cdot d\mathbf{r}_2$$

$$= -\oint_{C_2} \nabla^{(21)}\left(\frac{1}{r_{21}}\right) \cdot d\mathbf{r}_{21}$$

$$= -\oint_{C_2} d\left(\frac{1}{r_{21}}\right) = 0 \quad .$$

In the above, $\nabla^{(21)}$ denotes the gradient with respect to the components of $\mathbf{r}_{21}$, etc., and we have made use of the relation (1.5) keeping in mind $\mathbf{r}_{21} = \mathbf{r}_2 - \mathbf{r}_1$, where $\mathbf{r}_1$, $\mathbf{r}_2$ are the position vectors of $d\mathbf{r}_1$ and $d\mathbf{r}_2$ respectively. It follows that

$$\mathbf{F}_2 = -\frac{\mu_0}{4\pi} I_1 I_2 \oint_{C_1} \oint_{C_2} \frac{\mathbf{r}_{21}}{r_{21}^3} d\mathbf{r}_1 \cdot d\mathbf{r}_2 \quad ,$$

which is now anti-symmetrical with respect to the indices 1 and 2 so that $\mathbf{F}_1 = -\mathbf{F}_2$.

Equation (1.25) can be written as two equations:

$$\mathbf{B}_2 = \frac{\mu_0}{4\pi} I_1 \oint_{C_1} \frac{d\mathbf{r}_1 \times \mathbf{r}_{21}}{r_{21}^3}$$

and

$$\mathbf{F}_2 = I_2 \oint_{C_2} d\mathbf{r}_2 \times \mathbf{B}_2 \quad . \tag{1.26}$$

These equations can be interpreted as indicating that circuit C_1 sets up a magnetic field which in turn exerts a force on circuit C_2. $\mathbf{B}_2$ is a measure of the strength of the magnetic field produced by circuit C_1 at the position of the circuit element $I_2 d\mathbf{r}_2$ and is called the *magnetic induction* or *magnetic flux density*. We further interpret

$$d\mathbf{B} = \frac{\mu_0}{4\pi} \frac{I d\mathbf{r}' \times \bar{\mathbf{r}}}{\bar{r}^3} \tag{1.27}$$

as the contribution of Id**r**' to the magnetic induction at the field point at $\bar{\mathbf{r}}$. This expression is known as the Biot-Savart law. Note that according to this interpretation an element of a circuit can set up a field at another part of the same circuit and interact with the element there. This clearly has gone one step beyond the empirical result of Ampère.

According to Eq. (1.26), to find the strength of the magnetic field at a point a current element Id**r** may be introduced as the test body, similar to the use of a test charge in an electric field. However, the idea of an isolated current element obviously cannot be realized physically. The concept of current element is nevertheless useful as the interaction between complete circuits can be broken down into elementary interactions among circuit elements. Conceptually **B** is analogous to **E** as both specify the strength of the forces acting on a test body.

The above may be generalized to a volume distribution of currents. For a continuous current distribution with current density **J** whose magnitude is defined as the current per unit cross-sectional area normal to the direction of flow and whose direction is that of the current, we have by definition

$$\mathbf{J}\, dV = I d\mathbf{r} \quad ,$$

and

$$\mathbf{B} = \frac{\mu_0}{4\pi} \int_{V'} \frac{\mathbf{J} \times \bar{\mathbf{r}}}{\bar{r}^3}\, dV' \quad , \tag{1.28}$$

$$\mathbf{F} = \int_V \mathbf{J} \times \mathbf{B}\, dV \quad . \tag{1.29}$$

In the above, we have replaced the position vector $\mathbf{r}_1$ of the source point by $\mathbf{r}'$, $\mathbf{r}_2$ of the field point by $\mathbf{r}$, and $\mathbf{r}_{21}$ by $\bar{\mathbf{r}}$. The dash denotes integration with respect to the source coordinates. V' and V are respectively the volumes containing all source points and all field points.

Currents can of course be considered as charges in motion. The definition of current density gives

$$\mathbf{J} = \rho \mathbf{u} \quad ,$$

where ρ is the volume density and $\mathbf{u}$ the velocity of the moving charges in the elementary volume. This expression can be substituted in the above equations. In particular, in the limit of a point charge of magnitude q, Eq. (1.29) gives the magnetic force acting on it as

$$\mathbf{F} = \int_V \rho \mathbf{u} \times \mathbf{B} dV = q\mathbf{u} \times \mathbf{B} \quad . \tag{1.30}$$

The properties of magnetic induction $\mathbf{B}$ can be specified by its divergence and curl. Taking the divergence of $\mathbf{B}$ of a continuous current distribution contained in a volume V', we have

$$\nabla \cdot \mathbf{B} = \frac{\mu_0}{4\pi} \nabla \cdot \int_{V'} \frac{\mathbf{J} \times \bar{\mathbf{r}}}{\bar{r}^3} dV' \quad .$$

As the integration is to be performed over the source coordinates and the differentiation is with respect to the field coordinates, the divergence operator can be shifted inside the integral sign. Consider the new integrand

$$\nabla \cdot \left\{ \frac{\mathbf{J} \times \bar{\mathbf{r}}}{\bar{r}^3} \right\} = - \nabla \cdot \left\{ \mathbf{J} \times \bar{\nabla} \left(\frac{1}{\bar{r}}\right) \right\}$$

$$= - \nabla \times \mathbf{J} \cdot \bar{\nabla} \left(\frac{1}{\bar{r}}\right) + \mathbf{J} \cdot \nabla \times \bar{\nabla} \left(\frac{1}{\bar{r}}\right) \quad ,$$

where $\bar{\nabla}$ denotes operation with respect to the coordinates $\bar{x}_i = x_i - x_i'$. As $\mathbf{J}$ is a function of the source coordinates alone and ∇ operates with respect to the field coordinates, $\nabla \times \mathbf{J}$ vanishes. Also, by virtue of Eq. (1.5),

$$\mathbf{J} \cdot \nabla \times \bar{\nabla} \left(\frac{1}{\bar{r}}\right) = \mathbf{J} \cdot \bar{\nabla} \times \bar{\nabla} \left(\frac{1}{\bar{r}}\right) = 0 \quad .$$

we conclude therefore

$$\nabla \cdot \mathbf{B} = 0 \quad . \tag{1.31}$$

This shows that the lines of **B** form closed loops and **B** is said to be solenoidal.

As the evaluation of $\nabla \times \mathbf{B}$ is rather complicated, we shall in the first instance restrict ourselves to the field of stationary currents. A current is said to be *stationary* if there is no accumulation of charge at any point of its path. The conservation of charge requires that the net amount leaving the region bounded by an arbitrary closed surface per unit time equals the time rate of decrease of the charge enclosed therein, i.e.

$$\oint_S \mathbf{J} \cdot d\mathbf{S} = -\frac{d}{dt}\int_V \rho \, dV = -\int_V \frac{\partial \rho}{\partial t} dV$$

for an arbitrary volume V with boundary surface S. By the divergence theorem

$$\oint_S \mathbf{J} \cdot d\mathbf{S} = \int_V \nabla \cdot \mathbf{J} \, dV \quad .$$

As V is arbitrary, we have

$$\nabla \cdot \mathbf{J} + \frac{\partial \rho}{\partial t} = 0 \quad . \tag{1.32}$$

This equation is known as the *continuity equation of electric charge* and expresses charge conservation. Applying to stationary currents for which $\frac{\partial \rho}{\partial t} = 0$ everywhere, it requires that

$$\nabla \cdot \mathbf{J} = 0 \quad . \tag{1.33}$$

Hence the density of stationary currents is solenoidal. We shall also assume that **J** and its first and second derivatives are continuous and bounded in the region of interest.

Making use of Eq. (1.5) we have

$$\nabla \times \mathbf{B} = -\frac{\mu_0}{4\pi}\int_{V'} \nabla \times \{ \mathbf{J} \times \bar{\nabla} (\frac{1}{r})\} \, dV'$$

$$= -\frac{\mu_0}{4\pi}\int_{V'} \nabla \times \{ \mathbf{J} \times \nabla (\frac{1}{r})\} \, dV' \quad .$$

The integrand can be transformed:

$$\nabla \times \{ \mathbf{J} \times \nabla (\frac{1}{r}) \} = \mathbf{J}\nabla^2 (\frac{1}{r}) - (\mathbf{J} \cdot \nabla)\nabla (\frac{1}{r})$$

$$+ \{\nabla (\frac{1}{r}) \cdot \nabla\} \mathbf{J} - (\nabla \cdot \mathbf{J})\nabla (\frac{1}{r}) \quad .$$

Since the partial differentiations are to be taken with respect to the field coordinates and $\mathbf{J}$ depends on the source coordinates alone, the last two terms of the right-hand side vanish. Then, making use of Eq. (1.5) again, we obtain

$$\nabla \times \mathbf{B} = -\frac{\mu_0}{4\pi}\int_{V'} \mathbf{J}\nabla'^2 (\frac{1}{r}) dV' + \frac{\mu_0}{4\pi}\int_{V'} (\mathbf{J} \cdot \nabla')\nabla' (\frac{1}{r}) dV' \quad .$$

The second integral on the right-hand side can be integrated by parts. Consider its i-th component, whose integrand is

$$(\mathbf{J} \cdot \nabla') \frac{\partial}{\partial x_i'} (\frac{1}{r}) = \nabla' \cdot \{ \mathbf{J} \frac{\partial}{\partial x_i'} (\frac{1}{r}) \} - \{ \frac{\partial}{\partial x_i'} (\frac{1}{r}) \} \nabla' \cdot \mathbf{J} \quad .$$

The second term on the right-hand side vanishes on account of the condition $\nabla' \cdot \mathbf{J} = 0$ for stationary currents. Hence

$$\int_{V'} (\mathbf{J} \cdot \nabla') \frac{\partial}{\partial x_i'} (\frac{1}{r}) dV' = \int_{V'} \nabla' \cdot \{ \mathbf{J} \frac{\partial}{\partial x_i'} (\frac{1}{r}) \} \, dV'$$

$$= \oint_{S'} \{ \frac{\partial}{\partial x_i'} (\frac{1}{r}) \} \mathbf{J} \cdot d\mathbf{S}' \quad ,$$

where we have used the divergence theorem and S' is the boundary surface of V'. The surface integral will vanish if the volume V' is taken large enough so that $\mathbf{J}$ is everywhere zero on its boundary surface. We are then left with

$$\nabla \times \mathbf{B} = -\frac{\mu_0}{4\pi}\int_{V'} \mathbf{J}\,\nabla'^2\left(\frac{1}{\bar{r}}\right)dV' \quad . \tag{1.34}$$

Although the integration is to be taken over all space where $\mathbf{J}$ does not vanish, in practice the function $\nabla'^2\left(\frac{1}{\bar{r}}\right)$ and hence the integrand is zero everywhere except at the field point for which $\bar{r} = 0$. (Note that the field point is kept fixed while integration is taken over all source points). It suffices therefore to integrate over an arbitrary small volume V_0 containing the field point. We shall consider the i-th component of the above equation.

The integration of Eq. (1.34) is effected by applying Green's formula to the volume V'. The field point for which $\bar{r} = 0$, however, is a singular point and must be excluded. This is done by excluding a small sphere of radius ε with the field point as centre. Let its volume be V_0, its surface be S_0, and apply Green's formula to the volume $V' - V_0$:

$$\int_{V'-V_0} \left\{ J_i \nabla'^2 \left(\frac{1}{\bar{r}}\right) - \frac{1}{\bar{r}}\nabla'^2 J_i \right\} dV'$$

$$= \oint_{S'+S_0} \left\{ J_i \nabla' \left(\frac{1}{\bar{r}}\right) - \frac{1}{\bar{r}}\nabla' J_i \right\} \cdot d\mathbf{S}' \tag{1.35}$$

The function $\nabla'^2\left(\frac{1}{\bar{r}}\right)$ now vanishes everywhere in the domain of integration. Of the remaining terms consider the volume integral over V_0,

$$\int_{V_0} \frac{1}{\bar{r}}\nabla'^2 J_i dV' = \langle \nabla'^2 J_i \rangle \int_0^{\varepsilon} \frac{1}{\bar{r}}\, 4\pi\, \bar{r}^2 d\bar{r}$$

$$= 2\pi\varepsilon^2 \langle \nabla'^2 J_i \rangle$$

by the mean value theorem, $\langle \nabla'^2 J_i \rangle$ being the mean value of $\nabla'^2 J_i$ in the sphere V_0. Then, as $\nabla'^2 J_i$ is bounded, this expression tends to zero as $\varepsilon \to 0$.

Consider next the surface integrals over S_0. As the normal to an element of the inner surface S_0 of the volume $V' - V_0$ is directed towards the centre of the sphere, it is parallel to the unit vector $\hat{\mathbf{r}}$ drawn towards the field point, i.e.

$$\hat{\mathbf{r}} \cdot d\mathbf{S}' = dS'$$

on S_0. The integral $\oint_{S_0} J_i \nabla' \left(\frac{1}{r}\right) \cdot d\mathbf{S}'$

$$= - \oint_{S_0} J_i \bar{\nabla} \left(\frac{1}{r}\right) \cdot d\mathbf{S}' = - \langle J_i \rangle \int_{S_0} \frac{d}{dr} \left(\frac{1}{r}\right) \hat{\mathbf{r}} \cdot d\mathbf{S}'$$

$$= \frac{\langle J_i \rangle}{\varepsilon^2} \oint_{S_0} dS' = 4\pi \langle J_i \rangle \qquad ,$$

using the mean value theorem. If we now take the limit $\varepsilon \to 0$, $\langle J_i \rangle \to J_i(0)$, $J_i(0)$ being the value of J_i at the centre of the sphere, i.e. at the field point. The other surface integral over S_0

$$- \oint_{S_0} \frac{1}{r} \nabla' J_i \cdot d\mathbf{S}' = - \frac{1}{\varepsilon} \oint_{S_0} \frac{\partial J_i}{\partial x_j'} dS_j'$$

$$= - \frac{1}{\varepsilon} \left\langle \frac{\partial J_i}{\partial x_j'} \right\rangle \oint_{S_0} dS_j'$$

$$= - \frac{1}{\varepsilon} \left\langle \frac{\partial J_i}{\partial x_j'} \right\rangle \left(\oint_{S_0} d\mathbf{S}' \right)_j = 0 \qquad ,$$

since $\oint d\mathbf{S}' = 0$ over a closed surface.

Using the above results we see that in the limit $\varepsilon \to 0$ Eq. (1.35) gives

$$-\int_{V'} \frac{1}{r} \nabla'^2 J_i dV' = \oint_{S'} \left\{ J_i \nabla' \left(\frac{1}{r}\right) - \frac{1}{r} \nabla' J_i \right\} \cdot d\mathbf{S}' + 4\pi J_i(0) \quad .$$

On the other hand, taking the limit $\varepsilon \to 0$ in Eq. (1.35) without first performing the integration over V_0 and S_0 we would obtain

$$\int_{V'} \left\{ J_i \nabla'^2 \left(\frac{1}{r}\right) - \frac{1}{r} \nabla'^2 J_i \right\} dV' = \oint_{S'} \left\{ J_i \nabla' \left(\frac{1}{r}\right) - \frac{1}{r} \nabla' J_i \right\} \cdot d\mathbf{S}' \quad .$$

A comparison of these expressions yields

$$\int_{V'} J_i \nabla'^2 \left(\frac{1}{r}\right) dV' = -4\pi J_i(0) \quad . \tag{1.36}$$

When this is substituted in Eq. (1.34) we find for stationary currents

$$\nabla \times \mathbf{B} = \mu_0 J \quad . \qquad \text{(stationary currents)} \tag{1.37}$$

Here **B** and **J** refer to the same point.

The function $\nabla^2 \left(\frac{1}{r}\right)$ has the property that it vanishes everywhere except at $\bar{r} = 0$, i.e. $x_i' = x_i$. Furthermore, in the derivation of Eq. (1.36), J_i was treated as an arbitrary scalar function of x_i' subject only to the mathematical requirement that this function and its derivatives are continuous, single-valued and bounded. The physical properties of current density simply did not come in. Thus for an arbitrary function $f(x_i')$ satisfying the same mathematical conditions we have in general

$$\int_V f(x_i') \nabla'^2 \left(\frac{1}{r}\right) dV' = -4\pi f(x_i) \quad ,$$

if the point $\bar{r} = 0$, i.e. the point of position vector **r** is enclosed in V. The integral vanishes otherwise. These properties are similar to those of the Dirac delta function $\delta(\bar{r})$. In fact, we can relate the function $\nabla'^2 (\frac{1}{\bar{r}})$ to $\delta(\mathbf{r})$ by

$$\nabla'^2 (\frac{1}{\bar{r}}) = - 4\pi\delta(\bar{\mathbf{r}}) \quad . \tag{1.38}$$

Writing $\bar{\mathbf{r}} = \mathbf{r} - \mathbf{r}'$ the last integral can be written as

$$\int_V f(\mathbf{r}')\delta(\mathbf{r} - \mathbf{r}')dV' = f(\mathbf{r}) \quad . \tag{1.39}$$

Furthermore, if we interchange $\mathbf{r}'$ and $\mathbf{r}$, then as $\nabla'^2 (\frac{1}{\bar{r}}) = \nabla^2 (\frac{1}{\bar{r}})$, i.e. $\delta(\mathbf{r} - \mathbf{r}') = \delta(\mathbf{r}' - \mathbf{r})$, we also have

$$\int_V f(\mathbf{r})\delta(\mathbf{r} - \mathbf{r}')dV = f(\mathbf{r}') \quad ,$$

where V is assumed to contain the point $\mathbf{r}'$. We shall have occasion to use relation (1.38) in a later chapter.

Consider next the case of non-stationary currents in vacuum. The time derivative $\frac{\partial\rho}{\partial t}$ in the continuity equation no longer vanishes. We can however make use of Eq. (1.14)[b] and replace ρ by $\varepsilon_0\nabla \cdot \mathbf{E}$ in the continuity Eq. (1.32) which then becomes

$$\nabla \cdot (\mathbf{J} + \varepsilon_0\dot{\mathbf{E}}) = 0 \quad . \tag{1.40}$$

where the dot above $\mathbf{E}$ represents partial time derivation. This equation has the form of the characteristic equation of stationary currents, Eq. (1.33). All currents may therefore be considered stationary provided an additional quantity $\varepsilon_0\dot{\mathbf{E}}$ is added to the current

---

b Equation (1.14) was shown to be true for electrostatic fields. It is usually assumed to hold for electric fields in general. Actually it expresses charge conservation (cf. Eq. (2.15)) which is assumed hold under all conditions.

density **J** arising from the physical transportation of true charges. This additional quantity is called the *vacuum displacement current density* and was first introduced by Maxwell[c]. As will be seen later, the introduction of the displacement current serves to resolve the inconsistency encountered in the empirical Ampère circuital law, allowing it to be applied generally and at the same time giving it wider physical significance.

1.4 Magnetic Intensity H

In a material medium, not only have we to consider the conduction currents arising from the motion of the "true" charges and the displacement current identifiable with the time variation of the electric field, but also currents that are associated with the material medium itself. This is similar to the case of electrostatics where polarization charges as well as "true" charges must be considered in a material medium. If we confine ourselves to a stationary medium, two additional currents, polarization and magnetization currents, must also be included in the analysis.

Time variation of polarization **P** represents the process of charge rearrangement inside the molecules or atoms of a medium. These molecular currents will in general add up to give observable effects. Let ρ be the charge density and **u** the velocity of the charges at a fixed point in a molecule with position vector **r** with respect to the molecular centre. The current density at that point is then $\rho\mathbf{u}$, which is related to the change of the charge density through the continuity equation

$$\frac{\partial\rho}{\partial t} + \nabla \cdot (\rho\mathbf{u}) = 0 \quad .$$

[c] The interpretation of $\varepsilon_0\dot{\mathbf{E}}$ as a current density is tantamount to asserting that a time-varying electric field will generate a magnetic field.

The polarized molecule has an electric dipole moment

$$\mathbf{p} = \int_{V_0} \rho \mathbf{r} dV \quad ,$$

where V_0 is the molecular volume. The process of charge redistribution is accompanied by a time variation of $\mathbf{p}$:

$$\frac{d\mathbf{p}}{dt} = \int_{V_0} \mathbf{r} \frac{\partial \rho}{\partial t} dV = - \int_{V_0} \mathbf{r} \, \nabla \cdot (\rho \mathbf{u}) dV \quad .$$

The i-th component of the integrand of the last integral is

$$x_i \nabla \cdot (\rho \mathbf{u}) = \nabla \cdot (\rho \mathbf{u} x_i) - \rho \mathbf{u} \cdot \nabla x_i = \nabla \cdot (\rho \mathbf{u} x_i) - \rho u_i$$

since ∇x_i is simply the unit vector along the x_i-axis. Then, making use of the divergence theorem we have

$$\frac{d\mathbf{p}}{dt} = - \oint_{S_0} \mathbf{r} \, \rho \mathbf{u} \cdot d\mathbf{S} + \int_{V_0} \rho \mathbf{u} \, dV \quad ,$$

where S_0 is the boundary of V_0. No error would be introduced if we choose V_0 to be just slightly larger than the molecular volume. The charge density will then vanish on its boundary and the surface integral can be dropped. From its definition, the rate of change of polarization $\mathbf{P}$ is just the rate of change of the electric dipole moment in a unit volume, i.e.

$$\frac{\partial \mathbf{P}}{\partial t} = \frac{1}{V_0} \frac{d\mathbf{p}}{dt} = \frac{1}{V_0} \int_{V_0} \rho \mathbf{u} \, dV = \langle \rho \mathbf{u} \rangle \quad . \qquad (1.41)$$

This shows that the average molecular current density is simply the partial time derivative of $\mathbf{P}$. This is called the *polarization current density*.

The magnetic properties of a substance can be accounted for by the small permanently circulating current loops of the orbiting atomic electrons in its interior. Each loop acts as a magnetic dipole and the total dipole moment per unit volume is represented by a macroscopic vector, magnetization $\mathbf{M}$. If the magnetization is uniform, the neighbouring atomic current loops cancel without producing any net effective current across an arbitrary macroscopic surface drawn inside the substance. If the magnetization is non-uniform, currents of neighbouring loops do not cancel completely and a net effective current density results. This current is called the *magnetization current* and its density will be shown in Sec. 3.9 to be

$$\mathbf{J}_m = \nabla \times \mathbf{M} \quad . \tag{1.42}$$

In a stationary material medium the resultant current density is therefore the sum of three contributions: conduction current density $\mathbf{J}$, polarization current density $\partial\mathbf{P}/\partial t$ and magnetization current density $\nabla\times\mathbf{M}$. The principle of conservation of electric charge still holds and the continuity equation which must be satisfied now becomes

$$\nabla \cdot \left(\mathbf{J} + \frac{\partial \mathbf{P}}{\partial t} + \nabla\times\mathbf{M}\right) + \frac{\partial(\rho+\rho')}{\partial t} = 0 \quad , \tag{1.43}$$

where ρ and ρ' are the "true" and polarization charge densities respectively. Again, as in the case of non-stationary currents in vacuum, the total current can be made solenoidal by replacing $\rho+\rho'$ with $\varepsilon_0\nabla\cdot\mathbf{E}$ in accordance with Eq. (1.20).

Then after rearranging the terms, Eq. (1.43) becomes

$$\nabla \cdot \left\{ \mathbf{J} + \nabla\times\mathbf{M} + \frac{\partial}{\partial t}(\mathbf{P}+\varepsilon_0\mathbf{E}) \right\} = 0$$

or

$$\nabla \cdot \left\{ \mathbf{J} + \nabla\times\mathbf{M} + \frac{\partial \mathbf{D}}{\partial t} \right\} = 0 \quad . \tag{1.44}$$

The component represented by $\partial\mathbf{D}/\partial t$ is called the *displacement current density*. In vacuum it coincides with the vacuum displacement current density $\varepsilon_0\dot{\mathbf{E}}$.

According to Eq. (1.44), if the total current density is defined by

$$\mathbf{J}_t = \mathbf{J} + \nabla\times\mathbf{M} + \frac{\partial \mathbf{D}}{\partial t} \quad ,$$

it will always be solenoidal or stationary. This is true whether the medium is vacuum or material, and whether or not the conduction current $\mathbf{J}$ is stationary by itself. Consider, for instance, the flow of charges in a closed circuit which includes a condenser. While charges accumulate on the plates of the condenser and none crosses the space between them, the current may still be considered continuous across the plates, but in the form of the time rate of change of the displacement vector $\mathbf{D}$. In general, therefore, the magnetic effects of currents can always be interpreted in terms of stationary currents, provided all relevant currents are included.

It then follows that in a material medium Eq. (1.37) is to be replaced by

$$\nabla\times\mathbf{B} = \mu_0\left(\mathbf{J} + \nabla\times\mathbf{M} + \frac{\partial \mathbf{D}}{\partial t}\right) \quad , \tag{1.45}$$

or

$$\nabla\times(\mathbf{B} - \mu_0\mathbf{M}) = \mu_0\left(\mathbf{J} + \frac{\partial \mathbf{D}}{\partial t}\right) \quad .$$

This again has the form of Eq. (1.37) and may be interpreted as showing that the magnetic field produced by the sum of the conduction and displacement currents is $\mathbf{B} - \mu_0\mathbf{M}$. It is however more convenient to introduce a new magnetic vector $\mathbf{H}$ the *magnetic intensity*, defined by

$$\mathbf{H} = \frac{1}{\mu_0}(\mathbf{B} - \mu_0\mathbf{M}) \quad , \tag{1.46}$$

which is to be used when material media are involved. The field-current relationship now takes the simple form

$$\nabla\times\mathbf{H} = \mathbf{J} + \frac{\partial \mathbf{D}}{\partial t} \quad . \tag{1.47}$$

Integrating **H** around an arbitrary closed loop C, we obtain by Stokes' theorem

$$\oint_C \mathbf{H}\cdot d\mathbf{r} = \int_S \nabla\times\mathbf{H}\cdot d\mathbf{S}$$

$$= \int_S \left(\mathbf{J}+\frac{\partial \mathbf{D}}{\partial t}\right)\cdot d\mathbf{S} = I + \frac{d}{dt}\int_S \mathbf{D}\cdot d\mathbf{S} \quad , \qquad (1.48)$$

where S is an arbitrary open, two-sided surface with C as boundary. This relation is called the *general circuital law*, whose simplicity is one of the reasons for introducing **H**.

For a large class of substances an approximate linear relationship exists between magnetization **M** and magnetic intensity **H**:

$$\mathbf{M} = \chi_m \mathbf{H} \quad .$$

from which follows

$$\mathbf{B} = \mu_0(1+\chi_m)\mathbf{H} = \mu\mathbf{H} \quad .$$

The constants of proportionality χ_m and μ are called the *magnetic susceptibility* and the *permeability* of the substance respectively. For paramagnetic substances $\mu > \mu_0$ and for diamagnetic substances $\mu < \mu_0$. For these substances generally, $|\chi_m| \ll 1$ and it is a good approximation to take $\mu \simeq \mu_0$. For ferromagnetic materials hysteresis occurs and the definition of permeability as given above loses much of its significance even under normal conditions. Furthermore the relationship between **B** and **H** will in general depend on the absolute temperature. It is clear that compared with the linear relationship between **D** and **E** the linear relationship between **B** and **H** has a much more limited validity.

From the above discussion we have seen that to describe electromagnetic phenomena adequately, two field vectors, **E** and **B**, are sufficient if matter is not involved, whereas in material media, four field vectors are in general necessary. The additional field vectors required are **P** or **D** and **M** or **H**. We have noted that more simplified relations are obtained when the fields are described in terms of **E, B, D** and **H**. Furthermore, these vectors are more directly related to quantities which are readily measured experimentally, such as force, charge and current. They are therefore commonly employed for the description of electromagnetic effects. According to the nature of these vectors they may be separated into two groups: **E** and **B** expressing intensity, **D** and **H** expressing quantity.

1.5 Experimental Laws

Among the empirical electromagnetic laws, Faraday's law of induction and Ampère's circuital law are outstanding in that they are able to relate the spatial and temporal derivatives of the field vectors **E, B, D** and **H**. These laws constitute the foundation of Maxwell's electromagnetic theory. A brief discussion of them is given below.

A) Faraday's law of induction

Experimental results on the generation of electromotive forces (emf) by varying magnetic fields are summarized in Faraday's law of induction:

$$\oint_C \mathbf{E} \cdot d\mathbf{r} = -\frac{d}{dt} \int_S \mathbf{B} \cdot d\mathbf{S} \tag{1.49}$$

where S is an arbitrary open surface bounded by the closed circuit C. Physically the left-hand side is the emf generated in C and the right-hand side represents the rate of change of the number of magnetic lines of force crossing S. The negative sign is introduced to conform to Lenz's rule. It may be remarked that this law cannot be derived from any other empirical law.

It was first recognized by Maxwell that Faraday's law may have wider significance than merely giving the emf of a circuit, as the characteristics of the circuit play no part in the relation. Furthermore, the relation is independent of the way in which the flux crossing the circuit is changed. The circuit may be deformed or moved, the value of **B** may be varied by whatever means possible, but the law retains the same form. It appears plausible, therefore, that the validity of the law may not even require the physical presence of a conductor — the path of integration may cross conductors, dielectrics, free space or all of these — but that it expresses a general relationship between the electric and the magnetic field. It implies the creation of an electric field when the magnetic field changes with time.

Faraday's law therefore provides a quantitative basis for understanding the interactions between electric and magnetic fields.

B) Ampère's circuital law

From the empirical Biot-Savart law for the magnetic field of a stationary current distribution Eq. (1.37) was obtained:

$$\nabla \times \mathbf{B} = \mu_0 \mathbf{J} \quad .$$

Integrating both sides over an arbitrary surface S bounded by a closed loop C,

$$\int_S \nabla \times \mathbf{B} \cdot d\mathbf{S} = \mu_0 \int_S \mathbf{J} \cdot d\mathbf{S} \quad ,$$

we obtain by means of Stokes' theorem

$$\oint_C \mathbf{B} \cdot d\mathbf{r} = \mu_0 I \quad ,$$

where I is the current crossing the surface S. A similar relation for the magnetic field intensity **H** is obtained by dividing both sides of the above equation with μ_0:

$$\oint_C \mathbf{H} \cdot d\mathbf{r} = I \quad . \tag{1.50}$$

In analogy to the electromotive force for an electric circuit the closed loop line integral of $\mathbf{H}$ is called the *magnetomotive force*. Eq. (1.50) was the original Ampère's circuital law. While this law is useful in calculating the magnetic field intensity set up by a current, it leads to serious difficulties if accumulation of charge occurs anywhere on the path of the current as will be shown in the following example.

Consider the charging-up of a parallel-plate condenser with a steady current in a series circuit containing a suitable source of emf. Choose for C a large circle in a plane between and parallel to the condenser plates. If the surface S bounded by C is chosen to coincide with the plane, then as no current crosses it Eq. (1.50) asserts that the magnetomotive force round C is zero. On the other hand if a curved surface is chosen for S which cuts the charging circuit, the magnetomotive force as given by the same equation will be equal to the charging current I. Thus Ampère's circuital law in its original form leads to a contradiction, if applied to currents which are not stationary.

The difficulties encountered in Ampère's law would be resolved if an additional term, Maxwell's displacement current, is added to the conduction current. This would then give the general circuital law anticipated in the previous section:

$$\oint_C \mathbf{H} \cdot d\mathbf{r} = \int_S (\dot{\mathbf{D}} + \mathbf{J}) \cdot d\mathbf{S} \quad .$$

It can be seen immediately that the additional term $\varepsilon \int \dot{\mathbf{E}} \cdot d\mathbf{S}$ resolves the contradiction encountered in the above example. For as the electric field between the parallel plates is $\frac{\sigma}{\varepsilon}$, where σ is the surface charge density on either plate carrying charges of magnitude q, the surface integral of $\mathbf{D}$ over the plane surface bounded by C gives $\dot{q}$ which is just the charging current I in the circuit. On the other hand, as the current is steady, $\dot{\mathbf{E}}$ and hence $\dot{\mathbf{D}}$ is zero in the conductor of the circuit, so that the integral over the curved surface is not affected by the extra term introduced.

The success achieved by introducing the displacement current in resolving the difficulties encountered by Ampère's circuital law was itself a great triumph for Maxwell. He went on further to speculate that the general circuital law actually relates in general the magnetic intensity and the electric displacement, irrespective of whether or not a "true" current is present.

With Maxwell's general interpretation of Faraday's and Ampère's laws, electric and magnetic effects are no longer isolated phenomena; the variation in one must produce a change in the other. It is therefore more logical and realistic to speak of a unified *electromagnetic field*, with the electric and the magnetic fields representing different aspects of the same phenomenon. A theory formulated on such a basis is called an electromagnetic theory.

In this chapter we have briefly reviewed the concepts of the electric and the magnetic field as well as the empirical laws which relate the field vectors among themselves and to the sources: charges and currents. We have seen that in material media two additional vectors are required for a complete description of the fields. Two empirically based laws, Faraday's law of induction and the general circuital law, are of particular significance as they describe in a quantative way the interactions between the electric and the magnetic field. Their experimental basis and the generalization and reinterpretation proposed by Maxwell have also been discussed. These laws are to provide the basis of Maxwell's electromagnetic theory. In the following three chapters we shall assume that the field vectors **E**, **B**, **D** and **H** are well-defined physical quantities and using these laws as the fundamental axioms develop Maxwell's electromagnetic theory.

PROBLEMS

1. A conductor is a medium in which a charge can move freely. By means of Gauss' flux theorem show that (a) any excess charge on a conductor must lie on its surface, and (b) at the surface the electric field intensity is along the outward normal and has a

magnitude σ/ε where σ is the surface charge density and ε the permittivity of the medium in which the conductor is embedded. Investigate the case of a hollow conductor.

2. Two infinite parallel conducting plates carry uniform charges of surface densities σ and $-\sigma$ respectively on their inner surfaces. Using Gauss' flux theorem show that (a) the electric field between the plates is uniform and has a magnitude $\frac{\sigma}{\varepsilon}$, and (b) the electric field in the region external to the plates is zero.

3. Calculate the potential and the electrostatic field intensity for a sphere carrying a charge q which is distributed uniformly (a) over its surface, (b) over its volume.

4. If the electrostatic field intensity of a point charge q were $qr^{-3-\delta}\mathbf{r}$, where $\mathbf{r}$ is the radius vector of a field point with respect to the charge and δ a small positive constant, calculate $\nabla\cdot\mathbf{E}$ and $\nabla\times\mathbf{E}$ for $r\neq 0$. Would the field be conservative? Would Gauss' flux theorem still hold? How would an excess charge on a conductor be distributed?

5. Show that an electric dipole of moment $\mathbf{p}$ in an electrostatic field of intensity $\mathbf{E}$ suffers (a) a force $(\mathbf{p}\cdot\nabla)\mathbf{E}$, and (b) a torque $\mathbf{r}\times(\mathbf{p}\cdot\nabla)\mathbf{E}+\mathbf{p}\times\mathbf{E}$ about the origin, where $\mathbf{r}$ is the position vector of the dipole.

6. A dielectric of volume V and surface S has polarization $\mathbf{P}$. Show that if ρ' and σ' are the equivalent volume and surface charge densities the following relation holds:

$$\int_V \mathbf{P}\,dV = \int_V \rho'\mathbf{r}\,dV + \int_S \sigma'\mathbf{r}\,dS$$

where $\mathbf{r}$ is the position vector from any fixed origin. Give a physical interpretation of this result.

7. The *polarizability* α of an atom is defined as the ratio of the dipole moment of the atom to the local electric field $\mathbf{E}_\ell$ at the atom. Then if $\mathcal{N}_i$ and α_i are respectively the number per unit volume and polarizability of the i-th type atom, the polarization is $\mathbf{P} = \sum_i \mathcal{N}_i \alpha_i \mathbf{E}_\ell$. For a homogeneous and isotropic dense medium the effective local field may be taken to be (Sec. 6.6)

$$\mathbf{E}_\ell = \mathbf{E} + \frac{\mathbf{P}}{3\varepsilon_0} ,$$

where $\mathbf{E}$ is the intensity of the applied field.

From the above relations derive the Clausius-Mossotti relation

$$\frac{\varepsilon - \varepsilon_0}{\varepsilon + 2\varepsilon_0} = \sum_i \frac{\mathcal{N}_i \alpha_i}{3\varepsilon_0} ,$$

where ε is the permittivity of the medium.

8. Two long straight parallel conductors separated by a distance d carry currents I_1 and I_2 flowing in the same direction. Show that the forces between them are attractive. If the length of each conductor is 2ℓ, find the force per unit length acting near the middle portion of either conductor. Hence show that for $\ell \to \infty$ the result is the same as that obtained by means of the circuital law.

9. A sphere of radius R and uniform volume charge density ρ rotates with a constant angular velocity ω about a fixed diameter. Show that the magnetic induction at the centre of the sphere is

$$\frac{1}{3} \mu_0 \rho R^2 \omega .$$

10. Find the electric intensity at a distance r from a long straight filament carrying a uniform linear charge density τ, using (a) Coulomb's law, (b) Gauss' flux theorem.

11. Find the magnetic induction at a distance r from a long, straight conductor carrying a steady current I, using (a) the Biot-Savart law, (b) Ampère's circuital law.

12. A circular loop of wire of radius a and resistance per unit length ρ lies in the xy-plane with its centre at the origin. A uniform magnetic field of induction $\mathbf{B}$ is applied parallel to the z-axis. It is assumed that ρ is large so that the magnetic field of any induced current can be neglected. Find the induced current in the loop when (a) the loop is maintained stationary while $\mathbf{B} = \mathbf{B}_0 e^{-\beta t}$, (b) $\mathbf{B}$ is kept constant while the loop rotates about the y-axis at a constant angular velocity ω, and (c) the loop rotates as in (b) while $\mathbf{B} = \mathbf{B}_0 e^{-\beta t}$, β being a constant.

Chapter II

MAXWELL'S EQUATIONS

Maxwell's equations are based on Faraday's law of induction and Ampère's circuital law with the extension and re-interpretation introduced by Maxwell. Also incorporated are the physical ideas of the inseparability of magnetic poles and charge conservation. For the immediate applications, the integral Maxwell's equations are used to deduce the boundary conditions for the field vectors, and the differential Maxwell's equations are used to obtain the Poynting theorem and the wave equations. Scalar and vector potentials are introduced to facilitate the general solution of Maxwell's equations.

2.1 Maxwell's Equations in Integral Form

A consistent and logical description of the structure and properties of the electromagnetic field together with its relation to the sources can be formulated on the basis of the integral equations which express Faraday's law of induction and Ampère's circuital law. These laws, with the extension and generalized interpretation of Maxwell, serve as the fundamental axioms in Maxwell's formulation of an electromagnetic theory. These fundamental axioms are

$$\int_S \dot{\mathbf{B}} \cdot d\mathbf{S} = - \oint_C \mathbf{E} \cdot d\mathbf{r} \quad , \tag{2.1}$$

$$\int_S (\dot{\mathbf{D}} + \mathbf{J}) \cdot d\mathbf{S} = \oint_C \mathbf{H} \cdot d\mathbf{r} \quad , \tag{2.2}$$

where S is an open, two-sided surface bounded by a closed loop C. For convenience, partial derivatives with respect to time is indicated by an overhead dot. For mathematical manipulations to be possible it is assumed that the field vectors involved and their derivatives are continuous, single-valued and bounded in the domain of integration. It should be noted also that these relations are assumed to be valid only for stationary media, to which we shall confine ourselves throughout this book.

In addition to the fundamental axioms two supplementary axioms are required. They are obtained from the former by endowing the fields with certain physical properties which are found to be true from our empirical experience.

Consider the special case where S is a closed surface. Its boundary C is now contracted into a point with the result that the line integrals vanish. Equations (2.1) and (2.2) then become

$$\frac{d}{dt} \oint_S \mathbf{B} \cdot d\mathbf{S} = 0 \tag{2.3}$$

and

$$\frac{d}{dt} \oint_S \mathbf{D} \cdot d\mathbf{S} + \oint_S \mathbf{J} \cdot d\mathbf{S} = 0 \quad . \tag{2.4}$$

It follows from Eq. (2.3) that $\oint_S \mathbf{B} \cdot d\mathbf{S}$ is a constant. This integral represents the net number of magnetic lines of force crossing the closed surface S in the outward direction. It is an empirical fact that a magnetic pole cannot be isolated. Macroscopically, no matter how many times a magnet is sub-divided, north and south poles are always formed anew on every segment. More fundamentally, we may think of current loops. Each loop, however small, acts as a magnetic dipole or a short magnet. In either case each line of force leaves a magnet from the north pole and returns through the south pole forming a closed loop. This closure property of a magnetic line of force ensures that the number leaving a closed surface must equal the number entering. The constant of integration must therefore be zero, i.e.

$$\oint_S \mathbf{B} \cdot d\mathbf{S} = 0 \quad . \tag{2.5}$$

Equation (2.4) means that the conduction and the displacement current crossing a closed surface must compensate each other. By virtue of the definition of current density the surface integral of $\mathbf{J}$ represents the total conduction current leaving the volume enclosed by the closed surface. If the principle of charge conservation is to hold, this must equal the rate of decrease of the electric charge q contained in the volume, i.e.

$$\oint_S \mathbf{J} \cdot d\mathbf{S} = -\frac{dq}{dt} \quad .$$

A comparison of this equation with Eq. (2.4) shows that

$$\oint_S \mathbf{D} \cdot d\mathbf{S} = q \quad . \tag{2.6}$$

which may be recognized as Gauss' flux theorem. The constant of integration has been put equal to zero since in the static case there is no field in the absence of charge.

Equations (2.5) and (2.6) are the supplementary axioms required. Note that in these are incorporated the physical ideas of the inseparability of magnetic poles and of the conservation of electric charge.

The Four axioms presented above are known as *Maxwell's equations* in the integral form. From these the more familiar Maxwell's differential equations can be readily obtained. A general formulation of the theory must proceed from the differential equations, although for specific problems the integral equations may often turn out to be advantageous. One such problem is the determination of the boundary conditions for the field vectors, to be considered in the next section.

2.2 Boundary Conditions

The behaviour of the electromagnetic field vectors at the boundary between two media can be readily obtained using Maxwell's integral equations. These equations, however, have been postulated for regions where the vectors and their derivatives vary continuously. This condition requires the physical properties of matter also to vary continuously across the boundary from one medium to the other. To overcome the conceptual difficulty arising from the discontinuity that exists at the interface of two physical media we may imagine, as far as macroscopic phenomena are concerned, the boundary surface to be replaced by a transition layer, the thickness h of which may be made arbitrarily small, and within which the properties of matter such as permittivity and permeability vary continuously though rapidly from their values in one medium to those in the other. Consequently the field vectors and their derivatives may be assumed to be continuous. The boundary conditions are obtained by applying the integral equations to the field vectors across the transition layer, which is finally eliminated with the limiting process $h \to 0$.

A. *Magnetic Induction* **B**

Normal to the boundary between medium 1 and medium 2 draw a small right cylinder or "pill-box" of finite but small cross-sectional area A with its top and base on the surfaces of the transition layer as shown in Fig. 2.1. Apply Eq. (2.5) to the cylinder, integrating over its surfaces. Since the transition layer can be made arbitrarily thin, and as we are going to take the limit $h \to 0$ anyway, only the contribution of the end surfaces to the integral needs to be considered. Furthermore, A can be made sufficiently small so that the field over each end surface is essentially uniform. Equation (2.5) then gives

$$\mathbf{B}_2 \cdot \mathbf{A}_2 + \mathbf{B}_1 \cdot \mathbf{A}_1 = (B_{2n} - B_{1n})A = 0 \quad ,$$

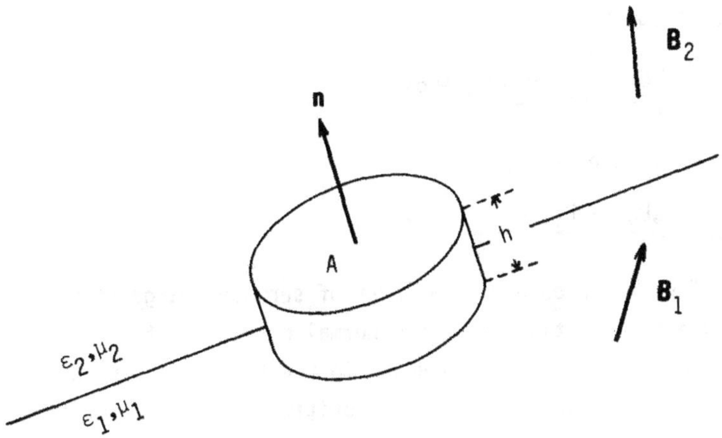

Fig. 2.1 A "pill-box" of height h equal to the thickness of the "transition layer" is drawn at the boundary between two media of constants ε_1, μ_1 and ε_2, μ_2 respectively.

where the subscript n signifies normal components, the normal being taken to direct from medium 1 to medium 2. As A is arbitrary we require that

$$B_{1n} = B_{2n} \quad . \tag{2.7}$$

The normal component of magnetic induction is therefore continuous across a surface of discontinuity.

B. *Electric Displacement* **D**

Apply Eq. (2.6) to the right cylinder of Fig. 2.1. As explained above, only the contribution of the end surfaces to the integral needs to be considered. In the limiting process of $h \to 0$, the charge q enclosed by the cylinder will become a thin layer of charge with a surface density of

$$\sigma = \frac{q}{A} \quad .$$

Equation (2.6) then gives

$$\mathbf{D}_2 \cdot \mathbf{A}_2 + \mathbf{D}_1 \cdot \mathbf{A}_1 = \sigma A$$

or, as A is arbitrary,

$$D_{2n} - D_{1n} = \sigma \quad . \tag{2.8}$$

Thus the presence of a layer of surface charge at the boundary will cause a discontinuity in the normal component of electric displacement to an extent numerically equal to the surface charge density. Note that if the surface charge is positive the normal component of the displacement vector will be greater in the medium where it directs into the medium. The normal component of **D** will of course be continuous if no surface charge is present at the boundary.

C. *Electric Intensity* **E**

Apply Eq. (2.1) to a rectangle drawn normal to the boundary with its long sides, each of length a, on the two surfaces of the transition layer, its short sides being each equal to the thickness h of the layer, as shown in Fig. 2.2(a). The length a is finite but arbitrary and can be made sufficiently small so that the field vectors do not vary appreciably along it. Furthermore, as the limiting process $h \to 0$ is to be taken in the final stage, we can neglect the contribution of the short sides to the closed line integral, which is therefore

$$\oint \mathbf{E} \cdot d\mathbf{r} = \mathbf{E}_2 \cdot \mathbf{a}_2 + \mathbf{E}_1 \cdot \mathbf{a}_1 = (E_{2t} - E_{1t})a$$

with the subscript t signifying the component tangential to the boundary surface. In the limiting process of $h \to 0$, provided $|\dot{\mathbf{B}}|$ near the boundary remains finite, the surface integral $\int \dot{\mathbf{B}} \cdot d\mathbf{S} = \dot{B}_n ha \to 0$, where the subscript n indicates the component normal to the rectangle, giving

$$E_{1t} = E_{2t} \quad . \tag{2.9}$$

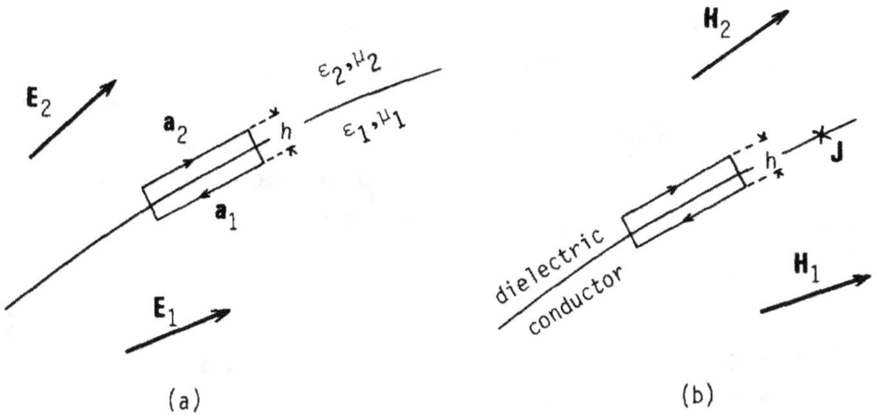

Fig. 2.2 A rectangle is drawn normal to the boundary between two media, its height h being equal to the thickness of the "transition layer". In (b), one of the media is a conductor in which a high frequency current of density **J** flows and the other is a dielectric. The rectangle is drawn normal to both the boundary surface and **J**.

Thus the tangential component of **E** is continuous across the boundary.

D. *Magnetic Intensity* **H**

Apply Eq. (2.2) to the rectangle considered above. If the conduction and displacement current densities near the boundary surface do not become infinitely large as $h \to 0$, the surface integral

$$\int (\dot{\mathbf{D}} + \mathbf{J}) \cdot d\mathbf{S} = (\dot{D}_n + J_n)ha \to 0 \quad ,$$

where n again indicates the component normal to the rectangle, giving

$$H_{1t} = H_{2t} \quad . \tag{2.10}$$

This is the case of low frequency fields. At high frequencies current becomes highly concentrated near the surface of a conducting medium and the surface integral of **J** does not tend to zero as $h \to 0$.

Consider for example the case where medium 1 is a conductor in which a high-frequency current is flowing. For a conductor, $|\dot{\mathbf{D}}|$ is much smaller than $|\mathbf{J}|$ and can be neglected (see Sec. 4.1). As shown in Fig. 2.2(b), suppose $\mathbf{J}$ is directed into the diagram. Draw a rectangle normal to the boundary and to $\mathbf{J}$ and carry out the line integration in a clockwise sense. Equation (2.2) gives

$$\mathbf{H}_2 \cdot \mathbf{a}_1 + \mathbf{H}_1 \cdot \mathbf{a}_1 = (H_{2t} - H_{1t})a = Jha \quad .$$

At high frequencies the current concentrates near the surface so that $Jh \to I_\ell$, where I_ℓ is finite, as $h \to 0$. Hence

$$H_{2t} - H_{1t} = I_\ell \quad . \tag{2.11}$$

I_ℓ measures the rate of flow of charge through a cross section of unit width and is known as the *linear current density*.

2.3 Maxwell's Equations in Differential Form

Applying Stokes' theorem to the right-hand side of Eq. (2.1) we obtain

$$\int_S \dot{\mathbf{B}} \cdot d\mathbf{S} = - \int_S \nabla \times \mathbf{E} \cdot d\mathbf{S} \quad .$$

Since S is arbitrary we require that

$$\dot{\mathbf{B}} = - \nabla \times \mathbf{E} \quad , \tag{2.12}$$

Similarly Eq. (2.2) gives

$$\dot{\mathbf{D}} + \mathbf{J} = \nabla \times \mathbf{H} \quad . \tag{2.13}$$

Applying the divergence theorem the supplementary axioms, Eqs. (2.5) and (2.6), give

$$\nabla \cdot \mathbf{B} = 0 \tag{2.14}$$

and $$\nabla \cdot \mathbf{D} = \rho \tag{2.15}$$

where ρ is the volume charge density.

The last four equations, (2.12)-(2.15), are Maxwell's electromagnetic field equations for a point in a stationary medium. The restriction to stationary media arises from the fact that convective currents and magnetic flux changes due to the motion of the medium have not been taken into account.

Maxwell's equations specify the relations among the four field vectors **E**, **B**, **D** and **H**, on the one hand and between these and the source quantities, charge density ρ and current density **J**, on the other. While the equations apply to a point in space and involve the source quantities at that point, the field vectors nevertheless represent the electric and magnetic fields due to the entire charge and current distributions. The equations are assumed to hold for every point of space, whether in vacuum or in a material medium, subject only to the usual mathematical conditions. Physically, Maxwell's equations incorporate the experimental laws of Faraday and Ampère in their general form, the inseparability of magnetic poles and the conservation of electric charge. These are the basic assumptions of Maxwell's electromagnetic theory.

Note that the choice of **E**, **B**, **D** and **H** as the basic field vectors is not a matter of necessity but of convenience only, for we can with every right replace Eqs. (2.13) and (2.15) with

$$\nabla \times \mathbf{B} = \mu_0(\varepsilon_0 \dot{\mathbf{E}} + \dot{\mathbf{P}} + \nabla \times \mathbf{M} + \mathbf{J})$$

and $$\nabla \cdot \mathbf{E} = \frac{1}{\varepsilon_0}(-\nabla \cdot \mathbf{P} + \rho)$$

in a formulation of the theory.

2.4 Electromagnetic Properties of a Material Medium

For further development of the theory it will be necessary to reduce the number of field vectors by imposing conditions on the electromagnetic properties of the medium. These conditions are called constitutive equations and involve constants which are characteristic of the medium. While in general the constitutive equations need not be simple, we shall restrict ourselves to media which have linear properties, and which are, furthermore, isotropic and homogeneous. While we would also reduce the generality of the solutions obtained, most of the materials commonly encountered do satisfy the conditions imposed to a good approximation.

The constitutive equations assumed are

$$\mathbf{D} = \varepsilon \mathbf{E} \tag{2.16}$$

and

$$\mathbf{B} = \mu \mathbf{H} \quad . \tag{2.17}$$

The parameters ε and μ are assumed to be constants and are respectively called the *permittivity* and *permeability* of the medium. Some mention of the validity of these equations has already been made in the last chapter. A useful point to note is that, electromagnetically, the atmosphere to a large extent acts as free space.

We shall also assume a linear relationship between **J** and **E** when the occasion arises:

$$\mathbf{J} = \sigma \mathbf{E} \tag{2.18}$$

where σ is a constant called the *conductivity* of the medium. This relation is known as Ohm's law and applies to macroscopic events alone, even then often approximately only. It fails completely in atomic phenomena such as Ampère's microscopic current loops or the orbiting electrons in an atom and in such quantum effects as superconductivity.

Unless otherwise stated we shall confine our attention to linear, isotropic and homogeneous media only, for which Eqs. (2.16)-(2.18) hold, throughout this book. It should be particularly stressed that the material constants are in general frequency-dependent and we shall often restrict ourselves to phenomena in free space to avoid any complication that may arise thereby.

2.5 Vector and Scalar Potentials

Maxwell's equations relate the derivatives of the field vectors to the source quantities, the charge and current densities. We are interested in a solution which expresses each field vector in terms of the charge and current distributions. To achieve this aim we first note that the solution for the magnetic induction **B** must satisfy Eq. (2.14), i.e. its divergence must vanish everywhere. Since the divergence of the curl of any vector function vanishes identically, the equation is satisfied automatically if **B** is expressed as the curl of a vector function $\mathbf{A}(\mathbf{r}, t)$:

$$\mathbf{B} = \nabla \times \mathbf{A} \quad . \tag{2.19}$$

A is called the *vector potential*. It is, however, not uniquely defined by Eq. (2.19) for a given magnetic induction **B**. For if $\mathbf{A}_0$ is a solution, then

$$\mathbf{A} = \mathbf{A}_0 - \nabla\psi \quad , \tag{2.20}$$

where $\psi(\mathbf{r}, t)$ is an arbitrary scalar function, is also a solution since

$$\nabla \times \nabla\psi = 0 \tag{2.21}$$

identically.

Expressing **B** in Eq. (2.12) in terms of the vector potentials **A** and $\mathbf{A}_0$, we obtain

$$\nabla \times (\mathbf{E} + \dot{\mathbf{A}}) = 0$$

and

$$\nabla \times (\mathbf{E} + \dot{\mathbf{A}}_0) = 0$$

respectively. These can be automatically satisfied by virtue of Eq. (2.21) if we express $\mathbf{E} + \dot{\mathbf{A}}$ and $\mathbf{E} + \dot{\mathbf{A}}_0$ as gradients of scalar functions $\Phi(\mathbf{r}, t)$ and $\Phi_0(\mathbf{r}, t)$. Hence, **E** can be expressed as

$$\mathbf{E} = -\dot{\mathbf{A}} - \nabla\Phi \tag{2.22}$$

or

$$\mathbf{E} = -\dot{\mathbf{A}}_o - \nabla\Phi_o \quad . \tag{2.23}$$

Subtracting, we obtain

$$\frac{\partial}{\partial t}(\mathbf{A} - \mathbf{A}_o) = \nabla(\Phi_o - \Phi)$$

or, using Eq. (2.20),

$$\nabla\dot{\psi} = \nabla(\Phi - \Phi_o)$$

giving

$$\Phi = \Phi_o + \dot{\psi} \quad . \tag{2.24}$$

The function Φ and Φ_o are called *scalar potentials.*

In the above it has been shown that the given electromagnetic field vectors $\mathbf{E}$ and $\mathbf{B}$ can be expressed in terms of a pair of vector and scalar potentials $\mathbf{A}$ and Φ. These potentials however are not unique. Another pair of potentials $\mathbf{A}_o$ and Φ_o will give rise to the same electromagnetic field provided they are related to $\mathbf{A}$ and Φ by Eq. (2.20) and (2.24) through an arbitrary scalar function ψ. The infinite number of possible choices of ψ thus corresponds to an infinite number of possible pairs of potentials, all leading to the same field. Once one such pair of potentials is found for the given charge and current configuration, the field vectors can be obtained by straight differentiation.

To obtain the differential equations for the vector and scalar potentials we shall confine ourselves to a linear, isotropic and homogeneous medium so that the number of independent field vectors may be reduced by two through the constitutive equations. Let the medium have permittivity ε and permeability μ, and write the remaining Maxwell's equations as

$$\mu\varepsilon\dot{\mathbf{E}} + \mu\mathbf{J} = \nabla \times \mathbf{B} \tag{2.25}$$

and
$$\nabla \cdot \mathbf{E} = \frac{\rho}{\varepsilon} \quad . \tag{2.26}$$

Substituting in the above the expressions for **E** and **B** in terms of the potentials we obtain

$$\nabla^2\mathbf{A} - \mu\varepsilon\ddot{\mathbf{A}} + \mu\mathbf{J} = \nabla(\nabla \cdot \mathbf{A} + \mu\varepsilon\dot{\Phi}) \tag{2.27}$$

and
$$\nabla \cdot \dot{\mathbf{A}} + \nabla^2\Phi = -\frac{\rho}{\varepsilon} \quad . \tag{2.28}$$

These are the differential equations for the vector and scalar potentials **A** and Φ for charge and current distributions specified by the densities $\rho(\mathbf{r}, t)$ and $\mathbf{J}(\mathbf{r}, t)$. All particular solutions of these equations lead to the same electromagnetic field when subject to identical boundary conditions. Thus, while an infinite number of such solutions are allowed, nothing more can be learned after a particular pair of potentials has been found. This being so, we are at liberty to choose from the possible solutions potentials for which the differential equations assume the simplest forms. This is done by imposing a suitable *gauge condition* to the potentials.

A gauge condition commonly adopted is the *Lorentz condition*

$$\nabla \cdot \mathbf{A} + \mu\varepsilon\dot{\Phi} = 0 \quad . \tag{2.29}$$

How potentials can be chosen to satisfy this condition can be seen in the following. Suppose Φ_0 and $\mathbf{A}_0$ are a pair of potentials satisfying the differential equations (2.27) and (2.28); choose for the arbitrary function ψ the solution of the following equation

$$\nabla \cdot (\mathbf{A}_0 - \nabla\psi) + \mu\varepsilon \frac{\partial}{\partial t}(\Phi_0 + \dot{\psi}) = 0 \quad ,$$

or
$$\nabla^2\psi - \mu\varepsilon\ddot{\psi} = \nabla \cdot \mathbf{A}_0 + \mu\varepsilon\dot{\Phi}_0 \quad . \tag{2.30}$$

The new pair of potentials $\mathbf{A}$ and Φ obtained from $\mathbf{A}_0$, Φ_0 and ψ will then satisfy the Lorentz condition. The imposition of the Lorentz condition reduces the differential equations for the potentials to the simpler inhomogeneous wave equations

$$\nabla^2\mathbf{A} - \mu\varepsilon\ddot{\mathbf{A}} = -\mu\mathbf{J} \tag{2.31}$$

and

$$\nabla^2\Phi - \mu\varepsilon\ddot{\Phi} = -\frac{\rho}{\varepsilon} \quad . \tag{2.32}$$

It should be noted that even with the restriction imposed by the Lorentz condition the remaining potentials, which are solutions of the inhomogeneous wave equations, still are not unique; for if we take the solution of the homogeneous wave equation

$$\nabla^2\chi - \mu\varepsilon\ddot{\chi} = 0$$

and replace $\mathbf{A}$ by $\mathbf{A} - \nabla\chi$, Φ by $\Phi + \dot{\chi}$, the new potentials will again satisfy Eqs. (2.31) and (2.32).

The different possible choices which we can make for the potentials leaving the electromagnetic field unchanged are called *gauges*. The transformation from one set of potentials to another by means of Eqs. (2.20) and (2.24) is called *gauge transformation*. The above discussion shows that the electromagnetic field is invariant under gauge transformation. The various gauges are specified by the gauge conditions imposed. The gauge specified by the Lorentz condition is called the *Lorentz gauge*. The Lorentz gauge is useful in that in this gauge the scalar and vector potentials satisfy separate differential equations.

Another important gauge is the *Coulomb gauge* specified by the condition

$$\nabla \cdot \mathbf{A} = 0 \quad ,$$

in which the potentials are determined by the differential equations

$$\nabla^2\mathbf{A} - \mu\varepsilon\ddot{\mathbf{A}} + \mu\mathbf{J} = \mu\varepsilon\nabla\dot{\Phi}$$

and
$$\nabla^2\Phi = -\frac{\rho}{\varepsilon} \quad .$$

The last equation is simply the Poisson equation in electrostatics (Sec. 3.2). This means that the scalar potential is determined from the charges as if they were stationary. Hence the Coulomb gauge is useful in problems in which electromagnetic interactions among discrete charges such as nuclear particles are involved. Throughout this book we shall however keep to the Lorentz gauge exclusively.

Returning to the Lorentz gauge, we note that while $\mathbf{A}$ is given in terms of the current distribution specified by $\mathbf{J}(\mathbf{r}, t)$ and Φ is given by the charge distribution specified by $\rho(\mathbf{r}, t)$ in the inhomogeneous wave equations (2.31) and (2.32), they are nevertheless related through the Lorentz condition Eq. (2.29). This is to be expected since $\mathbf{J}$ and ρ are themselves related through the continuity equation expressing charge conservation. In fact it can be readily shown that the gauge condition is itself merely a re-statement of the principle of charge conservation. For if we take the divergence of Eq. (2.31), apply the operator $\mu\varepsilon \frac{\partial}{\partial t}$ to Eq. (2.32) and add the resulting equations we shall obtain

$$\nabla^2(\nabla\cdot\mathbf{A} + \mu\varepsilon\dot{\Phi}) - \mu\varepsilon\frac{\partial^2}{\partial t^2}(\nabla\cdot\mathbf{A} + \mu\varepsilon\dot{\Phi}) = -\mu(\nabla\cdot\mathbf{J} + \dot{\rho}) \quad .$$

The assumption of the Lorentz condition in the above at once gives

$$\nabla\cdot\mathbf{J} + \dot{\rho} = 0 \quad .$$

Exactly the same situation exists in the Coulomb gauge. Since the vector and the scalar potential are not completely independent, we may express one in terms of the other, e.g.

$$\dot{\Phi} = -\frac{1}{\mu\varepsilon}\nabla\cdot\mathbf{A} \quad .$$

Expressions for the electric and the magnetic field can therefore be written in terms of one potential only. This is done, for instance, by introducing the *Hertz vector* or *polarization potential* $\boldsymbol{\pi}$, in terms of which we have

$$\mathbf{A} = \mu\varepsilon \frac{\partial \boldsymbol{\pi}}{\partial t} \qquad \text{and} \qquad \Phi = -\nabla \cdot \boldsymbol{\pi} \quad .$$

In the above, we have expressed the field vectors $\mathbf{E}$ and $\mathbf{B}$ in terms of the vector and scalar potentials $\mathbf{A}$ and Φ, which are respectively solutions of the inhomogeneous wave Eqs. (2.31) and (2.32). Usually, we are required to find the electric and magnetic fields set up by given distributions of charges and currents with certain boundary conditions on given surfaces. The problem is equivalent to solving the differential equations and selecting the particular solutions which satisfy the boundary conditions specified.

2.6 The Poynting Theorem and Energy Conservation

Adding the scalar products of $\mathbf{H}$ with Eq. (2.12) and of $\mathbf{E}$ with Eq. (2.13) we obtain

$$\mathbf{E}\cdot\dot{\mathbf{D}} + \mathbf{H}\cdot\dot{\mathbf{B}} + \mathbf{E}\cdot\mathbf{J} = \mathbf{E}\cdot\nabla\times\mathbf{H} - \mathbf{H}\cdot\nabla\times\mathbf{E} = -\nabla\cdot(\mathbf{E}\times\mathbf{H}) \quad .$$

If the medium is linear and isotropic, then

$$\mathbf{E}\cdot\dot{\mathbf{D}} = \frac{1}{2}\frac{\partial}{\partial t}(\mathbf{E}\cdot\mathbf{D}) \quad , \qquad \mathbf{H}\cdot\dot{\mathbf{B}} = \frac{1}{2}\frac{\partial}{\partial t}(\mathbf{H}\cdot\mathbf{B}) \quad ,$$

and the above becomes

$$\nabla\cdot(\mathbf{E}\times\mathbf{H}) + \frac{\partial}{\partial t}\left(\frac{1}{2}\,\mathbf{E}\cdot\mathbf{D} + \frac{1}{2}\,\mathbf{H}\cdot\mathbf{B}\right) + \mathbf{E}\cdot\mathbf{J} = 0 \quad . \qquad (2.33)$$

This relation is known as the *Poynting theorem* and the vector

$$\mathbf{N} = \mathbf{E}\times\mathbf{H} \qquad (2.34)$$

is called the *Poynting vector*.

For the interpretation of the Poynting theorem we consider first the case of $\mathbf{J} = 0$ in the region of interest, e.g. in a perfect dielectric. The theorem now takes the form of a continuity equation

$$\nabla \cdot (\mathbf{E} \times \mathbf{H}) + \frac{\partial}{\partial t} \left(\frac{1}{2} \mathbf{E} \cdot \mathbf{D} + \frac{1}{2} \mathbf{H} \cdot \mathbf{B}\right) = 0$$

and must therefore express the conservation of some physical quantity. Integrating the above over an arbitrary volume V of surface S and applying the divergence theorem, we obtain

$$-\frac{d}{dt}\left(\int_V \frac{1}{2} \mathbf{E} \cdot \mathbf{D}\, dV + \int_V \frac{1}{2} \mathbf{H} \cdot \mathbf{B}\, dV\right) = \oint_S \mathbf{E} \times \mathbf{H} \cdot d\mathbf{S} \quad .$$

In static fields it can be demonstrated (see Sec. 3.6 and Sec. 4.3) that the first and second volume integrals on the left-hand side represent respectively the electric and magnetic field energies stored in the volume V. Maxwell suggested that they can be considered to represent the electric and magnetic field energies in general. With acceptance of this view, the left-hand side then represents the rate of decrease of the electromagnetic field energy contained in the volume. It follows that the right-hand side must be interpreted as the rate at which the energy escapes through the boundary surface of the volume if the principle of energy conservation is to hold. The Poynting vector $\mathbf{N}$ must then be interpreted as the energy flux density, i.e. the energy crossing unit area normal to the direction of flow per unit time.

It was further postulated by Maxwell that electric and magnetic energies are in fact localized in an electromagnetic field, so that $\frac{1}{2}\mathbf{E} \cdot \mathbf{D}$ and $\frac{1}{2}\mathbf{H} \cdot \mathbf{B}$ respectively give the electric and magnetic energy densities for every volume element of the medium. This would allow us to interpret the continuity equation as expressing energy balance at every point of the field.

The remaining term $\mathbf{E} \cdot \mathbf{J}$ in the Poynting theorem can be written as

$$\mathbf{E}\cdot\mathbf{J} = \mathbf{E}\cdot\rho\mathbf{u} = \mathbf{F}\cdot\mathbf{u}$$

when current is considered as flow of charge. Since $\mathbf{F}$ is the electric force acting on the charge ρ contained in a unit volume and $\mathbf{u}$ is the velocity of the charge, the expression $\mathbf{E}\cdot\mathbf{J}$ represents the work done on the charge per unit time per unit volume by the field. It includes the work done against any back electromotive force which may be present as well as the irreversible Joule heat loss, and represents the dissipation of energy per unit time per unit volume of the field. The physical significance of $\mathbf{E}\cdot\mathbf{J}$ strengthens the belief that $\frac{1}{2}\mathbf{E}\cdot\mathbf{D}$ and $\frac{1}{2}\mathbf{H}\cdot\mathbf{B}$ represent field energy densities.

Thus the Poynting theorem expresses energy balance in an electromagnetic field, stating that, for a volume element, the outward flow of energy per unit time plus the power expended in moving the charge contained therein is exactly equal to the rate of decrease of the stored energy.

2.7 Wave Equations

Maxwell's equations for a linear, isotropic and homogeneous medium with permittivity ε and permeability μ are

$$\mu\dot{\mathbf{H}} = -\nabla\times\mathbf{E} \quad , \tag{2.35}$$

$$\varepsilon\dot{\mathbf{E}} + \mathbf{J} = \nabla\times\mathbf{H} \quad , \tag{2.36}$$

$$\nabla\cdot\mathbf{H} = 0 \tag{2.37}$$

and

$$\nabla\cdot\mathbf{E} = \frac{\rho}{\varepsilon} \quad . \tag{2.38}$$

Taking the curl of Eq. (2.35) we have

$$\mu\frac{\partial}{\partial t}(\nabla\times\mathbf{H}) = \nabla^2\mathbf{E} - \nabla(\nabla\cdot\mathbf{E}) \quad ,$$

or, making use of Eq. (2.36),

$$\nabla^2\mathbf{E} - \mu\varepsilon\ddot{\mathbf{E}} - \mu\dot{\mathbf{J}} - \nabla(\nabla\cdot\mathbf{E}) = 0 \quad .$$

If we confine ourselves to a charge-free region[a] where $\rho = 0$, then $\nabla\cdot\mathbf{E} = 0$ by virtue of Eq. (2.38). Eliminating $\mathbf{J}$ by means of Ohm's law, Eq. (2.18), the above equation becomes

$$\nabla^2\mathbf{E} - \mu\varepsilon\ddot{\mathbf{E}} - \mu\sigma\dot{\mathbf{E}} = 0 \quad . \tag{2.39}$$

Similar elimination of $\mathbf{E}$ from Eqs. (2.35) and (2.36) gives

$$\nabla^2\mathbf{H} - \mu\varepsilon\ddot{\mathbf{H}} - \mu\sigma\dot{\mathbf{H}} = 0 \quad . \tag{2.40}$$

The last two equations are known as the *general wave equation* and describe the propagation of an electromagnetic wave in a charge-free region. However, since the field vectors must satisfy Maxwell's equations their solutions are not independent. In practice, to obtain consistent solutions only one wave equation is solved and the other field vector found from the solution using the appropriate Maxwell's equation.

In a perfect dielectric the conductivity σ is zero and the general wave equation becomes the equation of wave propagation without absorption:

$$(\nabla^2 - \mu\varepsilon\frac{\partial^2}{\partial t^2})\mathbf{E} = 0$$

or

$$(\nabla^2 - \mu\varepsilon\frac{\partial^2}{\partial t^2})\mathbf{H} = 0 \quad .$$

On the other hand, if the medium is a good conductor, the second term of Eq. (2.39) can be neglected and the equation becomes a diffusion equation. Wave propagation now will not be possible and the penetration of the electromagnetic field into the medium is similar to the diffusion of heat in matter.

---

[a] Unless stated otherwise, ρ and $\mathbf{J}$ are to be treated formally as independent quantities in the formulation.

It may be noted that the second and the third term of the general wave equation arise respectively from the displacement and the conduction currents. Their relative importance will depend on the frequency of oscillation of the electromagnetic field. Consider for example a field whose time variation is sinusoidal:

$$\mathbf{E}(\mathbf{r}, t) = \mathbf{E}_0(\mathbf{r})e^{-i\omega t} \quad . \tag{2.41}$$

Substitution in the general wave equation gives

$$\nabla^2\mathbf{E}_0 + \left(1 + \frac{i}{\omega\tau}\right)\mu\varepsilon\omega^2\mathbf{E}_0 = 0 \quad ,$$

where $$\tau = \frac{\varepsilon}{\sigma} \quad . \tag{2.42}$$

The parameter τ has the dimensions of time and is called the *relaxation time* of the medium which may be considered as the time required for the medium to return to its normal state after an external excitation. If the relaxation time is much greater than the period of field oscillation, i.e. $\tau \gg 2\pi/\omega$, propagation conditions prevail. If $\tau \ll 2\pi/\omega$, diffusion conditions predominate. For pure metals, $\tau \sim 10^{-14}$s and diffusion conditions prevail for all frequencies lower than the optical frequencies. In an actual dielectric, a small conductivity is always present and the solution of the general wave equation is a propagating wave with a small attenuation factor in its amplitude. The absorption arises from the finite value of σ, indicating the conversion of energy into heat through Joule's effect.

2.8 Plane Electromagnetic Waves

An important example of the solution of the wave equation is one that represents plane electromagnetic waves, because radio waves at sufficient distances from the antenna and the ground and light waves at large distances from a point source are approximately plane waves. Furthermore any wave can be Fourier-analysed in terms of plane waves. An electromagnetic wave is said to be plane if at any instant its fields on each plane perpendicular to the direction of propagation are uniform and are functions only of the distance of the plane from some fixed origin on the path of propagation[b].

Take the x-axis of a Cartesian system along the direction of propagation. The conditions of plane waves require that $\mathbf{E} = \mathbf{E}(x,t)$ and $\mathbf{H} = \mathbf{H}(x, t)$, from which it follows that

$$\frac{\partial E_i}{\partial x_2} = \frac{\partial E_i}{\partial x_3} = 0 \quad , \quad \frac{\partial H_i}{\partial x_2} = \frac{\partial H_i}{\partial x_3} = 0 \quad ,$$

$$\text{for} \quad i = 1,2,3 \quad .$$

Maxwell's equations for a linear, isotropic, homogeneous and charge-free medium then give for the x-components of the field vectors

$$\mu \frac{\partial H_1}{\partial t} = - \varepsilon_{1jk} \frac{\partial E_k}{\partial x_j} = 0 \quad . \tag{2.43}$$

$$\varepsilon \frac{\partial E_1}{\partial t} + \sigma E_1 = \varepsilon_{1jk} \frac{\partial H_k}{\partial x_j} = 0 \quad . \tag{2.44}$$

$$\frac{\partial H_i}{\partial x_i} = \frac{\partial H_1}{\partial x_1} = 0 \tag{2.45}$$

and

$$\frac{\partial E_i}{\partial x_i} = \frac{\partial E_1}{\partial x_1} = 0 \quad . \tag{2.46}$$

Equations (2.43) and (2.45) show that H_1 is uniform and stationary. Equations (2.44) and (2.46) show that E_1 is uniform in spatial coordinates but may have an exponential time factor $\exp(-t/\tau)$, where τ is the relaxation time, if the medium is not a perfect

---

[b] Strictly speaking such a wave whose amplitude does not vary over a plane transverse to the direction of propagation should be called a uniform or homogeneous plane wave. An example of an inhomogeneous plane wave is the plane electromagnetic wave in a conductor (Sec. 5.4).

dielectric. In either case we see that the longitudinal components of the field are uniform and constant, with at most an exponential time-decay factor, and cannot be part of a wave motion. A plane electromagnetic wave must therefore be transverse.

Consider the wave equation

$$\nabla^2 \mathbf{E} - \mu\varepsilon \frac{\partial^2 \mathbf{E}}{\partial t^2} = 0 \quad . \tag{2.47}$$

For a plane wave solution, it has only two non-trivial components:

$$\frac{\partial^2 E_2}{\partial x^2} - \mu\varepsilon \frac{\partial^2 E_2}{\partial t^2} = 0 \tag{2.48}$$

$$\frac{\partial^2 E_3}{\partial x^2} - \mu\varepsilon \frac{\partial^2 E_3}{\partial t^2} = 0 \tag{2.49}$$

As E_2 and E_3 are completely independent of each other[c], these equations may be solved separately. Consider for example Eq. (2.48). It's general solution is

$$E_2 = f_2(x - vt) + g_2(x + vt) \tag{2.50}$$

where f_2 and g_2 are arbitrary functions of $(x - vt)$ and $(x + vt)$ respectively, and $v = (\mu\varepsilon)^{-\frac{1}{2}}$, as can be seen by substitution.

Consider the set of solutions $f_2(x - vt)$. The points on a plane at x perpendicular to the x-axis all have the same phase

$$x - vt = \text{constant} \quad .$$

Whatever value the field takes on this plane at time t, it will

---

[c] The components of $\mathbf{E}$ are in general related to one another through Eq. (2.38). However, for a plane wave $\mathbf{E} = \mathbf{E}(x, t)$, $\nabla \cdot \mathbf{E} = \partial E_1/\partial x_1$, so that E_2 and E_3 are completely unrelated.

take on the same value at time $t+dt$ on the plane $x+dx$, where $x - vt = x + dx - v(t+dt)$,

or
$$\frac{dx}{dt} = v = \frac{1}{\sqrt{\mu\varepsilon}} \quad . \tag{2.51}$$

f_2 therefore represents a disturbance which travels without change of form in the positive direction of the x-axis with a velocity v, called the *phase velocity*. Similarly, g_2 represents a wave travelling in the negative x-direction with a phase velocity of the same magnitude. The surface of all the points with the same phase is called a wave front. For plane waves the wave fronts are plane.

A wave equation identical to Eq. (2.47) exists for the associated magnetic intensity **H** and its solution will have the same form as for the electric intensity **E**. However, as remarked before, to find the associated magnetic intensity it is more convenient to make use of Maxwell's Eq. (2.36), noting that in a perfect dielectric **J** = 0. Its y-component equation is

$$\varepsilon \frac{\partial E_2}{\partial t} = \varepsilon_{2jk} \frac{\partial H_k}{\partial x_j} = \varepsilon_{213} \frac{\partial H_3}{\partial x_1} + \varepsilon_{231} \frac{\partial H_1}{\partial x_3} = -\frac{\partial H_3}{\partial x_1} \quad , \tag{2.52}$$

showing that the magnetic field associated with E_2 has only one component, the z-component. As both E_2 and H_3 have functional arguments $u = x \mp vt$, we find

$$\frac{\partial}{\partial t} = \frac{\partial u}{\partial t}\frac{d}{du} = \mp v \frac{d}{du} \quad ,$$

$$\frac{\partial}{\partial x} = \frac{\partial u}{\partial x}\frac{d}{du} = \frac{d}{du} \quad ,$$

and Eq. (2.52) becomes

$$\frac{dH_3}{du} = \pm \, v\varepsilon \, \frac{dE_2}{du} = \pm \sqrt{\frac{\varepsilon}{\mu}} \, \frac{dE_2}{du} \quad .$$

We integrate both sides, neglecting any possible constant and uniform fields, and obtain

$$H_3 = \pm \sqrt{\frac{\varepsilon}{\mu}} \, E_2 \quad .$$

This can be written in the vector form

$$\mathbf{H} = \sqrt{\frac{\varepsilon}{\mu}} \, \hat{\mathbf{k}} \times \mathbf{E} \quad , \tag{2.53}$$

where $\hat{\mathbf{k}}$ is a unit vector in the direction of propagation.

In a similar way, we solve Eq. (2.49) and obtain the plane wave solution with **E** in the direction of the z-axis:

$$E_3 = f_3(x - vt) + g_3(x + vt) \quad , \tag{2.54}$$

where f_3 and g_3 are arbitrary functions of the arguments indicated. The associated magnetic intensity is now in the y-direction, again given by Eq. (2.53). This relation which relates the relative magnitudes of **E** and **H** is characteristic of plane electromagnetic waves in general.

Each of the solutions Eq. (2.50) and (2.54) represents plane waves whose electric and magnetic fields have fixed directions. Such waves are said to be *linearly* or *plane polarized.* We shall adopt the convention of referring to the plane containing **E** as the *plane of polarization.* The direction of **H** is perpendicular to this plane. The plane wave solution of the wave equation (2.47) consists of two linearly polarized waves with mutually perpendicular planes of polarization. For each polarization, **E**, **H** and $\hat{\mathbf{k}}$ are mutually perpendicular, forming a right-handed set in the order given. The direction of propagation $\hat{\mathbf{k}}$ therefore coincides with the direction of the electromagnetic energy flux density represented by the Poynting vector $\mathbf{N} = \mathbf{E} \times \mathbf{H}$.

A plane wave is unpolarized or randomly polarized if the direction of $\mathbf{E}$, and hence of $\mathbf{H}$, at any instant at any point of the path of propagation is completely random. The wave Eq. (2.47) must still be satisfied and the analysis that follows still applies. Since $\mathbf{E}$ now has equal probability of being in any one direction in the yz-plane, its components E_2 and E_3 in the y- and z-directions respectively must have on the average the same magnitude. In other words, $\mathbf{E}$ is to be represented by components E_2 and E_3 having the same amplitude and a completely random phase relationship. Thus we can regard an unpolarized plane wave as being composed of two independent linearly polarized components with equal amplitudes and mutually perpendicular planes of polarization.

The most general solution of the wave equation (2.47) corresponding to plane waves travelling along the x-axis is a superposition of the two solutions given by Eqs. (2.50) and (2.54), each consisting of a wave travelling in the positive x-direction and one travelling in the negative x-direction. Whether any of these components will be present depends on the initial and boundary conditions.

The energy density of the propagating electromagnetic field is by definition

$$U = \frac{1}{2}(\mathbf{E}\cdot\mathbf{D} + \mathbf{H}\cdot\mathbf{B}) = \frac{1}{2}(\varepsilon E^2 + \mu H^2) = \varepsilon E^2 = \mu H^2 \tag{2.55}$$

by virtue of Eq. (2.53). Note that the energy carried by a plane wave is equally shared between the electric and the magnetic fields.

The electromagnetic energy flux density is

$$\mathbf{N} = \sqrt{\frac{\varepsilon}{\mu}}\,\mathbf{E}\times(\hat{\mathbf{k}}\times\mathbf{E}) = \sqrt{\frac{\varepsilon}{\mu}}\,E^2\hat{\mathbf{k}} = \varepsilon E^2\mathbf{v} = U\mathbf{v} \quad , \tag{2.56}$$

showing that the energy is propagated with the phase velocity of the waves. This is however not generally true since for a dispersive medium the phase velocity depends on frequency and it is not usually possible to generate waves of a single frequency only. Energy will then be propagated with the group velocity of the waves (Sec. 6.8).

It should be noted that the above expressions give the instantaneous energy density and energy flux density. Since in any measurement the experimental time is much greater than the time for one oscillation the relevant quantities are the mean values over one oscillation:

$$\langle U \rangle = \varepsilon \langle E^2 \rangle = \mu \langle H^2 \rangle \quad , \tag{2.57}$$

$$\langle \mathbf{N} \rangle = \langle U \rangle \, \mathbf{v} \quad . \tag{2.58}$$

Maxwell was the first to point out that the speed of propagation of electromagnetic waves in vacuum is the same as that of light. These may be approximated by their values in air. Recent measurements give 299792.4 ± 0.5km s^{-1} for the speed of optical light in air, 299795.1 ± 1.9km s^{-1} for the speed of electromagnetic radiation in vacuum[d], while the formula $c = (\mu_0 \varepsilon_0)^{-\frac{1}{2}}$ gives 299784 ± 10km s^{-1}. These values agree within the experimental errors indicated.

The ratio of the magnitude of **E** to that of **H**, $(\mu/\varepsilon)^{\frac{1}{2}}$, has the dimensions of an impedance and is known as the *intrinsic impedance* or the *wave resistance* of the medium. For vacuum the intrinsic impedance has the value $(\mu_0/\varepsilon_0)^{\frac{1}{2}} = 120\pi$ ohms.

---

[d] Actually the speed of light in vacuum is known to a few parts in 10^9. Its value depends on the definitions of units of length and time. Defining these separately in terms of two different atomic transitions, $c = 299{,}792{,}456.2 \pm 1.1$ ms^{-1}, as given by K. Evenson *et al.*, *Phys. Rev. Lett.* **29**, 1346 (1972). There is no evidence of a frequency dependence of the speed of light in vacuum.

PROBLEMS

1(a). The electric intensities $\mathbf{E}_1$, $\mathbf{E}_2$ in media 1 and 2 near the boundary make angles θ_1, θ_2 respectively with the normal to the boundary surface. Show that "Snell's law"

$$|\mathbf{E}_1|\sin\theta_1 = |\mathbf{E}_2|\sin\theta_2$$

applies.

(b). If one medium is a perfect conductor, find by means of the boundary conditions the electric intensity near its surface.

2. A stationary current of density $\mathbf{J}$ crosses a surface S separating two media of permittivities ε_1, ε_2 and conductivities σ_1, σ_2 respectively. Find the boundary condition for $\mathbf{J}$. Hence show that if the line of flow makes angles θ_1 and θ_2 with the normal to S in the respective media, then

$$\frac{\tan\theta_1}{\tan\theta_2} = \frac{\sigma_1}{\sigma_2} \quad .$$

Show also that a layer of surface charge will in general appear on S and find its density.

3. Show that the density of any excess charge introduced in a linear, isotropic and homogeneous medium of permittivity ε and conductivity σ will decay exponentially with a time factor $\exp(-\sigma t/\varepsilon)$.

If the medium is not homogeneous show that the charge density ρ is given by the equation

$$\varepsilon\frac{\partial\rho}{\partial t} + \sigma\rho = \sigma\,\mathbf{J}\cdot\nabla\left(\frac{\varepsilon}{\sigma}\right) \quad ,$$

where $\mathbf{J}$ is the current density. Find the steady-state charge density.

4. Show that when the magnetic induction changes at the rate $\dot{\mathbf{B}}(\mathbf{r}', t)$ an electric field is induced whose intensity is

$$\mathbf{E}(\mathbf{r}, t) = \frac{1}{4\pi} \int \frac{\bar{\mathbf{r}} \times \dot{\mathbf{B}}(\mathbf{r}', t)}{\bar{r}^3} dV' \quad ,$$

where $\bar{\mathbf{r}} = \mathbf{r} - \mathbf{r}'$ and the integral is over all space.
Hint: the left-hand side can be written as an integral

$$\int \mathbf{E}(\mathbf{r}', t)\delta(\bar{\mathbf{r}})dV' \quad .$$

5. In a linear but anisotropic medium the components of **E, B, D** and **H** are related through the following linear equations:

$$D_i = \varepsilon_{ij}E_j \quad , \qquad B_i = \mu_{ij}H_j \quad ,$$

where $\varepsilon_{ij} = \varepsilon_{ji}$, $\mu_{ij} = \mu_{ji}$, being components of symmetric tensors. Show that for such a medium

$$\mathbf{E} \cdot \dot{\mathbf{D}} = \frac{1}{2}\frac{\partial}{\partial t}(\mathbf{E} \cdot \mathbf{D}) \quad , \qquad \mathbf{H} \cdot \dot{\mathbf{B}} = \frac{1}{2}\frac{\partial}{\partial t}(\mathbf{H} \cdot \mathbf{B}) \quad .$$

6. A stationary field has uniform electric intensity **E** and uniform magnetic induction **B**. Show that the vector and scalar potentials at a point **r** may be expressed as

$$\mathbf{A} = \frac{1}{2}(\mathbf{B} \times \mathbf{r}) \quad , \qquad \nabla\Phi = -\mathbf{E} \cdot \mathbf{r} \quad .$$

7. A region of space has $\rho = 0$, $\mathbf{J} = 0$, $\mu = \mu_0$ and polarization $\mathbf{P}(\mathbf{r}, t)$. Show that the region can be considered as free space with charge density $\rho' = -\nabla \cdot \mathbf{P}$ and current density $\mathbf{J}' = \dot{\mathbf{P}}$. Hence show that the electric intensity **E** and magnetic induction **B** may be expressed in terms of a single vector, the Hertz vector $\boldsymbol{\pi}$, by

$$\mathbf{E} = \nabla \times (\nabla \times \boldsymbol{\pi}) - \frac{\mathbf{P}}{\varepsilon_0} \quad , \qquad \mathbf{B} = \mu_0\varepsilon_0 \nabla \times \dot{\boldsymbol{\pi}} \quad ,$$

where π is given by

$$\nabla^2\pi - \mu_0\varepsilon_0\ddot{\pi} = -\frac{\mathbf{P}}{\varepsilon_0} \quad ,$$

How are the vector and scalar potentials expressed in terms of π?

8. The charge and current distributions in a linear, isotropic and homogeneous medium of permittivity ε and permeability μ are given by the densities $\rho(\mathbf{r}, t)$ and $\mathbf{J}(\mathbf{r}, t)$. Show that a vector $\mathbf{P}(\mathbf{r}, t)$ can be found such that

$$\nabla \cdot \mathbf{P} = -\rho \quad , \quad \dot{\mathbf{P}} = \mathbf{J} \quad .$$

Hence show that Maxwell's equations are correctly obtained from a Hertz vector π satisfying

$$\nabla^2\pi - \mu\varepsilon\ddot{\pi} = -\frac{\mathbf{P}}{\varepsilon}$$

through

$$\mathbf{E} = \nabla\times(\nabla\times\pi) - \frac{\mathbf{P}}{\varepsilon} \quad , \quad \mathbf{B} = \mu\varepsilon\nabla\times\dot{\pi} \quad .$$

9. Show that in a linear, isotropic and homogeneous medium of permittivity ε and permeability μ Maxwell's equations can be written as two complex equations in terms of a complex vector

$$\mathbf{F} = \mathbf{B} + i\sqrt{\mu\varepsilon}\,\mathbf{E}$$

and find these equations. How can $\mathbf{F}$ be expressed in terms of the scalar and vector potentials?

If there is no charge or current present show that $\mathbf{F}$ can be expressed in terms of a Hertz vector π through

$$\mathbf{F} = \mu\varepsilon\nabla\times\dot{\pi} + i\sqrt{\mu\varepsilon}\,\nabla\times(\nabla\times\pi) \quad .$$

10. A sinusoidal plane electromagnetic wave in a linear, isotropic and homogeneous medium of permittivity ε and permeability μ can be represented by

$$\mathbf{E} = \mathbf{E}_0 \exp\{i(\mathbf{k}\cdot\mathbf{r} - \omega t)\}$$

where $\mathbf{E}$ is its electric field intensity at a point $\mathbf{r}$ on the path of propagation, ω the angular frequency, and $\mathbf{k}$ the propagation vector defined as a vector of magnitude $\omega\sqrt{\mu\varepsilon}$ along the direction of propagation. Show that this function satisfies the wave equation

$$\nabla^2\mathbf{E} - \mu\varepsilon\ddot{\mathbf{E}} = 0$$

and find the associated magnetic field intensity.

If the electric intensity is given by the real part of $\mathbf{E}$ find expressions for the average energy flux and energy density of the wave.

11. Show that the wave equations for $\mathbf{E}$ and $\mathbf{H}$ in a linear, isotropic but inhomogeneous medium of permittivity $\varepsilon = \varepsilon(\mathbf{r})$ and permeability $\mu = \mu_0$ are

$$\nabla^2\mathbf{E} - \mu_0\varepsilon\ddot{\mathbf{E}} = \nabla(\nabla\cdot\mathbf{E})$$

$$\nabla^2\mathbf{H} - \mu_0\varepsilon\ddot{\mathbf{H}} = \varepsilon\nabla\left(\frac{1}{\varepsilon}\right)\times(\nabla\times\mathbf{H}) \quad .$$

The medium is assumed to be free of charge and current.

12. To obtain a plane wave solution of the wave equation we may assume $\mathbf{E}$ to be harmonic in time and to depend on one coordinate only. If we use the cylindrical coordinate system and consider a wave whose electric field is parallel to the z-axis and is a function of ρ only apart from a harmonic time dependance $\exp(-i\omega t)$, show that for large ρ the wave equation admits of a

solution

$$E_z(\rho, t) = E_0 \rho^{-\frac{1}{2}} e^{i(k\rho - \omega t)} \quad ,$$

where $k = \omega\sqrt{\mu\varepsilon}$. Hence show that at large distances from the origin **H** has only one component

$$|H_\phi| = \sqrt{\frac{\varepsilon}{\mu}}\, |E_z| \quad .$$

The solution is uniform over cylinders $\rho = \text{constant}$, representing a cylindrical wave front expanding from the z-axis. The field components are transverse to the direction of propagation. The wave is known as a *cylindrical wave*. Show that at large distances it approximates to a uniform plane wave.

13. In the previous problem, if we use spherical coordinates and consider a wave whose electric vector is parallel to $\mathbf{i}_\theta$ and is a function of r and θ only apart from a harmonic time dependence, show that for large r the wave equation has a solution

$$E_\theta(r, \theta, t) = \frac{E_0}{r \sin\theta} e^{i(kr - \omega t)}$$

off the $\theta = 0$ plane. Hence show that **H** is perpendicular to **E** and has a magnitude

$$|\mathbf{H}| = \sqrt{\frac{\varepsilon}{\mu}}\, |\mathbf{E}| \quad .$$

The solution represents a wave propagating along the radial direction with transverse fields. The wave is known as a *spherical wave*. Show that at large distances it approximates to a uniform plane wave.

Chapter III

ELECTROMAGNETIC FIELDS — STATIC FIELDS

Under static conditions, Maxwell's equations separate naturally into two groups which may be solved independently. Thus, a pure electrostatic field is obtained if stationary charges alone are present, and a pure magnetostatic field is obtained if stationary currents or magnetized bodies alone are present. Magnetized matter may be treated by attributing its magnetic properties to the inaccessible Ampèrian currents in its interior, thus avoiding the necessity of introducing a separate postulation of magnetic charges.

3.1 Types of Electromagnetic Field

On the basis of time dependence electromagnetic fields may be classified as static, quasi-stationary and rapidly varying fields.

If all the quantities involved in Maxwell's equations are constant in time, the electromagnetic field in question is said to be static or stationary. Maxwell's equations for static fields are

$$\nabla \times \mathbf{E} = 0 \quad , \qquad \nabla \cdot \mathbf{D} = \rho \quad ,$$

$$\nabla \times \mathbf{H} = \mathbf{J} \quad , \qquad \nabla \cdot \mathbf{B} = 0 \quad .$$

Here the field equations separate naturally into two groups: those involving the electric vectors **E** and **D** only and those involving the magnetic vectors **H** and **B** only. Since the electric vectors and the magnetic vectors are separately related by constitutive equations, each group can be solved independently. Consequently there are two types of static fields: the *electrostatic field* involving electric vectors which are constant in time and stationary charges; the *magnetostatic field* involving magnetic vectors which are constant in time and stationary currents. Note that the separation of Maxwell's equations into independent groups is not possible if the fields vary with time, since then the electric and magnetic vectors are both involved in the same equations.

If the electromagnetic field varies but slowly with time so that its time dependence needs to be taken account of only in the first approximation, it is said to be *quasi-static* or *quasi-stationary*. In the case of *rapidly varying fields*, time dependence is important and must be taken into account fully.

In this and the next chapter we shall derive from the fundamental axioms, Maxwell's equations, the general structure and properties of the various types of the electromagnetic field. In all cases we shall assume the medium to be linear, isotropic and homogeneous, unless stated otherwise.

3.2 Electrostatics

Under static conditions and without moving charges we need only to consider the following Maxwell's equations

$$\nabla \times \mathbf{E} = 0 \quad , \tag{3.1}$$

$$\nabla \cdot \mathbf{E} = \frac{\rho}{\varepsilon} \quad . \tag{3.2}$$

Equation (3.1) shows that **E** can be expressed in terms of a scalar potential $\Phi(\mathbf{r})$:

$$\mathbf{E} = -\nabla\Phi \quad . \tag{3.3}$$

Substitution in Eq. (3.2) gives

$$\nabla^2\Phi = -\frac{\rho}{\varepsilon} \quad , \tag{3.4}$$

which is known as Poisson's equation. In a charge-free region it becomes

$$\nabla^2\Phi = 0 \tag{3.5}$$

which is known as Laplace's equation.

The fundamental problem in electrostatics is then to determine a scalar potential $\Phi(\mathbf{r})$ that satisfies Poisson's or Laplace's equation at every point of the region and takes on the prescribed values of Φ or of its normal derivative on the surfaces of discontinuity which may be present, for on such surfaces the boundary conditions Eqs. (2.9) and (2.8) must be satisfied, which in terms of Φ can be written respectively as

$$\Phi_2 = \Phi_1 \tag{3.6}$$

and
$$\varepsilon_1 \left(\frac{\partial\Phi}{\partial n}\right)_1 - \varepsilon_2 \left(\frac{\partial\Phi}{\partial n}\right)_2 = \sigma \quad , \tag{3.7}$$

where σ is the surface charge density, $\frac{\partial\Phi}{\partial n}$ the component of $\nabla\Phi$ normal to the boundary surface, and the indices refer to the two media separated by the interface. In a physical problem, either Φ or $\frac{\partial\Phi}{\partial n}$ is to be prescribed for the surfaces present. Once Φ is determined for the given charge distribution $\rho(\mathbf{r})$ and the given set of boundary values, the field intensity $\mathbf{E}(\mathbf{r})$ can be calculated by means of Eq. (3.3). Conversely, if the field $\mathbf{E}(\mathbf{r})$ is known, the charge distribution can be found by means of Eqs. (3.2) and (3.7).

Consider a region where the charge density is $\rho(\mathbf{r})$ and let Φ_1 and Φ_2 be the solutions of Laplace's and Poisson's equations respectively. The sum $\Phi_1 + \Phi_2 = \Phi$ must satisfy Poisson's equation since

$$\nabla^2\Phi = \nabla^2\Phi_1 + \nabla^2\Phi_2 = \nabla^2\Phi_2 = -\frac{\rho}{\varepsilon} \quad .$$

This means that if we find a particular solution of Poisson's equation for a region where the charge density is ρ we can obtain a family of solutions by adding to this particular solution the family of functions satisfying Laplace's equation. The correct solution is then chosen by applying the appropriate boundary conditions.

3.3 Laplace's Equation

The manner in which the solution of Laplace's equation is obtained depends on the geometry of the problem. Only the solution in spherical coordinates will be given here as example. We shall first consider two important theorems which govern its solutions in general.

If $\Phi_1, \Phi_2, \ldots, \Phi_n$ are each a solution of Laplace's equation, then on account of the linearity of the equation, the sum

$$\Phi = C_1\Phi_1 + C_2\Phi_2 + \ldots + C_n\Phi_n \quad ,$$

where $C_1, C_2, \ldots, C_n$ are arbitrary constants, is also a solution. This is the *theorem of superposition of solutions*. It means that potentials are additive. It enables us to superpose two or more solutions in such a way as to satisfy a given set of boundary conditions.

The second theorem concerns the uniqueness of the solution which satisfies a given set of boundary conditions. If two solutions Φ_1 and Φ_2 both satisfy the same boundary conditions they can differ at most by an additive constant. The boundary conditions are usually given for a closed surface (which may be a surface closed at infinity) in the region of interest by prescribing either the potential (the *Dirichlet boundary condition*) or the normal derivative of the potential (the *Neumann boundary condition*) on the surface.

To prove this theorem we consider a charge-free region V bounded externally by a closed surface S, which may be a real physical surface or a surface at infinity, and internally by closed surfaces $S_1, S_2, \ldots, S_n$. Define $\Phi_2 - \Phi_1 = \Phi$; then by the previous theorem, Φ is also a solution of Laplace's equation, i.e.

$$\nabla^2\Phi = 0$$

in V. Hence

$$(\nabla\Phi)^2 = \nabla \cdot (\Phi\nabla\Phi) - \Phi\nabla^2\Phi = \nabla \cdot (\Phi\nabla\Phi) \quad ,$$

and
$$\int_V (\nabla\Phi)^2 dV = \int_V \nabla \cdot (\Phi\nabla\Phi) dV = \oint_{S+S_1+S_2+\ldots+S_n} \Phi\nabla\Phi \cdot d\mathbf{S}$$

using the divergence theorem. Now the surface integrals vanish for $\Phi = \Phi_2 - \Phi_1 = 0$ if the potentials on the boundary surfaces are prescribed; or $\mathbf{n} \cdot \nabla\Phi = \frac{\partial\Phi_2}{\partial n} - \frac{\partial\Phi_1}{\partial n} = 0$ if the normal derivatives of the potentials are prescribed. Hence

$$\int_V (\nabla\Phi)^2 dV = 0 \quad .$$

As $(\nabla\Phi)^2$ is either positive or zero, this requires that

$$\nabla\Phi = 0 \quad , \quad \text{i.e.} \quad \frac{\partial\Phi}{\partial x} = \frac{\partial\Phi}{\partial y} = \frac{\partial\Phi}{\partial z} = 0 \quad .$$

Hence
$$\Phi = \Phi_2 - \Phi_1 = \text{constant} \tag{3.8}$$

in V and on $S, S_1, S_2, \ldots, S_n$.

If the Dirichlet boundary condition is prescribed, $\Phi_2 = \Phi_1$ on $S, S_1, S_2, \ldots, S_n$. The continuity of Φ requires that the constant in Eq. (3.8) be zero, i.e. $\Phi_2 = \Phi_1$, everywhere in V. If the Neumann boundary condition is given, $\mathbf{n} \cdot \nabla\Phi = \frac{\partial(\Phi_2 - \Phi_1)}{\partial n} = 0$ on $S, S_1, S_2, \ldots, S_n$. The constant in Eq. (3.8) need not be zero, i.e., Φ_2 and Φ_1 can at most differ by a constant throughout the whole region. As the field is not affected by the addition of a constant to the potential, Φ_2 and

Φ_1 do not differ in any significant way and the solution is still unique.

On account of the uniqueness theorem we may obtain a solution Φ of Laplace's equation by any means whatever, and if Φ satisfies all the boundary conditions specified, then it represents the unique solution to the problem. Many methods such as the method of images, which simulates a given boundary condition by replacing the boundary with suitably placed charges, have been devised for accomplishing this end without specifically solving the differential equation.

In a number of orthogonal coordinate systems Laplace's equation can be separated in its variables and readily solved. The coordinate system to be used depends on the geometry of the problem. We shall consider as example its solution in spherical coordinates (r, θ, ϕ), in which Laplace's equation is

$$\frac{1}{r^2}\frac{\partial}{\partial r}\left(r^2\frac{\partial\Phi}{\partial r}\right)+\frac{1}{r^2\sin\theta}\frac{\partial}{\partial\theta}\left(\sin\theta\,\frac{\partial\Phi}{\partial\theta}\right)+\frac{1}{r^2\sin^2\theta}\frac{\partial^2\Phi}{\partial\phi^2}=0 \quad . \tag{3.9}$$

Consider a solution of the form

$$\Phi(r, \theta, \phi) = R(r)P(\theta)Q(\phi) \quad .$$

Substituting this in Eq. (3.9) and multiplying both sides by $r^2\Phi^{-1}\sin^2\theta$ we have

$$\left\{\frac{1}{R}\frac{d}{dr}\left(r^2\frac{dR}{dr}\right)+\frac{1}{P\sin\theta}\frac{d}{d\theta}\left(\sin\theta\frac{dP}{d\theta}\right)\right\}\sin^2\theta+\frac{1}{Q}\frac{d^2Q}{d\phi^2}=0 \quad . \tag{3.10}$$

The first term on the left-hand side depends on r and θ only, while the second term depends on ϕ only. A change of ϕ cannot affect the first term and consequently does not affect the second term. Thus each term must be equal to a constant, say $\pm m^2$. Equation (3.10) is therefore equivalent to two equations

$$\frac{d^2Q}{d\phi^2} + m^2Q = 0 \quad , \tag{3.11}$$

$$\frac{1}{R}\frac{d}{dr}\left(r^2\frac{dR}{dr}\right) + \left\{\frac{1}{P\sin\theta}\frac{d}{d\theta}\left(\sin\theta\frac{dP}{d\theta}\right) - \frac{m^2}{\sin^2\theta}\right\} = 0 \quad . \tag{3.12}$$

On the left-hand side of the last equation the two terms respectively depend on r and θ alone. Again each must be equal to a constant, say $\pm n(n+1)$. Thus Eq. (3.12) is equivalent to the following equations

$$\frac{1}{\sin\theta}\frac{d}{d\theta}\left(\sin\theta\frac{dP}{d\theta}\right) + \left\{n(n+1) - \frac{m^2}{\sin^2\theta}\right\}P = 0 \quad , \tag{3.13}$$

$$\frac{d}{dr}\left(r^2\frac{dR}{dr}\right) - n(n+1)R = 0 \quad . \tag{3.14}$$

The solution Φ is obtained by solving Eqs. (3.11), (3.13) and (3.14).

The solutions of Eq. (3.11) are the harmonic functions $\exp(\pm im\phi)$. We require Φ and hence Q to be single-valued, i.e. $Q(\phi) = Q(\phi + 2\pi)$. m must therefore be either zero or a positive integer. (Negative integers simply repeat the same solutions).

The solutions of Eq. (3.13) are the *associated Legendre functions* $P_n^m(\cos\theta)$. These functions are finite over the range $0 \le \theta \le \pi$ only if n is an integer equal to or greater than m. When m is zero the functions are called the *Legendre polynomials* $P_n(\cos\theta)$, which are related to the associated Legendre functions through the equation

$$P_n^m(z) = (1 - z^2)^{\frac{m}{2}} \frac{d^m P_n(z)}{dz^m}$$

for $|z| \le 1$. As both types of polynomials can be found in standard mathematical tables, it suffices to quote the first few functions here:

$P_0(\cos\theta) = 1 \quad , \qquad P_1(\cos\theta) = \cos\theta \quad ,$

$P_2(\cos\theta) = \frac{1}{4}(3\cos 2\theta + 1) \quad ,$

$P_3(\cos\theta) = \frac{1}{8}(5\cos 3\theta + 3\cos\theta) \quad ,$

...

$P_1^1(\cos\theta) = \sin\theta \quad , \qquad P_2^1(\cos\theta) = \frac{3}{2}\sin 2\theta \quad ,$

$P_3^1(\cos\theta) = \frac{3}{8}(\sin\theta + 5\sin 3\theta) \quad ,$

...

$P_2^2(\cos\theta) = \frac{3}{2}(1 - \cos 2\theta) \quad ,$

$P_3^2(\cos\theta) = \frac{15}{4}(\cos\theta - \cos 3\theta) \quad ,$

...

$P_3^3(\cos\theta) = \frac{15}{4}(3\sin\theta - \sin 3\theta) \quad ,$

...

Equation (3.14) is known as *Cauchy's equation*. Assuming solutions of the type $R = r^s$, we find $s = n$ and $s = -n - 1$, i.e. its solutions are r^n and r^{-n-1}. The general solution of Laplace's equation in spherical coordinates is the superposition of all its particular solutions:

$$\Phi = \sum_{n=0}^{\infty} \sum_{m=0}^{n} (a_{nm} r^n + b_{nm} r^{-n-1}) P_n^m(\cos\theta) e^{\pm im\phi} \quad . \tag{3.15}$$

For specific problems the constant coefficients a_{nm} and b_{nm} are to be determined from the boundary conditions. It is interesting to note that the only spherically symmetric solution is the one for which $n = m = 0$, i.e.

$$\Phi = \frac{1}{r} \quad .$$

To illustrate the manner in which the coefficients of the expression (3.15) may be determined, consider the following example. A homogeneous sphere of radius a is introduced in a uniform electric field of intensity $\mathbf{E}_0$ and we wish to find the field in its interior and immediate neighbourhood.

Choose a Cartesian coordinate system with the origin at the centre of the sphere and the z-axis in the direction of the field. Since potentials are additive, the potential Φ_+ at a point (r, θ, ϕ) outside the sphere can be written as the sum of two parts, the potential if the sphere were absent, which by Eq. (3.3) is

$$\Phi_\infty = - E_0 z = - E_0 r \cos\theta = - E_0 r P_1(\cos\theta) \quad ,$$

and the potential Φ_1 due to the polarization of the sphere. Since the point (r, θ, ϕ) is in a charge-free region, Laplace's equation holds so that

$$\nabla^2\Phi_+ = \nabla^2\Phi_\infty + \nabla^2\Phi_1 = \nabla^2\Phi_1 = 0 \quad ,$$

i.e. Φ_1 also satisfies Laplace's equation and has the general form (3.15).

To determine Φ_1 we note that the geometry involved is cylindrically symmetric, which requires $m = 0$. Furthermore at large distances from the origin the effect of the presence of the sphere is negligible, hence $\Phi_+ = \Phi_\infty$, or $\Phi_1 = 0$, for $r \to \infty$. Under such conditions Eq. (3.15) gives

$$\Phi_1 = \sum_{n=0}^{\infty} b_n r^{-n-1} P_n(\cos\theta) \quad ,$$

and hence $$\Phi_+ = \sum_{n=0}^{\infty} b_n r^{-n-1} P_n(\cos\theta) - E_o r P_1(\cos\theta) \quad , \tag{3.16}$$

where the coefficients b_n are to be determined from the boundary conditions.

If the sphere is *conducting* its surface charge will be distributed in such a way that the field inside it is zero everywhere, i.e. the potential is constant throughout the sphere. Thus

$$\Phi_- = \text{constant} = \Phi_S \quad , \quad \text{say,} \quad \text{for} \quad r \leq a \quad .$$

This is the boundary condition for the surface and requires that

$$\Phi_S = \sum_{n=0}^{\infty} b_n a^{-n-1} P_n(\cos\theta) - E_o a P_1(\cos\theta) \quad .$$

But as the functions P_n are linearly independent, and Φ_S is constant over the sphere and hence independent of θ, we require that $b_n = 0$ for $n > 1$ and the coefficient of $P_1(\cos\theta)$ to vanish, i.e.

$$\frac{b_1}{a^2} - E_o a = 0 \quad , \quad \text{giving} \quad b_1 = a^3 E_o \quad .$$

Only one term now remains in the expression for Φ_S:

$$\Phi_S = \frac{b_o}{a} \quad , \quad \text{giving} \quad b_o = a\Phi_S \quad .$$

Substitution of these coefficients in Eq. (3.16) gives the potential at a point (r, θ, ϕ), where $r \geq a$, as

$$\Phi_+ = - E_o r \cos\theta + \frac{a\Phi_S}{r} + \frac{a^3 E_o \cos\theta}{r^2} \quad .$$

The parameter Φ_S which represents the potential of the sphere depends on the net charge carried by it as can be seen in the following.

Inside the conducting sphere the field is zero and the boundary condition for **D** becomes

$$\sigma = -\varepsilon_1 \left(\frac{\partial\Phi_+}{\partial r}\right)_{r=a} = 3\varepsilon_1 E_0 \cos\theta + \frac{\varepsilon_1 \Phi_S}{a} ,$$

where ε_1 is the permittivity of the medium in which the sphere is embedded. The surface charge density σ can be integrated to give the total charge

$$\int_0^{\pi} 2\pi\sigma \sin\theta\, a^2 d\theta = 4\pi\varepsilon_1 a \Phi_S .$$

If the sphere is uncharged before its introduction in the field the induced charges will cancel out and Φ_S vanishes, in which case

$$\Phi_+ = -E_0 r\cos\theta + \frac{a^3 E_0 \cos\theta}{r^2} .$$

The electric field on the spherical surface is normal everywhere as can be seen by differentiation. By comparing this equation with Eq. (1.16) it is noted that Φ_+ consists of a dipole potential superimposed on the potential of the primal uniform field.

Consider next the case where the sphere is an uncharged *dielectric* of permittivity ε_2. On its surface, the boundary condition Eq. (3.7) becomes

$$\varepsilon_2 \left(\frac{\partial\Phi_-}{\partial r}\right)_{r=a} = \varepsilon_1 \left(\frac{\partial\Phi_+}{\partial r}\right)_{r=a} , \tag{3.17}$$

where Φ_- is the potential at a point in the interior of the sphere and Φ_+ the potential at an external point given by Eq. (3.16). The boundary condition Eq. (3.6) becomes

$$\Phi_-(a, \theta, \phi) = \Phi_+(a, \theta, \phi) \tag{3.18}$$

At an interior point, as there is no net charge anywhere

(macroscopically speaking!), the potential is a solution of Laplace's equation which is finite at $r=0$. As it is furthermore cylindrically symmetric, it will have the form

$$\Phi_- = \sum_{n=0}^{\infty} a_n r^n P_n(\cos\theta) \quad . \tag{3.19}$$

Equation (3.18) therefore gives

$$\sum_{n=0}^{\infty} (b_n a^{-n-1} - a_n a^n) P_n(\cos\theta) - E_o a P_1(\cos\theta) = 0 \quad .$$

This condition is to be satisfied by every point (a, θ, ϕ) on the surface and must therefore hold for all values of θ. This requires the coefficient of each Legendre polynomial to vanish separately. Thus for $n=0$ and $n\geq 1$ this implies that a_n and b_n are proportional. The same argument applied to the boundary condition Eq. (3.17) requires that, for each such n, a_n and b_n are again proportional, but with a different constant of proportionality. To satisfy both requirements, we must have for $n=0$ and $n\geq 1$, $a_n = b_n = 0$, and for $n=1$,

$$\frac{b_1}{a^3} - a_1 = E_o$$

and $$\frac{2b_1}{a^3} + \frac{\varepsilon_2 a_1}{\varepsilon_1} = -E_o \quad .$$

Solving the last two equations we find that

$$a_1 = -\left(\frac{3\varepsilon_1}{\varepsilon_2 + 2\varepsilon_1}\right)E_o$$

and $$b_1 = \left(\frac{\varepsilon_2 - \varepsilon_1}{\varepsilon_2 + 2\varepsilon_1}\right)a^3 E_o \quad .$$

The potential is therefore

$$\Phi_- = -\left(\frac{3\varepsilon_1}{\varepsilon_2 + 2\varepsilon_1}\right)E_o r \cos\theta \qquad \text{for} \qquad r \le a$$

and $$\Phi_+ = \left(\frac{\varepsilon_2 - \varepsilon_1}{\varepsilon_2 + 2\varepsilon_1}\right)\frac{a^3}{r^2}E_o \cos\theta - E_o r\cos\theta \qquad \text{for} \qquad r \ge a \quad , \tag{3.20}$$

It can be seen by differentiation that the electric field inside the sphere is uniform and parallel to $\mathbf{E}_o$.

A comparison of Eq. (3.20) with Eq. (1.16) shows that the induced field of the sphere in the outside region is that of a dipole of moment

$$\mathbf{p} = 4\pi\varepsilon_1 a^3 \left(\frac{\varepsilon_2 - \varepsilon_1}{\varepsilon_2 + 2\varepsilon_1}\right)\mathbf{E}_o \quad .$$

3.4 Poisson's Equation

A particular solution of Poisson's equation

$$\nabla^2\Phi = -\frac{\rho}{\varepsilon}$$

for a linear, isotropic and homogeneous medium of permittivity ε can be obtained by integration making use of Green's formula and the Dirac delta function.

Let $\mathbf{r}$ be the field point where the potential $\Phi(\mathbf{r})$ due to a charge distribution of density $\rho(\mathbf{r}')$ is to be evaluated and denote the radius vector from a source point $\mathbf{r}'$ to the field point by $\bar{\mathbf{r}}$, i.e. $\bar{\mathbf{r}} = \mathbf{r} - \mathbf{r}'$. Enclose the field point with an arbitrary closed surface S and apply Green's formula to the enclosed volume V:

$$\int_V \{ \psi(\mathbf{r}')\nabla'^2\Phi(\mathbf{r}') - \Phi(\mathbf{r}')\nabla'^2\psi(\mathbf{r}') \} dV'$$

$$= \oint_S \{ \psi(\mathbf{r}')\nabla'\Phi(\mathbf{r}') - \Phi(\mathbf{r}')\nabla'\psi(\mathbf{r}') \} \cdot d\mathbf{S}' \quad ,$$

where Φ and ψ are arbitrary scalar functions. If we choose for Φ the solution of Poisson's equation and

$$\psi(\mathbf{r}') = \frac{1}{\bar{r}} \quad ,$$

Green's formula becomes

$$- \int_V \Phi(\mathbf{r}')\nabla'^2(\frac{1}{\bar{r}})dV' = \frac{1}{\varepsilon}\int_V \frac{\rho(\mathbf{r}')}{\bar{r}} dV' + \oint_S \{ \frac{1}{\bar{r}} \nabla'\Phi(\mathbf{r}')$$

$$- \Phi(\mathbf{r}')\nabla' (\frac{1}{\bar{r}}) \} \cdot d\mathbf{S}'$$

Using Eqs. (1.38) and (1.39) we have

$$\int_V \Phi(\mathbf{r}')\nabla'^2(\frac{1}{\bar{r}})dV' = - 4\pi \int_V \Phi(\mathbf{r}')\delta(\mathbf{r} - \mathbf{r}')dV' = - 4\pi\Phi(\mathbf{r}) \quad .$$

The solution of Poisson's equation is therefore

$$\Phi(\mathbf{r}) = \frac{1}{4\pi\varepsilon}\int_V \frac{\rho}{\bar{r}} dV' + \frac{1}{4\pi}\oint_S \{ \frac{1}{\bar{r}} \nabla'\Phi - \Phi\nabla' (\frac{1}{\bar{r}}) \} \cdot d\mathbf{S}' \quad . \tag{3.21}$$

For the interpretation of the above, consider first the case where S does not enclose any charge, i.e. the case of $\rho = 0$ in V. The above becomes

$$\Phi(\mathbf{r}) = \frac{1}{4\pi}\oint_S \{ \frac{1}{\bar{r}} \nabla'\Phi - \Phi\nabla' (\frac{1}{\bar{r}}) \} \cdot d\mathbf{S}' \quad . \tag{3.22}$$

This obviously represents the contribution to the potential at $\mathbf{r}$ of all charges on or external to S. The fact that it satisfies Laplace's equation can be readily seen by direct differentiation. The density of the external charge does not appear in the expression; instead, the potential $\Phi(\mathbf{r})$ is given by the potential Φ_S and its normal derivative $\left(\frac{\partial\Phi}{\partial n}\right)_S$ on the boundary surface.

It should be noted however that the potential and its normal derivative are related and cannot be independently assigned for the same surface. On the other hand, the knowledge of one for a surface does not allow us to write down the other without first obtaining the solution. This difficulty is overcome by the use of Green's functions.

The choice we have made for ψ in the above is just one of an infinite number of possibilities. The analysis still holds if we choose for ψ any function which satisfies the requirement

$$\nabla^2\psi = -4\pi\delta(\mathbf{r} - \mathbf{r}') \quad . \tag{3.23}$$

In fact Eq. (3.23) defines a family of functions called *Green's functions* which have the form

$$\psi = \frac{1}{r} + \chi(\mathbf{r}, \mathbf{r}') \equiv G \quad , \tag{3.24}$$

where $\chi(\mathbf{r}, \mathbf{r}')$ is a symmetric function of $\mathbf{r}$ and $\mathbf{r}'$ that satisfies Laplace's equation in V, i.e.

$$\nabla^2\chi = 0 \quad , \tag{3.25}$$

but is otherwise arbitrary.

The freedom in the choice of χ enables us to select a Green's function G that eliminates one of the two boundary functions, Φ and $\frac{\partial\Phi}{\partial n}$, appearing on the right-hand side of Eq. (3.21). A detailed discussion of Green's functions is beyond the scope of this book and we shall be satisfied with some remarks for the case where the Dirichlet boundary condition is prescribed. Here we set $G = 0$ on S so that Eq. (3.21) becomes

$$\Phi(\mathbf{r}) = \frac{1}{4\pi\varepsilon}\int_V G\rho dV' - \frac{1}{4\pi}\oint_S \Phi(\mathbf{r}')\nabla' G \cdot d\mathbf{S}' \quad .$$

Equations (3.24) and (3.25) show that G may be considered as the sum of two potentials, one due to a point charge $4\pi\varepsilon$ located at $\mathbf{r}$, the other satisfying Laplace's equation in V and arising therefore from external charges. The result of superposition of the two potentials is to make G vanish on the surface S. This suggests that Green's function G can often be obtained by the method of images (e.g. Prob. 3.11). Note also that if S is a conducting surface, χ may be interpreted as being due to the charges induced on the surface by the point charge located at $\mathbf{r}$.

Next consider the case where every element of the charge distribution is located within a finite distance of some fixed origin O. Such a distribution is said to be *finite* or *closed* and can be circumscribed by a sphere Q of finite radius R centred at O. Suppose the field point P is at some finite distance r from O. It may be inside or outside Q. In any case we can take S so large as to include both the sphere of charge distribution and the field point as shown in Fig. 3.1. Consider the surface integral over S in Eq. (3.21). As there is no charge outside S, the potential Φ_S on the surface is due entirely to the charge distribution. This means that if S is taken sufficiently large Φ_S and $\nabla\Phi_S$ can be made to fall to zero by taking the potential at infinity as the reference level since the charge distribution is finite. Hence if the integral is taken over the whole space we have

$$\Phi(\mathbf{r}) = \frac{1}{4\pi\varepsilon}\int \frac{\rho(\mathbf{r}')}{\bar{r}} dV' \quad . \tag{3.26}$$

In practice, of course, the integrand vanishes outside the charge distribution and the integral is really over the charge distribution. This shows that the surface integral in Eq. (3.21) vanishes if all the

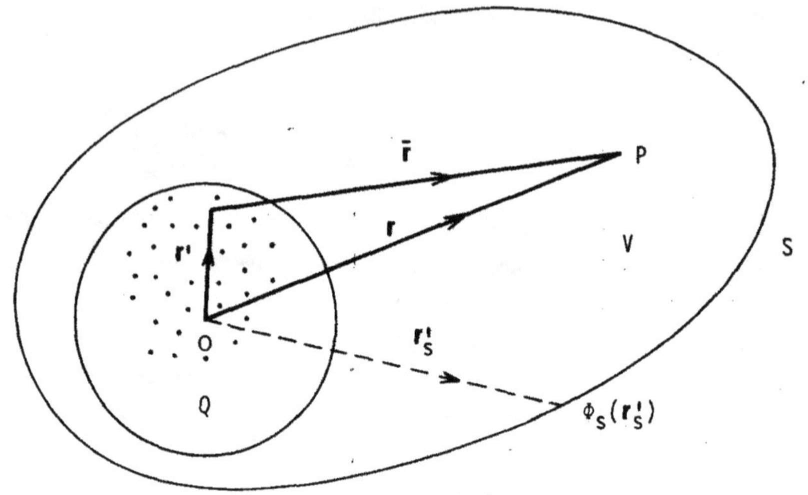

Fig. 3.1 The closed surface S is taken so large as to enclose both the sphere of charge distribution and the field point P, which may be either outside or inside the sphere.

charges are enclosed in S.

Suppose now that the field point $\mathbf{r}$ is situated at a large distance from the centre O of the sphere containing the charge distribution. As $r \gg r'$, we can expand $\bar{r}^{-1}$ as a power series in r'/r:

$$\bar{r}^{-1} = |\mathbf{r} - \mathbf{r}'|^{-1} = \frac{1}{r}\left\{1 - \left[\frac{2\mathbf{r}\cdot\mathbf{r}'}{r^2} - \left(\frac{r'}{r}\right)^2\right]\right\}^{-\frac{1}{2}}$$

$$= \frac{1}{r}\left\{1 + \frac{1}{2}\left[\frac{2\mathbf{r}\cdot\mathbf{r}'}{r^2} - \left(\frac{r'}{r}\right)^2\right] + \frac{3}{8}\left[\frac{2\mathbf{r}\cdot\mathbf{r}'}{r^2} - \left(\frac{r'}{r}\right)^2\right]^2 + \ldots\right\}$$

$$= \frac{1}{r}\left\{1 + \frac{\mathbf{r}\cdot\mathbf{r}'}{r^2} + \frac{1}{2}\left[3\left(\frac{\mathbf{r}\cdot\mathbf{r}'}{r^2}\right)^2 - \left(\frac{r'}{r}\right)^2\right] + \ldots\right\} \tag{3.27}$$

Substitution in Eq. (3.26) gives

$$\Phi(\mathbf{r}) = \frac{1}{4\pi\varepsilon r}\int \rho(\mathbf{r}')dV' + \frac{1}{4\pi\varepsilon r^3}\int \mathbf{r}\cdot\mathbf{r}'\,\rho(\mathbf{r}')dV'$$

$$+ \frac{1}{4\pi\varepsilon r^5}\int \frac{\rho(\mathbf{r}')}{2}\left\{ 3(\mathbf{r}\cdot\mathbf{r}')^2 - r^2 r'^2 \right\} dV'$$

$$+ \ldots$$

$$= \Phi_1 + \Phi_2 + \Phi_3 \ldots \quad .$$

In the above expansion, the first term on the right-hand side,

$$\Phi_1 = \frac{1}{4\pi\varepsilon}\frac{q}{r} \quad ,$$

corresponds to the potential of a point charge at 0 of magnitude q equal to the net total charge of the distribution. This part of the potential is called the *Coulomb potential*. The second term,

$$\Phi_2 = \frac{1}{4\pi\varepsilon}\frac{\mathbf{r}\cdot\mathbf{p}}{r^3} = -\frac{1}{4\pi\varepsilon}\mathbf{p}\cdot\nabla\left(\frac{1}{r}\right) \quad ,$$

has the form of Eq. (1.16) and corresponds to the potential of a dipole located at 0 of moment

$$\mathbf{p} = \int \mathbf{r}'\,\rho(\mathbf{r}')dV'$$

and is therefore called the *dipole potential*. The third term can be written as

$$\Phi_3 = \frac{1}{4\pi\varepsilon}\frac{x_i x_j}{2r^5} Q_{ij}$$

and represents the *quadrupole potential*. The quantities

$$Q_{ij} = \int (3x_i'x_j' - r'^2\delta_{ij})\rho(\mathbf{r}')dV' \qquad (3.28)$$

form the quadrupole moment tensor. The remaining terms represent the potentials of multipoles of still higher orders. Beyond the quadrupole, however, the expansion becomes extremely tedious to handle.

A closed charge distribution at a large distance may therefore be approximated by a point charge plus multipole moments. Thus for an unknown charge distribution we may try separately to measure the moments of the various multipole components and thereby get some idea about the entire charge distribution. Such a method is particularly useful in the investigation of the internal charge structure of an atomic particle.

The ratio of the successive multipole potentials is of the order r'/r. Hence at a sufficiently large distance the potential of a finite charge distribution is given essentially by the first non-vanishing multipole term in the expansion. In particular, a distant finite charge distribution having a non-vanishing net charge will behave like a point charge.

Equation (3.26) may be generalized to include regions where there are surface charges of density σ on surfaces ΣS in addition to the volume charges:

$$\Phi = \frac{1}{4\pi\varepsilon}\int_V \frac{\rho}{r}\,dV' + \frac{1}{4\pi\varepsilon}\int_{\Sigma S} \frac{\sigma}{r}\,dS' \quad . \tag{3.29}$$

3.5 Electrostatic Field in Several Dielectric Media

The treatment presented above applies to a linear, isotropic and homogeneous medium extending over the entire space. We shall now consider the field of a charge distribution in several rigid, stationary dielectric media, each of them linear, isotropic and homogeneous. As surfaces of discontinuity are now involved we shall have to consider, in addition to a volume charge distribution, surface charges of density $\sigma(\mathbf{r}')$ on these surfaces.

As several media are now involved Eq. (2.15) should be used instead of Eq. (3.2). However, for a material medium it is often convenient to use instead of the electric displacement $\mathbf{D}$ the *polarization* $\mathbf{P}$ as the additional field vector which is defined as

$$\mathbf{P} = \mathbf{D} - \varepsilon_0 \mathbf{E} \quad .$$

Note that $\mathbf{P}$ vanishes outside a material medium. Writing Eq. (2.15) as

$$\nabla \cdot \mathbf{E} = \frac{1}{\varepsilon_0} (\rho - \nabla \cdot \mathbf{P})$$

and comparing it with the corresponding equation in vacuum

$$\nabla \cdot \mathbf{E} = \frac{\rho}{\varepsilon_0} \quad ,$$

we see that the presence of a dielectric is accounted for by introducing an equivalent charge density, the polarization charge density,

$$\rho' = -\nabla \cdot \mathbf{P} \tag{3.30}$$

in the space occupied by the dielectric. At the interface of two dielectrics, the boundary condition for the electric displacement $\mathbf{D}$ is given by Eq. (2.8)

$$D_{2n} - D_{1n} = \sigma \quad ,$$

where the normal $\mathbf{n}$ is directed from medium 1 to medium 2. This, together with the definition of the polarization $\mathbf{P}$, gives

$$E_{2n} - E_{1n} = \frac{1}{\varepsilon_0} \{ \sigma - (P_{2n} - P_{1n}) \} = \frac{1}{\varepsilon_0} (\sigma + \sigma') \quad ,$$

say. On the other hand, if the media were both vacuum but separated by a surface carrying a surface charge density σ, the boundary condition would have been

$$E_{2n} - E_{1n} = \frac{\sigma}{\varepsilon_0} \quad .$$

A comparison of the two equations shows that

$$\sigma' = -(P_{2n} - P_{1n}) \tag{3.31}$$

may be considered as the equivalent or polarization surface charge density resulting from the presence of the dielectrics.

The expression for the potential due to a charge distribution as given by Eq. (3.29) is now to be modified to

$$\Phi(\mathbf{r}) = \frac{1}{4\pi\varepsilon_0}\int_{V'} \frac{1}{r}(\rho - \nabla' \cdot \mathbf{P})dV' + \frac{1}{4\pi\varepsilon_0}\int_{\Sigma S'} \frac{1}{r}\left\{\sigma - (P_{2n}-P_{1n})\right\} dS' \tag{3.32}$$

where the volume integral is to extend over the entire region V' occupied by the dielectrics and the "true" charges, and the surface integral extends over all surfaces of discontinuity $\Sigma S'$. Note that the permittivity of vacuum is used in the expression because the presence of matter is taken account of by including in the polarization charge densities. The above expression can be simplified through the following consideration.

If the region V' occupied by the entire system of charge distribution and dielectrics is assumed to be finite, we may enclose it with a larger but arbitrary closed surface S_0 of volume V. We may also surround the surfaces of discontinuity with closely-fitting closed surfaces S_m'' as shown in Fig. 3.2. The volume V_m'' enclosed by S_m'' can be made as nearly equal as possible to the volume of the m-th dielectric. Similarly, the volume V_0'' bounded internally by S_0'' and externally by S_0 approximates the portion of the volume enclosed by S_0 that is not occupied by the system. Thus we may write $V = V_0'' + \sum_{m=1} V_m''$. Since both ρ and $-\nabla' \cdot \mathbf{P}$ vanish outside V', the volume integral in Eq. (3.32) over V' can be replaced by the integral over V, or by the sum of the integrals over V_0'', V_1'', V_2'',...

The surface integral in Eq. (3.32) involving the normal component of $\mathbf{P}$ over the interfaces S' can be replaced by integrals

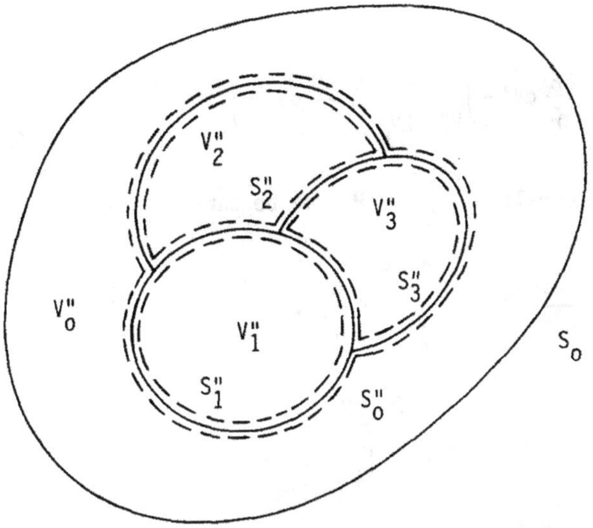

Fig. 3.2 The interfaces between different dielectrics are enclosed by closely-fitting closed surfaces S''_m. S_0 is an arbitrary surface enclosing the entire system of charges and dielectrics.

over S''. Consider the integral over the interface S'_{12} between media 1 and 2. Adopting the convention that the normal to the surface S''_m is directed outwards from the volume V''_m and bearing in mind that the normal to the interface S'_{12} has been taken to direct from medium 1 to medium 2, we can write

$$\int_{S'_{12}} \frac{(P_{2n} - P_{1n})dS'}{\bar{r}} = \int_{S'_{12}} \frac{P_{2n}dS'}{\bar{r}} - \int_{S'_{12}} \frac{P_{1n}dS'}{\bar{r}}$$

$$= - \int_{S''_{21}} \frac{\mathbf{P}\cdot d\mathbf{S}'}{\bar{r}} - \int_{S''_{12}} \frac{\mathbf{P}\cdot d\mathbf{S}'}{\bar{r}} \quad ,$$

where S''_{mn} is the portion of S''_m adjacent to the n-th dielectric. The same transformation applies to other interfaces so that Eq. (3.32) can be written as

$$4\pi\varepsilon_0\Phi = \int_{V'} \frac{\rho}{r}\,dV' - \int_{V''_0 + \Sigma V''_m} \frac{\nabla'\cdot\mathbf{P}}{r}\,dV' + \int_{\Sigma S'} \frac{\sigma dS'}{r} + \int_{S''_0 + \Sigma S''_m} \frac{\mathbf{P}\cdot d\mathbf{S}'}{r} \quad .$$

The volume integrals of $\frac{1}{r}\nabla'\cdot\mathbf{P}$ can be partially integrated:

$$\int_{V''_m} \frac{\nabla'\cdot\mathbf{P}}{r}\,dV' = \int_{V''_m} \nabla'\cdot\left(\frac{\mathbf{P}}{r}\right)dV' - \int_{V''_m} \mathbf{P}\cdot\nabla'\left(\frac{1}{r}\right)dV'$$

$$= \oint_{S''_m} \frac{\mathbf{P}\cdot d\mathbf{S}'}{r} - \int_{V''_m} \mathbf{P}\cdot\nabla'\left(\frac{1}{r}\right)dV' \quad ,$$

$$\int_{V''_0} \frac{\nabla'\cdot\mathbf{P}}{r}\,dV' = \oint_{S''_0} \frac{\mathbf{P}\cdot d\mathbf{S}'}{r} + \oint_{S_0} \frac{\mathbf{P}\cdot d\mathbf{S}'}{r} - \int_{V''_0} \mathbf{P}\cdot\nabla'\left(\frac{1}{r}\right)dV' \quad ,$$

where use has been made of the divergence theorem. The integrals over the surfaces S_0 and S''_0 vanish as $\mathbf{P}$ is zero on these surfaces which lie outside all the dielectrics. Then as the surface integrals over $\Sigma S''_m$ cancel out we have

$$\Phi = \frac{1}{4\pi\varepsilon_0}\int_{V'} \frac{\rho}{r}\,dV' + \frac{1}{4\pi\varepsilon_0}\int_{V'} \mathbf{P}\cdot\nabla'\left(\frac{1}{r}\right)dV' + \frac{1}{4\pi\varepsilon_0}\int_{\Sigma S'} \frac{\sigma}{r}\,dS' \quad , \tag{3.33}$$

where the volume integrals are to be taken over all regions in which ρ and $\mathbf{P}$ do not vanish.

A comparison with Eq. (1.16) shows that the contribution of the dielectrics to the field, namely the second volume integral of Eq. (3.33) is simply the potential produced by a continuous distribution of dipoles with dipole moment per unit volume **P**. This is exactly what we would have expected from the manner the polarization vector was introduced in Chap. I.

3.6 Electrostatic Field Energy

The electrostatic field is conservative since $\nabla \times \mathbf{E} = 0$. Thus the work done *against* electrical force in bringing a point charge q slowly from a point of potential Φ_1 to a point of potential Φ_2 is stored as the potential energy

$$W_e = -\int_1^2 q\mathbf{E} \cdot d\mathbf{r} = q \int_1^2 \nabla\Phi \cdot d\mathbf{r} = q \int_1^2 d\Phi = q(\Phi_2 - \Phi_1) \quad .$$

If we take as the initial position a point at infinity and use its potential as the level of reference, then the potential energy of a charge at a point of potential Φ is

$$W_e = q\Phi \quad . \tag{3.34}$$

Consider a system of two point charges. The introduction of a charge q_2 slowly to a point B in a field-free space requires no work. The work done in bringing a charge q_1 from infinity to a point A in the field of charge q_2 is

$$W_e = q_1\Phi_{12} \quad .$$

where Φ_{12} is the potential at A due to the charge q_2 at B. This work is stored as the potential energy of the two charges and is a result of their mutual Coulomb interaction. Now, suppose q_1 is first located at A and q_2 is then introduced from infinity to B; the potential energy of the system would be

$$W_e = q_2\Phi_{21} \quad ,$$

where Φ_{21} is the potential at B due to the charge q_1 at A. Since the configuration of the charges remains unchanged we require that

$$W_e = q_1\Phi_{12} = q_2\Phi_{21} = \frac{1}{2}(q_1\Phi_{12} + q_2\Phi_{21}) \quad . \tag{3.35}$$

We can extend the system to include a third charge q_3. If first q_2 and then q_3 are introduced in the field of q_1, the potential energy of the system will be

$$W_e = q_2\Phi_{21} + q_3(\Phi_{31} + \Phi_{32}) \quad ,$$

since potentials are additive. Then, by virtue of the reciprocal relation, Eq. (3.35), between a pair of charges, this can be rewritten as

$$W_e = \frac{1}{2} q_1(\Phi_{12} + \Phi_{13}) + \frac{1}{2} q_2(\Phi_{23} + \Phi_{21}) + \frac{1}{2} q_3(\Phi_{31} + \Phi_{32})$$

$$= \frac{1}{2} q_1\Phi_1 + \frac{1}{2} q_2\Phi_2 + \frac{1}{2} q_3\Phi_3 = \frac{1}{2} \sum_{s=1}^{3} q_s\Phi_s \quad ,$$

where Φ_s is the potential at the location of q_s due to all the other charges. By induction, the result

$$W_e = \frac{1}{2} \sum_{s=1}^{n} q_s\Phi_s \tag{3.36}$$

applies in general to a system of n point charges located at finite distances from one another (i.e. a closed system). W_e is called the *electrostatic field energy* of the system.

For a closed system of continuous charge distribution of density ρ in a volume V' and surface density σ on surfaces of discontinuity $\Sigma S'$, the electrostatic field energy is obtained by generalizing Eq. (3.36):

$$W_e = \frac{1}{2} \int_{V'} \rho\Phi dV + \frac{1}{2} \int_{\Sigma S'} \sigma\Phi dS \quad . \tag{3.37}$$

Using Maxwell's equation (2.15) and the boundary condition for $\mathbf{D}$, the above can be written as

$$W_e = \frac{1}{2} \int_{V'} \Phi\nabla \cdot \mathbf{D} dV + \frac{1}{2} \int_{\Sigma S'} \Phi(\mathbf{D}_2 - \mathbf{D}_1) \cdot d\mathbf{S} \quad . \tag{3.38}$$

We can enclose the entire system with a large but arbitrary closed surface S_0 of volume V which contains the entire volume V' of the charge distribution and surround the boundary surfaces $\Sigma S'_m$ with closely-fitting closed surfaces $\Sigma S''_m$ as shown in Fig. 3.2. An argument similar to that given in the last section shows that the surface integral in Eq. (3.38) may be replaced by

$$-\frac{1}{2}\oint_{S''_0+\Sigma S''_m} \Phi\, \mathbf{D}\cdot d\mathbf{S} \quad .$$

At the same time, the volume integral in the equation may be taken as over the volume V since ρ, and hence the integrand, is zero outside V'. It is then partially integrated using the divergence theorem:

$$\frac{1}{2}\int_V \Phi\nabla\cdot\mathbf{D}dV = \frac{1}{2}\int_{V''_0+\Sigma V''_m} \nabla\cdot(\Phi\mathbf{D})dV - \frac{1}{2}\int_{V_0} \mathbf{D}\cdot\nabla\Phi dV$$

$$= \frac{1}{2}\oint_{S_0+S''_0+\Sigma S''_m} \Phi\mathbf{D}\cdot d\mathbf{S} - \frac{1}{2}\int_{V_0} \mathbf{D}\cdot\nabla\Phi dV \quad .$$

By making the arbitrary closed surface S_0 sufficiently large, the surface integral over S_0 can be made negligible as the charge distribution is closed[a]. Also, the surface integral over $S''_0 + \Sigma S''_m$ cancels with the second term on the right-hand side of Eq. (3.38). We therefore have

$$W_e = -\frac{1}{2}\int \mathbf{D}\cdot\nabla\Phi dV = \frac{1}{2}\int \mathbf{E}\cdot\mathbf{D}\, dV \quad . \tag{3.39}$$

The integral is now to be taken over all space.

---

[a] Note that for a closed distribution as $\Phi\sim r^{-1}$, $D\sim r^{-2}$ and $dS\sim r^2$, $\oint_{S_0} \sim r^{-1}$.

Thus the electrostatic energy of a closed system of stationary charges can be expressed in terms of the field intensity. If it is hypothesized that this energy is localised in every volume element of the field, we may define the *electrostatic energy density* as

$$U_e = \frac{1}{2}\mathbf{E}\cdot\mathbf{D} = \frac{1}{2}\varepsilon E^2 \quad . \tag{3.40}$$

This expression can neither be justified nor be disproved but must be regarded as a hypothesis which gives the correct total energy of the electrostatic field. Maxwell further postulated that the electrostatic field energy density U_e as defined by Eq. (3.40) can in fact be used to represent the energy density of an electric field in general, not necessarily static.

The energy density U_e and hence the total field energy given by Eq. (3.39) are non-negative while the interaction energy given by Eq. (3.36) can have either sign depending on the relative signs of the interacting charges. The apparent contradiction is due to the fact that in the latter expression we do not include the energy of the interactions between different elements of the same charge, the *self-energy*, which is always positive, and which becomes infinitely large for a point charge. These two expressions are therefore actually different, a fact which serves to emphasize the hypothetical nature of Eq. (3.40).

Consider now a system of n insulated conductors $G_1, G_2, \ldots, G_n$ which carry charges $q_1, q_2, \ldots, q_n$ respectively in a linear, isotropic and homogeneous medium of permittivity ε. As the charge on a conductor is distributed over its surface [Prob. (1.1)] we can write

$$q_\ell = \oint_{G_\ell} \sigma_\ell \, dS_\ell \quad , \tag{3.41}$$

where σ_ℓ is the surface charge density of G_ℓ and the integration is over its surface. The potential at a point $\mathbf{r}_\ell$ on G_ℓ is given by the equivalent of Eq. (3.29):

$$\Phi_\ell = \frac{1}{4\pi\varepsilon} \oint_{\sum_{m=1}^{n} G_m} \frac{\sigma'_m}{\bar{r}_{\ell m}} dS'_m \quad ,$$

where $\bar{r}_{\ell m} = |\mathbf{r}_\ell - \mathbf{r}'_m|$, $\mathbf{r}'_m$ being a point of G_m, and the integral is to be taken over all the conductors of the system. Using Eq. (3.41) we can write

$$\Phi_\ell = \frac{1}{4\pi\varepsilon q_\ell} \oint_{G_\ell} \oint_{\sum_{m=1}^{n} G_m} \frac{\sigma_\ell \sigma'_m dS_\ell dS'_m}{\bar{r}_{\ell m}}$$

$$= \frac{1}{4\pi\varepsilon} \sum_{m=1}^{n} \frac{1}{q_\ell} \oint_{G_\ell} \oint_{G_m} \frac{\sigma_\ell \sigma'_m}{\bar{r}_{\ell m}} dS_\ell \, dS'_m$$

$$= \sum_{m=1}^{n} p_{\ell m} q_m \quad , \tag{3.42}$$

where

$$p_{\ell m} = \frac{1}{q_\ell q_m} \oint_{G_\ell} \oint_{G_m} \frac{\sigma_\ell \sigma'_m}{4\pi\varepsilon \bar{r}_{\ell m}} dS_\ell \, dS'_m \quad .$$

The parameters $p_{\ell m}$ are called *coefficients of potential*. For a given conductor we may assume that the relative charge distribution σ/q does not depend on the total charge q on it. It follows then that $p_{\ell m}$ depends only on the geometrical configuration of the system. Equation (3.42) therefore shows that the potential of a conductor in a system is a linear function of the charges carried by all the conductors of the system. Furthermore in the expression for $p_{\ell m}$ the field point and the source point are interchangeable whether they are on the same conductor or on different conductors. Hence we have the reciprocal relation: $p_{\ell m} = p_{m\ell}$. It may also be noted that, as a positive charge on any one conductor will give rise to positive potentials everywhere if all other conductors are uncharged, $p_{\ell m} \geq 0$.

As $\ell = 1,2,\ldots,n$, Eq. (3.42) represents n linear algebraic equations which can be solved for the charges:

$$q_\ell = \sum_m C_{\ell m}\Phi_m \quad , \tag{3.43}$$

where the parameters $C_{\ell m}$ are called *coefficients of capacitance*. Again these coefficients depend only on the geometrical configuration of the conductors and satisfy the reciprocal relation $C_{\ell m} = C_{m\ell}$.

The electric field energy of a system of conductors can be expressed in terms of either the charges or the potentials:

$$\begin{aligned} W_e &= \frac{1}{2}\sum_\ell q_\ell \Phi_\ell \\ &= \frac{1}{2}\sum_\ell \sum_m p_{\ell m} q_\ell q_m \\ &= \frac{1}{2}\sum_\ell \sum_m C_{\ell m}\Phi_\ell \Phi_m \quad . \end{aligned} \tag{3.44}$$

As a special case, the energy stored in a capacitor or condenser, which is a system of two conductors carrying equal and opposite charges q and $-q$, is

$$W_e = \frac{1}{2}(p_{11} + p_{22} - p_{12} - p_{21})q^2 = \frac{q^2}{2C} = \frac{C}{2}(\Phi_2 - \Phi_1)^2 \quad , \tag{3.45}$$

where Φ_1 and Φ_2 are the potentials of the conductors and $C = (p_{11} + p_{22} - p_{12} - p_{21})^{-1}$ is called the *capacitance* of the capacitor, use having been made of Eq. (3.42).

3.7 Magnetostatic Field of Magnetized Matter

The magnetostatic field is described by the time-independent Maxwell's equations

$$\nabla \times \mathbf{H} = \mathbf{J} \quad , \tag{3.46}$$

$$\nabla \cdot \mathbf{B} = 0 \quad . \tag{3.47}$$

It may be produced either by stationary currents or by stationary magnetized or ferromagnetic bodies. We shall first consider the case where current is everywhere zero and where the magnetic field is produced entirely by stationary and rigid magnetized matter. The first equation now becomes

$$\nabla \times \mathbf{H} = 0 \quad , \tag{3.48}$$

which means that $\mathbf{H}$ can be expressed in terms of a *magnetic scalar potential* Φ_m:

$$\mathbf{H} = - \nabla\Phi_m \quad . \tag{3.49}$$

In a material medium the magnetic intensity is related to the induction through Eq. (1.46), which may be taken here as the definition of the *magnetization* $\mathbf{M}$:

$$\mathbf{M} = \frac{\mathbf{B}}{\mu_o} - \mathbf{H} \quad . \tag{3.50}$$

While $\mathbf{M}$ is usually related to $\mathbf{H}$, for ferromagnetic materials, however, magnetization may exist even in the absence of external excitation. To include such a possibility we shall assume that the magnetization $\mathbf{M}$ may contain a permanent part, which has no relation to $\mathbf{H}$. Substituting the expression (3.49) for $\mathbf{H}$ in Eq. (3.50) we obtain

$$\mathbf{B} = - \mu_o\nabla\Phi_m + \mu_o\mathbf{M} \quad .$$

Equation (3.47) then gives

$$\nabla^2\Phi_m = - \rho_m \quad , \tag{3.51}$$

with $$\rho_m = - \nabla \cdot \mathbf{M} \quad . \tag{3.52}$$

Equation (3.51) has the form of Poisson's equation in electrostatics and we may by analogy define ρ_m as the *"magnetic charge" density*. From Eq. (3.49) we also have the relation

$$\nabla \cdot \mathbf{H} = \rho_m \quad , \tag{3.53}$$

analogous to Eq. (3.2).

Across a surface of discontinuity the boundary conditions for **H** is

$$H_{2t} = H_{1t} \quad . \tag{3.54}$$

Another equation is obtained from the boundary condition for **B**,

$$B_{2n} = B_{1n} \quad ,$$

which gives

$$H_{2n} - H_{1n} = -(M_{2n} - M_{1n}) = \sigma_m \quad , \tag{3.55}$$

where σ_m may be called the *surface magnetic charge density*.

The similarity between the forms of Eqs. (3.52) and (3.30) and Eqs. (3.55) and (3.31) shows that the so-called magnetic charges are "induced" or "equivalent" in nature, similar to the equivalent polarization charges. Their introduction is a matter of convenience rather than from any fundamental necessity, since no single magnetic charges, or monopoles, have been found to exist in nature, as the results obtained in electrostatics can then be applied directly to magnetostatics.

The magnetic scalar potential Φ_m is therefore given by

$$\Phi_m = \frac{1}{4\pi}\int \frac{\rho_m}{\bar{r}}\,dV' + \frac{1}{4\pi}\int \frac{\sigma_m}{\bar{r}}\,dS' = -\frac{1}{4\pi}\int \frac{\nabla'\cdot\mathbf{M}}{\bar{r}}\,dV'$$

$$-\frac{1}{4\pi}\int \frac{1}{\bar{r}}(\mathbf{M}_2 - \mathbf{M}_1)\cdot d\mathbf{S}' \quad ,$$

where the volume integral is to be taken over all regions of space where ρ_m does not vanish and the surface integral over all surfaces of discontinuity, and $\bar{r} = |\mathbf{r} - \mathbf{r}'|$, being the distance of a source point from the field point. By analogy with electrostatics, this can be rewritten as

$$\Phi_m = \frac{1}{4\pi}\int \mathbf{M}\cdot\nabla'\left(\frac{1}{\bar{r}}\right)dV' \qquad . \tag{3.56}$$

integrating over all magnetized bodies. The magnetization $\mathbf{M}$ can now be interpreted as magnetic dipole moment per unit volume.

If the system of magnetized bodies has dimensions small compared with its distance from the field point, i.e. $r' \ll r$, taking the origin within or near the region occupied by the system, then $\bar{r} \simeq r$ and Eq. (3.56) can be written as

$$\Phi_m = -\frac{1}{4\pi}\int \mathbf{M}\cdot\nabla\left(\frac{1}{r}\right)dV' \approx -\frac{1}{4\pi}\nabla\left(\frac{1}{r}\right)\cdot\int \mathbf{M}dV'$$

$$= -\frac{1}{4\pi}\nabla\left(\frac{1}{r}\right)\cdot\mathbf{m} \tag{3.57}$$

where

$$\mathbf{m} = \int \mathbf{M}dV'$$

is the total magnetic dipole moment of the system. Note that by keeping only the first term in the expansion of $\bar{r}^{-1} = |\mathbf{r}-\mathbf{r}'|^{-1}$ we retain the dipole potential only and neglect all other multipole components. The magnetic field intensity under the same approximation is

$$\mathbf{H} = -\nabla\Phi_m = \frac{1}{4\pi}\nabla\left\{\mathbf{m}\cdot\nabla\left(\frac{1}{r}\right)\right\} \qquad . \tag{3.58}$$

3.8 Magnetostatic Field of Stationary Currents

Magnetostatic fields can be produced also by stationary currents. Consider a closed region where no magnetized matter is present, while stationary currents of density $\mathbf{J}$ are prescribed. The full time-independent equations

$$\nabla\times\mathbf{H} = \mathbf{J}$$

and

$$\nabla\cdot\mathbf{B} = 0$$

are to be used. The second equation means that $\mathbf{B}$ can be expressed in terms of a vector potential $\mathbf{A}$:

$$\mathbf{B} = \nabla \times \mathbf{A} \quad . \tag{3.59}$$

If we assume the medium to be linear, isotropic and homogeneous with permeability μ, then Eq. (2.17) applies. Under static conditions, the Lorentz gauge condition becomes $\nabla \cdot \mathbf{A} = 0$. Thus, substituting this and Eq. (3.59) in the first Maxwell's equation above, we find the differential equation for the vector potential in the Lorentz gauge:

$$\nabla^2 \mathbf{A} = - \mu \mathbf{J} \quad . \tag{3.60}$$

Each component of Eq. (3.60) has the form of Poisson's equation and can be integrated separately. The result can again be expressed in vector form

$$\mathbf{A}(\mathbf{r}) = \frac{\mu}{4\pi} \int_V \frac{\mathbf{J}(\mathbf{r}')}{\bar{r}} \, dV' \quad , \tag{3.61}$$

where $\bar{r} = |\mathbf{r} - \mathbf{r}'|$ and the integration is over the entire current field of volume V.

It is convenient to introduce the concept of the *current element.* For currents flowing in linear conductors, i.e., conductors whose normal cross section has dimensions small compared with the distance $\bar{r}$ from the field point, the current density may be assumed to be uniform over its cross section a and furthermore parallel to the conductor. We can write

$$\mathbf{J}\, dV' = \mathbf{J}\, a\, d\mathbf{r}' = I d\mathbf{r}' \quad , \tag{3.62}$$

where I is the current flowing in the conductor and $d\mathbf{r}'$ a length element. If the cross sectional area of the conductor is appreciable it is possible to imagine the conductor as consisting of a bundle of tubes of small cross sections parallel to the current density $\mathbf{J}$, in each of which the current density is uniform. Equation (3.62) can then be applied to each current tube. In either case, the element of a current distribution is the current element $Id\mathbf{r}'$.

The contribution of a current element $Id\mathbf{r}'$ at $\mathbf{r}'$ to the vector potential $\mathbf{A}(\mathbf{r})$, in accordance with Eq. (3.61), is

$$d\mathbf{A} = \frac{\mu}{4\pi} I \frac{d\mathbf{r}'}{\bar{r}} \quad . \tag{3.63}$$

The magnetic induction $\mathbf{B}(\mathbf{r})$ due to the current distribution is therefore

$$\mathbf{B} = \frac{\mu}{4\pi} \nabla \times \int \frac{Id\mathbf{r}'}{\bar{r}} = \frac{\mu}{4\pi} \int I\nabla \times \left(\frac{d\mathbf{r}'}{\bar{r}}\right) = \int d\mathbf{B} \quad ,$$

where the integration is to be taken over all the currents in the distribution. Note that the curl, which is to be taken with respect to the field coordinates, may be moved over under the integral sign which signifies integration over the source coordinates. It follows that

$$d\mathbf{B} = \frac{\mu I}{4\pi} \nabla \times \left(\frac{d\mathbf{r}'}{\bar{r}}\right) \quad ,$$

which assumes that the contributions to the magnetic induction of the current elements are vectorially additive. As

$$\left\{ \nabla \times \left(\frac{d\mathbf{r}'}{\bar{r}}\right) \right\}_i = \varepsilon_{ijk} \frac{\partial}{\partial x_j} \left(\frac{dx'_k}{\bar{r}}\right) = \varepsilon_{ijk} \frac{\partial}{\partial x_j} \left(\frac{1}{\bar{r}}\right) dk'_k$$

$$= \left\{ \nabla \left(\frac{1}{\bar{r}}\right) \times d\mathbf{r}' \right\}_i \quad ,$$

the above can be written as

$$d\mathbf{B} = \frac{\mu I}{4\pi} \bar{\nabla} \left(\frac{1}{\bar{r}}\right) \times d\mathbf{r}' = \frac{\mu}{4\pi} \frac{Id\mathbf{r}' \times \bar{\mathbf{r}}}{\bar{r}^3} \quad , \tag{3.64}$$

making use of Eq. (1.5). This equation is called the *Biot-Savart law.*

3.9 Magnetic Dipole Moment of a Closed Current Filament

The assumption of a system of stationary currents means that all currents therein form closed loops. Consider a loop of current I. As shown in Fig. 3.3, it can be decomposed into a number of arbitrarily small loops, each of current I. Such small current loops are called *closed current filaments*. The magnetic effects of a current distribution can then be considered as the sum of the effects of all such closed current filaments.

Consider a closed current filament C situated near some origin O_s at a large distance r from the field point P as shown in Fig. 3.4. The reciprocal of the distance from P to a current element $I d\mathbf{r}'$ at $\mathbf{r}'$, $\bar{r}^{-1} = |\mathbf{r} - \mathbf{r}'|^{-1}$, can be expanded as a power series in r'/r as was done in Eq. (3.27). The vector potential $d\mathbf{A}(\mathbf{r})$ due to C can then be written as a series. Thus,

$$d\mathbf{A} = \frac{\mu}{4\pi}\oint_C \frac{I d\mathbf{r}'}{\bar{r}}$$

$$= \frac{\mu I}{4\pi r}\oint_C d\mathbf{r}' - \frac{\mu I}{4\pi}\oint_C \left\{\nabla\left(\frac{1}{r}\right)\cdot \mathbf{r}'\right\} d\mathbf{r}' + \text{higher order terms}$$

$$\approx -\frac{\mu I}{4\pi}\oint_C \left\{\nabla\left(\frac{1}{r}\right)\cdot \mathbf{r}'\right\} d\mathbf{r}' \quad , \qquad (3.65)$$

where we have retained the dominant non-vanishing term only. Since

$$\left\{\nabla\left(\frac{1}{r}\right)\cdot \mathbf{r}'\right\} d\mathbf{r}' = \nabla\left(\frac{1}{r}\right)\times(d\mathbf{r}'\times\mathbf{r}') + \left\{\nabla\left(\frac{1}{r}\right)\cdot d\mathbf{r}'\right\}\mathbf{r}'$$

$$= \nabla\left(\frac{1}{r}\right)\times(d\mathbf{r}'\times\mathbf{r}') + d\left[\left\{\nabla\left(\frac{1}{r}\right)\cdot \mathbf{r}'\right\}\mathbf{r}'\right]$$

$$- \left\{\nabla\left(\frac{1}{r}\right)\cdot \mathbf{r}'\right\} d\mathbf{r}' \quad ,$$

we have

$$\left\{\nabla\left(\frac{1}{r}\right)\cdot \mathbf{r}'\right\} d\mathbf{r}' = \frac{1}{2}\nabla\left(\frac{1}{r}\right)\times(d\mathbf{r}'\times\mathbf{r}') + \frac{1}{2} d\left[\left\{\nabla\left(\frac{1}{r}\right)\cdot \mathbf{r}'\right\}\mathbf{r}'\right] \quad .$$

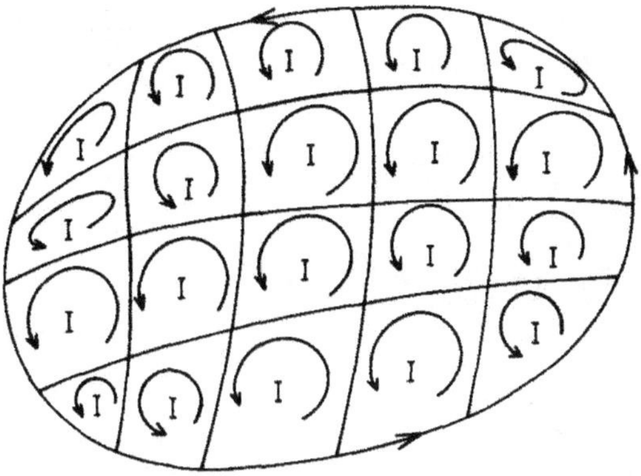

Fig. 3.3 The decomposition of a closed loop carrying a current I into arbitrarily small current filaments, each carrying a current I. Note that the currents carried by neighbouring filaments cancel out, leaving only the current in the original loop.

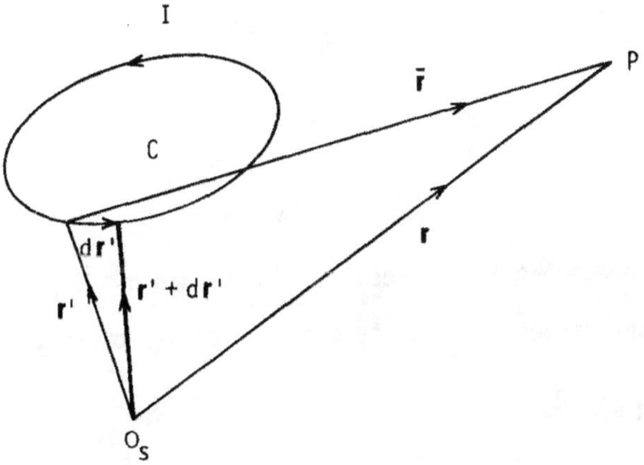

Fig. 3.4 A closed current filament C situated near some origin O_s at a large distance $\mathbf{r}$ from the field point P.

The second term on the right-hand side being a full differential, its integral over a closed loop vanishes so that Eq. (3.65) becomes

$$d\mathbf{A} = -\frac{\mu}{4\pi}\mathbf{m}\times\nabla\left(\frac{1}{r}\right) \quad , \tag{3.66}$$

where

$$\mathbf{m} = I\oint_C \frac{1}{2}\,\mathbf{r}'\times d\mathbf{r}' \quad . \tag{3.67}$$

The vector potential of the entire current distribution is therefore, writing r_s for r for the s-th filament,

$$\mathbf{A} = -\frac{\mu}{4\pi}\sum_s \mathbf{m}_s \times\nabla^{(s)}\left(\frac{1}{r_s}\right) \quad , \tag{3.68}$$

where the summation extends over all closed current filaments in the distribution. The corresponding magnetic intensity is

$$\mathbf{H} = \frac{1}{\mu}\nabla\times\mathbf{A} = -\frac{1}{4\pi}\sum_s \nabla\times\left\{\mathbf{m}_s\times\nabla^{(s)}\left(\frac{1}{r_s}\right)\right\}$$

$$= -\frac{1}{4\pi}\sum_s \nabla^{(s)}\times\left\{\mathbf{m}_s\times\nabla^{(s)}\left(\frac{1}{r_s}\right)\right\} \quad ,$$

as a small displacement of P results in a change of dr_s in r_s.

Thus the magnetic intensity due to a closed current filament is

$$\delta\mathbf{H} = -\frac{1}{4\pi}\nabla\times\left\{\mathbf{m}\times\nabla\left(\frac{1}{r}\right)\right\}$$

$$= -\frac{1}{4\pi}\mathbf{m}\nabla^2\left(\frac{1}{r}\right) + \frac{1}{4\pi}\nabla\left\{\mathbf{m}\cdot\nabla\left(\frac{1}{r}\right)\right\}$$

$$= \frac{1}{4\pi}\nabla\left\{\mathbf{m}\cdot\nabla\left(\frac{1}{r}\right)\right\} \quad , \tag{3.69}$$

for as r is finite $\nabla^2\left(\frac{1}{r}\right)$ vanishes.

A comparison of Eq. (3.69) with Eq. (3.58) shows that at a sufficiently large distance the field of a closed circuit carrying a stationary current I reduces to that of a magnetic dipole of moment $\mathbf{m}$ given by Eq. (3.67). In this derivation we have omitted terms of magnitude r'^2/r^3 in Eq. (3.65). Physically we have neglected the contribution of magnetic multipoles of orders higher than the dipole. Note that the same approximation has been used in the derivation of Eq. (3.58). Equation (3.69) also suggests that to the same approximation the magnetic field of a closed current filament can likewise be expressed in terms of a magnetic scalar potential Φ_m given by Eq. (3.57).

The integrand in Eq. (3.67) represents the vector area of the triangle formed by $\mathbf{r}'$, $\mathbf{r}'+d\mathbf{r}'$ and $d\mathbf{r}'$, so that $\mathbf{m}$ is equal to the product of the current in the closed filament C and the vector area of the cone formed with C as base and O_s as vertex. Let O_s' be a point on the other side of C and form a second cone with C as base and O_s' as vertex. Denote the two cones by K and K'. Since the vector area of a closed surface is equal to zero, we have

$$\oint_{K+K'} d\mathbf{S} = \int_K d\mathbf{S} + \int_{K'} d\mathbf{S} = 0 \quad ,$$

hence
$$\int_K d\mathbf{S} = -\int_{K'} d\mathbf{S} \quad .$$

This means that irrespective of the exact location of O_s and hence the exact shape of the cone K, the integral $\int_K d\mathbf{S}$ has a definite value as long as C forms the base of the cone. Denoting this quantity by $\mathbf{S}$, the dipole moment of the closed current filament is

$$\mathbf{m} = I\,\mathbf{S} \quad . \tag{3.70}$$

It should be emphasized that under the condition $r \gg r'$ the position of O_s, i.e., the location of the equivalent magnetic dipole, is arbitrary insofar as it is in the neighbourhood of the loop C.

It is sometimes useful to write the total dipole moment of a closed current distribution making use of Eqs. (3.67) and (3.62) as

$$\sum_s \mathbf{m}_s = \frac{1}{2}\int_V \mathbf{r}'\times \mathbf{J}dV' = \frac{1}{2}\int_V \rho\mathbf{r}'\times \mathbf{u}dV' \quad , \tag{3.71}$$

where $\mathbf{u}$ is the velocity of the charge in the volume element dV', the integration being over the entire distribution.

In a similar way, the force on a closed current filament can be expressed in terms of $\mathbf{m}$. Using the geometry of Fig. 3.4, the force on the closed current filament is by definition

$$\mathbf{F} = I\oint_C d\mathbf{r}'\times\mathbf{B} \quad .$$

As the dimensions of the loop C is assumed to be small, the i-th component of $\mathbf{B}$ at $\mathbf{r}'$ can be expanded as a *Taylor series:*

$$B_i(\mathbf{r}') = B_i(0) + \frac{\partial B_i}{\partial x_j}x_j' + \ldots \quad ,$$

where $B_i(0)$ is the magnetic induction at $\mathbf{r}'=0$, i.e. the origin O_s, and the derivatives are to be taken at the same point, whence

$$\mathbf{F} = I\left(\oint_C d\mathbf{r}'\right)\times\mathbf{B}(0) + I\oint_C d\mathbf{r}'\times(\mathbf{r}'\cdot\nabla)\mathbf{B} + \ldots \quad .$$

The first term in the series vanishes because its integrand is a full differential. The second term is then the dominant term. If higher order terms are neglected, the i-th component of the force is

$$F_i = \varepsilon_{ijk}\left\{I\oint_C (dx_j')x_\ell'\right\}\frac{\partial B_k}{\partial x_\ell} \quad .$$

The integral in the above equation can be transformed [Prob. (3.35)] into

$$I \oint_C (dx'_j)x'_\ell = I \oint_C d(x'_j x'_\ell) - I \oint_C x'_j dx'_\ell = - I \oint_C x'_j dx'_\ell$$

$$= \frac{1}{2} I \oint_C \{(dx'_j)x'_\ell - x'_j dx'_\ell\} = \varepsilon_{s\ell j} m_s \quad ,$$

where $\mathbf{m}$ is the dipole moment of the filament as given by Eq. (3.67). Hence

$$F_i = \varepsilon_{jki}\varepsilon_{js\ell} m_s \frac{\partial B_k}{\partial x_\ell} = (\delta_{ks}\delta_{i\ell} - \delta_{k\ell}\delta_{is}) m_s \frac{\partial B_k}{\partial x_\ell}$$

$$= m_k \frac{\partial B_k}{\partial x_i} - m_i \frac{\partial B_k}{\partial x_k} = \mathbf{m} \cdot \frac{\partial \mathbf{B}}{\partial x_i}$$

where we have made use of the relation $\nabla \cdot \mathbf{B} = 0$. In the present case we can also write

$$\mathbf{F} = \nabla(\mathbf{m} \cdot \mathbf{B}) \quad . \tag{3.72}$$

Note that $\mathbf{m}$ is given by Eq. (3.70) and does not depend on the exact location of O_s. It is seen that the force increases with the inhomogenuity of the magnetic field and vanishes if the field is uniform. Furthermore, Eq. (3.72) shows that if we interpret the force as the negative gradient of a potential function U, then

$$U = - \mathbf{m} \cdot \mathbf{B} \tag{3.73}$$

is the potential energy of a closed filament of dipole moment $\mathbf{m}$ in the magnetic field.

3.10 Magnetization Current

In the above discussion we have seen that, if we choose, we may describe the magnetic effects of a closed stationary current distribution in terms of equivalent magnetic dipoles. We can therefore assign a magnetization $\mathbf{M}$, which is simply the magnetic dipole moment per unit volume, to a current distribution. Equation (3.71) suggests that the magnetization due to the current density $\mathbf{J}$ is

$$\mathbf{M} = \frac{1}{2}\,\mathbf{r}' \times \mathbf{J} = \frac{\rho}{2}\,\mathbf{r}' \times \mathbf{u} \qquad . \tag{3.74}$$

This means that magnetization with respect to a point will occur if the current has a component transverse to the radius vector $\mathbf{r}'$ drawn from that point, or equivalently, if the charges have a net rotational motion about that point.

Equation (3.74) also suggests that conversely, we may describe the magnetic effects of magnetized matter in terms of a distribution of equivalent currents called the *magnetization currents* of density $\mathbf{J}_m$ as follows. Consider a magnetized body with magnetization $\mathbf{M}(\mathbf{r}')$. Its vector potential at a distant field point $\mathbf{r}$ is given by the integral form of Eq. (3.68):

$$A(\mathbf{r}) = -\frac{\mu}{4\pi}\int_V \mathbf{M}(\mathbf{r}') \times \bar{\nabla}\left(\frac{1}{\bar{r}}\right) dV' \qquad , \tag{3.75}$$

where the integration is over the volume V of the entire magnetized body. Note that while Eq. (3.68) was derived for a distribution of stationary currents, the identification of $\mathbf{M}$ with the magnetic dipole moment per unit volume makes it applicable to a distribution of magnetic dipoles. By Eq. (1.5),

$$-\mathbf{M} \times \bar{\nabla}\left(\frac{1}{\bar{r}}\right) = \mathbf{M} \times \nabla'\left(\frac{1}{\bar{r}}\right) = -\nabla' \times \left(\frac{\mathbf{M}}{\bar{r}}\right) + \frac{1}{\bar{r}}\,\nabla' \times \mathbf{M} \qquad .$$

Hence

$$\mathbf{A} = -\frac{\mu}{4\pi}\int_V \nabla' \times \left(\frac{\mathbf{M}}{\bar{r}}\right) dV' + \frac{\mu}{4\pi}\int_V \frac{\nabla' \times \mathbf{M}}{\bar{r}}\, dV' \qquad .$$

The first integral on the right-hand side can be integrated partially, the i-th component being

$$\varepsilon_{ijk}\int_V \frac{\partial}{\partial x_j'}\left(\frac{M_k}{\bar{r}}\right) dV' = \varepsilon_{ijk}\oint_S \frac{M_k}{\bar{r}}\, dS_j' = -\oint_S \left(\frac{\mathbf{M}}{\bar{r}} \times d\mathbf{S}'\right)_i$$

using the corollary of the divergence theorem (A11). The volume of integration V can be taken slightly larger than the volume of the magnetized body since **M** vanishes outside the body. The integral over its boundary surface S will then vanish. Hence

$$\mathbf{A} = \frac{\mu}{4\pi}\int_V \frac{\nabla' \times \mathbf{M}}{\bar{r}}\, dV' \tag{3.76}$$

integrating over the volume of the magnetized body. A comparison of Eq. (3.76) with Eq. (3.61) shows that volume magnetization can be expressed in terms of an equivalent current, the *magnetization current* or the *Ampèrian current* which was a density of

$$\mathbf{J}_m(\mathbf{r}') = \nabla' \times \mathbf{M}(\mathbf{r}') \quad . \tag{3.77}$$

The magnetization current, which was first suggested by Ampère to explain the magnetic properties of matter may be interpreted as representing the inaccessible microscopic currents of the orbiting atomic electrons. According to Eq. (3.77), the microscopic currents cancel in a region of uniform magnetization. $\mathbf{J}_m$ is the net current density arising from incomplete cancellation of the microscopic currents when magnetization is non-uniform.

Note that, although the integrals in both Eqs. (3.61) and (3.76) are over the entire distribution and not an arbitrary volume, and Eq. (3.77) strictly speaking does not follow therefore, it may be derived for every volume element of the distribution of magnetized matter directly from Eq. (3.70). This is shown in the following.

Consider adjacent volume elements as shown in Fig. 3.5. The difference in the magnetization of adjacent volumes may be obtained by Taylor's expansion similar to the expansion of **B** in the previous section:

$$\mathbf{M}(\mathbf{r}+d\mathbf{r}) - \mathbf{M}(\mathbf{r}) = (d\mathbf{r}\cdot\nabla)\mathbf{M} + \ldots \quad .$$

Consider the z-component of the magnetic moments of two adjacent volume elements along the x-axis. Eq. (3.70) may be written for the

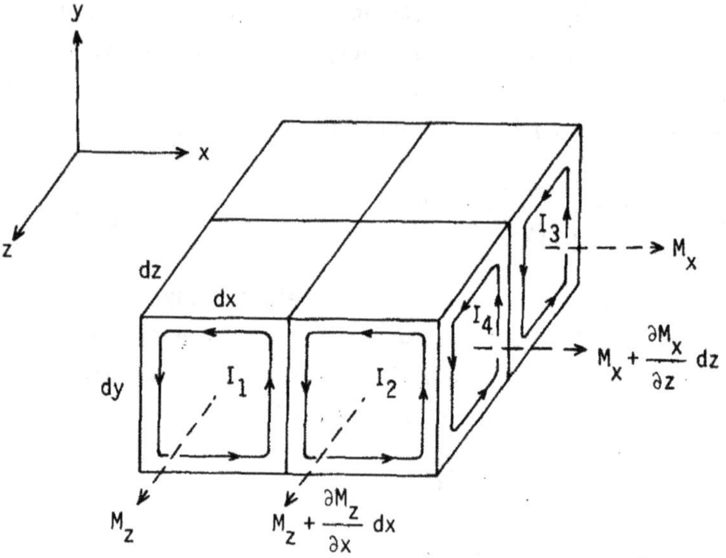

Fig. 3.5 Adjacent volume elements of magnetized matter of magnetization **M**. I_1 and I_2 are Ampèrian currents circulating about axes in the z-direction, and I_3, I_4 are Ampèrian currents circulating about axes in the x-direction.

first volume as

$$M_z dxdydz = I_1 dxdy \quad ,$$

where I_1 is the Ampèrian or magnetization current circulating in this volume about an axis in the z-direction. Similarly, for the second element we have

$$(M_z + \frac{\partial M_z}{\partial x} dx)dxdydz = I_2 dxdy \quad ,$$

neglecting the higher order terms in Taylor's expansion. Subtracting we find

$$-\frac{\partial M_z}{\partial x} dxdz = I_1 - I_2 \quad .$$

Similarly, for adjacent volumes along the z-axis we have

$$\frac{\partial M_x}{\partial z} dxdz = I_4 - I_3 \qquad .$$

The net current in the y-direction for an effective area of dxdz arising from incomplete cancellation of Ampèrian currents is therefore

$$I_1 - I_2 + I_4 - I_3 = (\frac{\partial M_x}{\partial z} - \frac{\partial M_z}{\partial x})dxdz \qquad ,$$

i.e.

$$J_y = (\nabla \times \mathbf{M})_y \qquad .$$

It follows that, in general, the Ampèrian or magnetization current is

$$\mathbf{J}_m = \nabla \times \mathbf{M} \qquad .$$

In conclusion it can be said that the magnetic field of magnetized matter is identical with that of a distribution of stationary currents. In analysing the field of a magnetized body, we can consider it either as the manifest of the magnetic charges induced by magnetization, or as arising from the inaccessible closed current filaments, the Ampèrian currents, in its interior. In either case, a magnetic scalar potential or a vector potential may be used. Although the two treatments are mathematically equivalent, the former treatment is purely phenomenological while the latter is fundamental as it can be incorporated in the general frame work of an electromagnetic theory of charge-field interactions without having to postulate the separate existence of magnetic charges, real or otherwise. This view is also consistent with the modern atomic theory.

PROBLEMS

1. Laplace's equation in cylindrical coordinates is

$$\frac{1}{\rho}\frac{\partial}{\partial\rho}\left(\rho\frac{\partial\Phi}{\partial\rho}\right)+\frac{1}{\rho^2}\frac{\partial^2\Phi}{\partial\phi^2}+\frac{\partial^2\Phi}{\partial z^2}=0 \quad .$$

Show that by assuming $\Phi(\rho, \phi, z) = R(\rho)Q(\phi)Z(z)$ it can be separated into three equations:

$$\frac{d^2Q}{d\phi^2} + m^2Q = 0 \quad ,$$

$$\frac{d^2Z}{dz^2} - k^2Z = 0 \quad ,$$

and
$$\frac{d^2R}{d\rho^2} + \frac{1}{\rho}\frac{dR}{d\rho} + \left(k^2 - \frac{m^2}{\rho^2}\right)R = 0 \quad ,$$

where m and k are constants.

The R-equation is *Bessel's equation* in $k\rho$ and its solutions, finite at $\rho = 0$, are Bessel's functions of the first kind of order m, $\mathcal{J}_m(k\rho)$. Show that the general solution of Laplace's equation which is single-valued and finite everywhere, and which, furthermore, vanishes at $\rho = a$ is

$$\Phi(\rho, \phi, z) = \sum_{m=0}^{\infty}\sum_{n=1}^{\infty}\mathcal{J}_m(k_{mn}\rho)(A_{mn}\sin m\phi + B_{mn}\cos m\phi)e^{-k_{mn}|z|} \quad ,$$

where k_{mn} is the n-th positive root of $\mathcal{J}_m(k_{mn}a) = 0$ divided by a, and A_{mn} and B_{mn} are constants.

2. For problems where the z coordinate is not involved, Laplace's equation in cylindrical coordinates is reduced to

$$\frac{1}{\rho}\frac{\partial}{\partial\rho}\left(\rho\frac{\partial\Phi}{\partial\rho}\right)+\frac{1}{\rho^2}\frac{\partial^2\Phi}{\partial\phi^2}=0 \quad .$$

Separate the variables by considering solutions of the form $\Phi = R(\rho)Q(\phi)$ and obtain the general solution

$$\phi = A \ln \rho + B + \sum_{m=1}^{\infty} (a_m \rho^m + b_m \rho^{-m}) \cos(m\phi + \delta_m) \quad ,$$

where A, B, a_m, b_m and δ_m are constants. (Hint: Integrate the R-equation for $m = 0$ and $m \geq 1$ separately).

A long coaxial line consists of a cylindrical conductor of radius a and a hollow conducting cylinder of inner radius $b > a$. The region $a < \rho < b$ is filled with a linear, isotropic and homogeneous dielectric of permittivity ε. The inner conductor is earthed while the outer conductor is charged to a fixed potential V. Using the above general solution to show that the potential distribution in the dielectric is

$$\phi = \frac{V}{\ln(b/a)} \ln \left(\frac{\rho}{a}\right)$$

and find the electric intensity there. From the boundary condition for **D** find the surface charge density for both conductors. Hence show that the capacitance per unit length of the coaxial line is

$$C = \frac{2\pi\varepsilon}{\ln(b/a)} \quad .$$

3. A long hollow cylindrical conductor of radius a is cut into two halves by a plane through its axis which is taken to be z-axis. The two halves are electrically insulated from each other and kept at potentials Φ_1 and Φ_2. Taking the x-axis along a direction so that the gaps are at $\phi = \pm\pi/2$, show that the potential at a point of cylindrical coordinates (ρ, ϕ, z) within the cylinder is

$$\Phi = \frac{1}{2}(\Phi_1 + \Phi_2) + \frac{2}{\pi}(\Phi_1 - \Phi_2) \sum_{m=1}^{\infty} \frac{(-1)^{m-1}}{(2m-1)} \left(\frac{\rho}{a}\right)^{2m-1} \cos(2m-1)\phi \quad .$$

Hint: The solution of Laplace's equation in cylindrical coordinates which is independent of the z-coordinate and which satisfies the conditions that Φ is finite at $\rho = 0$ and that $\Phi(-\phi) = \Phi(\phi)$ may be written as

$$\Phi = \sum_{n=0}^{\infty} a_n \rho^n \cos n\phi \quad .$$

To evaluate a_m multiply both sides with $\cos m\phi$ and integrate from 0 to 2π, noting that $\Phi(\rho = a) = \Phi_1$ for $-\frac{\pi}{2} \leq \phi \leq \frac{\pi}{2}$ and that $\Phi(\rho = a) = \Phi_2$ for $\frac{\pi}{2} \leq \phi \leq \frac{3\pi}{2}$.

4. A long conducting cylinder of circular cross section of radius a is placed in an electric field with its axis coinciding with the z-axis. In the absence of the cylinder, the field is uniform with intensity E_o and is parallel to the x-axis. If the cylinder is maintained at a potential Φ_S, show that the potential at an external point is

$$\Phi = \Phi_S + \left(\frac{a^2}{\rho} - \rho\right) E_o \cos \phi$$

and find the electric intensity.

5. Show that the two-dimensional Laplace's equation in Cartesian coordinates

$$\frac{\partial^2 \Phi}{\partial x^2} + \frac{\partial^2 \Phi}{\partial y^2} = 0$$

has a general solution $\Phi = (A \sin kx + B \cos kx)(C \sinh ky + D \cosh ky)$, where A, B, C, D and k are constants. Note that k may be real or imaginary.

A long hollow conducting tube has a rectangular cross section of internal dimensions a and b. The three conducting walls $x = 0$, $x = a$ and $y = 0$ are connected and are earthed. The fourth wall, $y = b$, is insulated from the others and maintains a potential distribution $V \sin(\pi x/a)$, where V is a constant. Show that the potential distribution inside the tube is given by

$$\Phi = \frac{V}{\sinh(\pi b/a)} \sin\frac{\pi x}{a} \sinh\frac{\pi y}{a} \quad .$$

6. Surface currents flow in two parallel infinite planes $y = b$ and $y = -b$. The linear current densities are parallel to the z-axis and are $I_0 \sin(\frac{\pi}{a}x)$ and $-I_0 \sin(\frac{\pi}{a}x)$ respectively. Solve Laplace's equation for the magnetic potential Φ_m between the planes and show that

$$\Phi_m = \frac{a}{\pi} I_0 \operatorname{csch}\frac{\pi b}{a} \cos\frac{\pi}{a}x \sinh\frac{\pi}{a}y \quad .$$

Hence find the magnetic intensity.

7. The conjugate of a point $\mathbf{r}$ with respect to a sphere of radius a centred at the origin is defined as the point $\mathbf{r}' = (a^2/r^2)\mathbf{r}$. The transformation which transforms every point into its conjugate is called an *inversion transformation*. Show that Laplace's equation is invariant with respect to an inversion transformation, i.e. if $\Phi(r, \theta, \phi)$ is a solution of Laplace's equation, then $\Phi'(r', \theta, \phi) = \frac{a}{r'} \Phi(\frac{a^2}{r'}, \theta, \phi)$ is a solution of Laplace's equation, $\nabla'^2\Phi' = 0$, in the inverted space. This is the basis for the method of inversion in boundary-value problems.

8. A point charge q_s at $\mathbf{r}_s$ can be represented by a charge distribution in space of density $\rho_s(\mathbf{r}') = q_s\delta(\mathbf{r}' - \mathbf{r}_s)$. By means of Eq. (3.26) show that the potential of a system of point charges $q_1, q_2, \ldots, q_n$ located at $\mathbf{r}_1, \mathbf{r}_2, \ldots, \mathbf{r}_n$ is

$$\Phi(\mathbf{r}) = \frac{1}{4\pi\varepsilon} \sum_{s=1}^{n} \frac{q_s}{|\mathbf{r} - \mathbf{r}_s|} \quad .$$

9. An infinite cylinder of radius a is uniformly charged with a volume density ρ. Show by direct integration of Poisson's equation that the potential of the cylinder at a point r from the axis is

$$\Phi(r) = -\frac{\rho r^2}{4\varepsilon} \qquad \text{for} \qquad 0 \le r \le a \quad ,$$

$$\Phi(r) = \frac{a^2}{4}\frac{\rho}{\varepsilon}\left\{2\ln\left(\frac{a}{r}\right) - 1\right\} \qquad \text{for} \qquad r \ge a \quad ,$$

assuming that $\Phi(0) = 0$.

10. For the purpose of field computations the presence of a cavity in a volume distribution of density ρ is equivalent to the superposition of additional charges of density $-\rho$ in the cavity. Show that the field in an infinite cylindrical cavity of circular cross section in an infinite uniformly charged circular cylinder of density ρ is uniform, being

$$\mathbf{E} = \frac{1}{2}\frac{\rho}{\varepsilon_0}\mathbf{r}_0 \quad ,$$

where $\mathbf{r}_0$ is the displacement of the axis of the cavity from that of the cylinder. What is the geometry for the field in the cavity to vanish?

11. A spherical surface of radius a and potential $\Phi(a, \theta, \phi)$ encloses a charge distribution of volume density $\rho(r, \theta, \phi)$. Show by the method of images that for the potential $\Phi(\mathbf{r})$ of an interior point the Green's function to be used is

$$G(\mathbf{r}', \mathbf{r}) = \frac{1}{|\mathbf{r}' - \mathbf{r}|} - \frac{a}{r\left|\mathbf{r}' - \frac{a^2}{r^2}\mathbf{r}\right|}$$

and write down $\Phi(\mathbf{r})$ as the sum of volume and surface integrals.

12. Show that Eq. (3.22) is a solution of Laplace's equation.

13. Show that for a charge-free region the electrostatic potential at a point is equal to the average potential of an arbitrary spherical surface centred at that point. Hint: use Eq. (3.22) and Gauss' flux theorem.

14. An *equipotential surface* in an electric field is defined as the surface at all points of which the potential is equal to some prescribed constant. Thus the equation of an equipotential surface is given by

$$\Phi(\mathbf{r}) = \Phi_0$$

where Φ_0 is a constant. The different values of Φ_0 give rise to a family of equipotential surfaces. Find the equation of equipotential surfaces for (a) a system of two charges, q situated at $x = a$, and $-q$ at $x = -a$, and (b) an electric field $\mathbf{E} = x^2\mathbf{i} + yz^2\mathbf{j} + y^2z\mathbf{k}$.

15. Two straight lines, each of length 2ℓ, are parallel to the z-axis and carry charges per unit length ρ and $-\rho$. Their mid-points are in the xy plane and have coordinates $(d, 0, 0)$ and $(-d, 0, 0)$ respectively. Find the potential Φ at a point $(x, y, 0)$ and show that in the limit of $\ell \to \infty$, $\Phi = \frac{\rho}{2\pi\varepsilon_0} \ln k$, where $k = \left\{\frac{(x+d)^2 + y^2}{(x-d)^2 + y^2}\right\}^{\frac{1}{2}}$. Hence show that the equipotential surfaces due to the infinite line charges are the family of surfaces given by

$$\left[x - \frac{d(k^2+1)}{k^2-1}\right]^2 + y^2 = \left(\frac{2dk}{k^2-1}\right)^2$$

for different constant values of k. Note that their intercepts on the xy-plane give rise to non-concentric circles of radii $\frac{2dk}{k^2-1}$ with centres at $x = \frac{d(k^2+1)}{k^2-1}$.

Now if the lines are replaced by infinitely long cylindrical conductors of radius a with axes at $(b, 0, 0)$ and $(-b, 0, 0)$

respectively their surfaces must themselves be equipotential surfaces. Show that their surfaces correspond to

$$k = \frac{b \pm \sqrt{b^2 - a^2}}{a} .$$

Hence find the potentials of the conductors and show that the capacity per unit length of the parallel conductors is

$$C = \frac{\pi\varepsilon_0}{\ln\left\{(b + \sqrt{b^2 - a^2})/a\right\}} .$$

16. In a vector field lines of force may be drawn which are directed curves such that the forward drawn tangent at any point has the direction of the vector there. Thus in an electric field the element $d\boldsymbol{\ell}$ of a line of force is related to $\mathbf{E}$ by $d\boldsymbol{\ell} = \lambda\mathbf{E}$, where λ is a scalar factor. The differential equation of the lines of force is therefore

$$\frac{dx}{E_x} = \frac{dy}{E_y} = \frac{dz}{E_z}$$

in Cartesian coordinates, or

$$\frac{dr}{E_r} = \frac{rd\theta}{E_\theta} = \frac{r\sin\theta\, d\phi}{E_\phi}$$

in spherical coordinates.

Find the equations of the lines of force for (a) a system of two charges, q situated at $x = a$, and $-q$ at $x = -a$; (b) the field given by the potential

$$\Phi = \frac{\cos\theta \sin\phi}{r^2} .$$

17. A system consists of two conductors carrying equal and opposite charges. Show that its capacitance is

$$C = \frac{C_{11}C_{22} - C_{12}^2}{C_{11} + C_{22} + 2C_{12}}$$

where C_{11}, C_{22} and C_{12} are the coefficients of capacitance. What is the capacitance in terms of the coefficients of potential?

18. The capacitance of an isolated conductor is defined as the ratio of the charge on the conductor to its potential. Two isolated conductors have capacitances C_1 and C_2. A system is formed with these two conductors by placing them in vacuum at a distance r from each other, where r is large compared with the linear dimensions of either conductor. Show that the coefficients of potential of the system are

$$P_{11} = C_1^{-1} \quad , \qquad P_{22} = C_2^{-1} \quad ,$$

$$P_{12} = P_{21} = \frac{1}{4\pi\varepsilon_o r} \quad .$$

Hence show that the coefficients of capacitance are approximately

$$C_{11} \simeq C_1\left\{1 + \frac{C_1C_2}{(4\pi\varepsilon_o)^2r^2}\right\} \quad ,$$

$$C_{12} = C_{21} \simeq -\frac{C_1C_2}{4\pi\varepsilon_o r} \quad ,$$

$$C_{22} \simeq C_2\left\{1 + \frac{C_1C_2}{(4\pi\varepsilon_o)^2r^2}\right\} \quad .$$

19. An isolated system consists of two conductors such that conductor 1 is placed inside conductor 2 which is hollow. By means of Gauss' flux theorem show that (a) if q_1 is the charge on

conductor 1, then the inner surface of conductor 2 carries a charge $-q$, and (b) the field in the external region, as well as the potential Φ_2 of conductor 2, depends only on the charge q on the outer surface of conductor 2.

If the capacitance C of a capacitor is defined as the ratio of the magnitude of the charge on one of its conductors to the potential difference between the conductors, show that C is equal to the coefficient of capacitance $C_{11} = -C_{12}$.

20. Prove *Green's reciprocity theorem;* if charges $q_1, q_2, \ldots, q_n$ on n conductors produce potentials $\Phi_1, \Phi_2, \ldots, \Phi_n$ respectively on the conductors, and if charges $q_1', q_2', \ldots, q_n'$ on the same conductors produce potentials $\Phi_1', \Phi_2', \ldots, \Phi_n'$ respectively, then

$$\sum_{s=1}^{n} q_s' \Phi_s = \sum_{s=1}^{n} q_s \Phi_s' \quad .$$

21. A sphere of radius a is uniformly charged with a volume density ρ. Show by means of Eqs. (3.37) and (3.39) that the electrostatic energy of the distribution is $4\pi a^5 \rho^2/(15\varepsilon_0)$. How is the energy shared between the regions inside and outside the sphere?

22(a). The configuration of a system of stationary charges with f degrees of freedom can be specified by generalized coordinates $q_1, q_2, \ldots, q_f$. Show that the generalized forces Q_s are given by

$$Q_s = -\frac{\partial W}{\partial q_s} \quad ,$$

where $W(q_1, \ldots, q_f)$ is the energy of the system.

(b) Show that the energy of a dipole $\mathbf{p}$ in an electrostatic field of intensity $\mathbf{E}$ is

$$W = -\mathbf{p} \cdot \mathbf{E} \quad .$$

Hence show that the force on the dipole is $(\mathbf{p} \cdot \nabla)\mathbf{E}$.

(c) Show that the force per unit volume on a dielectric of permittivity ε in an electrostatic field of intensity $\mathbf{E}$ is

$$\mathbf{f} = \frac{\varepsilon - \varepsilon_0}{2} \nabla(E^2) \quad .$$

23. A right circular dielectric cylinder of length ℓ and radius a is placed with its axis along the z-axis and one of its ends on the xy-plane. If the polarization $\mathbf{P}$ is uniform and is parallel to the axis, find the electric potential at a point $z > \ell$ along the axis. Hence show that the electric field along the axis at that point is

$$E_z = \frac{P}{2\varepsilon_0} \left\{ \frac{\ell - z}{\sqrt{a^2 + (z - \ell)^2}} + \frac{z}{\sqrt{a^2 + z^2}} \right\} \quad .$$

24. An electret is a substance which exhibits polarization in the absence of external electric field. Show that for the field of a uniform electret the integral $\frac{1}{2}\int \mathbf{E} \cdot \mathbf{D}\, dV$ taken over all space vanishes. (Hint: Poisson's equation for an electret is $\nabla^2\Phi = \nabla \cdot \mathbf{P}/\varepsilon_0$).

25. Show that the magnetic induction at a distant point P due to a closed loop carrying a current I is

$$\mathbf{B} = -\frac{\mu_0 I}{4\pi} \nabla\Omega \quad ,$$

where Ω is the solid angle subtended at P by the loop.

26. Two long, straight parallel wires carry the same current I in opposite directions. Show that at a point distances r_1 and r_2 from the wires the vector potential is

$$\mathbf{A} = \frac{\mu_0 I}{2\pi} \ln \left(\frac{r_2}{r_1}\right)\mathbf{n}$$

where $\mathbf{n}$ is a unit vector along the direction of the wires.

Find the magnetic induction if the wires are separated by a distance d.

27. An axially symmetric magnetic field is represented by a vector potential with components $A_\phi(\rho, z)$ and $A_\rho = A_z = 0$ in cylindrical coordinates. Show that the equation for the lines of magnetic induction is

$$\rho A_\phi(\rho, z) = \text{constant} \quad .$$

28. A circular loop of radius a carrying a steady current I has its axis coinciding with the z-axis. Show that at a point (r, θ, ϕ) the vector potential has only one component, the ϕ-component, given by

$$A_\phi = \frac{\mu_0}{4\pi} Ia \int_0^{2\pi} \frac{\cos(\phi' - \phi)d\phi'}{\{r^2 - 2ra \sin\theta \cos(\phi' - \phi) + a^2\}^{\frac{1}{2}}} \quad .$$

Hence show that for $r \gg a$, A_ϕ is the same as that given by Eq. (3.66).

29. Show that at a point near the middle of a long straight wire of length ℓ along the z-axis carrying a current I the vector potential is

$$\mathbf{A} = - \mathbf{i}_z \frac{\mu_0 I}{2\pi} \ln\left(\frac{r}{\ell}\right) \quad ,$$

where r is the distance from the wire. Hence derive an expression for the magnetic induction **B** at that point and compare it with that obtained by means of the circuital law. (Hint: use cylindrical coordinates)

30. In the previous problem, if the current density **J** is calculated from $\nabla \times \mathbf{B} = \mu_0 \mathbf{J}$, a zero value is obtained except for $r \to 0$, which is a singular point. To obtain the current, it is necessary to assume the wire to have a finite cross section and to integrate **J** over the cross sectional area. Let it be a circle of area a

and radius r_o. Then using

$$\int_a J dS = \frac{1}{\mu_o} \int_a \nabla \times \mathbf{B} \cdot d\mathbf{S}$$

and Stokes' theorem evaluate the current flowing in the wire.

31. A uniform sphere of radius a and magnetic permeability μ is placed in a uniform magnetic field $\mathbf{H}_o$. By analogy with the case of a dielectric sphere in a uniform electric field, show that inside the sphere the magnetic field is

$$\mathbf{H} = \frac{3\mu_o}{\mu + 2\mu_o} \mathbf{H}_o$$

while outside the sphere the magnetic field is equal to the sum of $\mathbf{H}_o$ and the field due to a magnetic dipole of moment

$$\mathbf{m} = 4\pi a^3 \left(\frac{\mu - \mu_o}{\mu + 2\mu_o}\right) \mathbf{H}_o$$

32. A magnetic cylinder of infinite length has a uniform circular cross section of radius a and a constant permeability μ_1. It is introduced into a magnetic field in free space, which in the absence of the cylinder is uniform and has intensity $\mathbf{H}_o$, such that $\mathbf{H}_o$ is perpendicular to the axis of the cylinder. Show that in cylindrical coordinates with the z-axis coinciding with the axis of the cylinder and the x-axis parallel to $\mathbf{H}_o$ the magnetic scalar potentials inside and outside the cylinder are respectively

$$\Phi_m^- = -\frac{2\mu_o}{\mu_1 + \mu_o} H_o \rho \cos\phi$$

and $$\Phi_m^+ = \left(\frac{\mu_1 - \mu_o}{\mu_1 + \mu_o}\frac{a^2}{\rho^2} - 1\right) H_o \rho \cos\phi \quad .$$

Hence find the corresponding magnetic fields and the magnetization vector in the cylinder.

33. A spherical shell of radius a and having a uniform surface charge density σ rotates with angular velocity ω about a fixed axis through its centre which is taken as the z-axis. To calculate the vector potential at a distant point of radius vector $\mathbf{r}$ which makes an angle θ with the z-axis, the shell may be considered as consisting of parallel layers each having the form of a circular loop. Then using Eq. (3.70) show that the shell has a total magnetic dipole moment of

$$\mathbf{m} = \frac{4}{3}\pi\sigma\omega a^4 \mathbf{i}_z \quad .$$

Hence show that the vector potential at $\mathbf{r}$ is $\dfrac{\mu_0\sigma\omega a^4 \sin\theta}{3r^2}\mathbf{i}_\phi$ and calculate the magnetic induction at the point. (Hint: First use the equation $\int \mathbf{J}dV = \int I d\boldsymbol{\ell}$ to show that a charge q distributed uniformly over a circular ring rotating with angular velocity ω is equivalent to a circular loop carrying a current $\frac{q\omega}{2\pi}$.)

34. Show that the magnetic force per unit volume on a magnetized body of permeability μ in a magnetic field of induction $\mathbf{B}$ is

$$\mathbf{f} = \frac{1}{2}\left(\frac{1}{\mu_0} - \frac{1}{\mu}\right)\nabla(B^2) \quad .$$

The body is assumed to have linear properties.

35. Show that from the definition $\mathbf{m} = \frac{1}{2} I \oint_C \mathbf{r}\times d\mathbf{r}$ we can write

$$\frac{1}{2} I \oint_C (x_i dx_j - x_j dx_i) = \varepsilon_{ijk} m_k \quad .$$

Hence show that the torque $\mathbf{L}$ on a small closed current filament in a magnetic field of induction $\mathbf{B}$ is approximately

$$\mathbf{L} = \mathbf{m}\times\mathbf{B} \quad .$$

36. Show that for the magnetostatic field of a finite distribution of uniform permanent magnetization in a linear, isotropic and homogeneous medium

$$\frac{1}{2}\int \mathbf{B}\cdot\mathbf{H}\, dV = 0 \quad ,$$

where the integration is over all space.

37(a). The magnetic scalar potential of a closed loop carrying current I can be evaluated by first decomposing it into small loops of magnetic dipole moment $I\mathbf{S}$, where $\mathbf{S}$ is its vector area. A circular loop of radius a carrying a current I lies in the xy plane with its centre at the origin. Show that the magnetic scalar potential at an axial point is

$$\Phi_m(z) = \frac{I}{2}\left(1 - \frac{z}{\sqrt{a^2+z^2}}\right) \quad .$$

Expand Φ_m into a series valid for $z < a$.

(b) The magnetic scalar potential can also be determined from Laplace's equation. Write $\Phi_m(r, \theta, \phi)$ as a series in spherical coordinates and determine the coefficients from the requirement that for an axial point near the origin the expression should reduce to the series obtained in (a). Hence show that the components of the magnetic intensity at an off-axial point are

$$H_r = \frac{I}{2a}\left\{\cos\theta - \frac{3r^2}{16a^2}(5\cos 3\theta + 3\cos\theta) + \ldots\right\} \quad ,$$

$$H_\theta = \frac{I}{2a}\left\{-\sin\theta + \frac{3r^2}{16a^2}(5\sin 3\theta + \sin\theta) + \ldots\right\} \quad .$$

Chapter IV

ELECTROMAGNETIC FIELDS — TIME-DEPENDENT FIELDS

If time variation is not negligible a solution of Maxwell's equations will involve both the electric and magnetic fields. Small time variations which need to be taken into account in the first approximation only give rise to quasi-stationary fields, for which the effect of radiation may be neglected. Rapidly varying fields are characterized by the presence of a radiation field showing the dissipation of energy through electromagnetic radiation. Simple examples of radiating systems are the accelerated charge, the Hertzian dipole and the loop antenna. In general, a radiation field may be broken down into the various multipole components.

4.1 Maxwell's Equations for Quasi-Stationary Fields

If an electromagnetic field varies but slowly with time it is said to be *quasi-static* or *quasi-stationary*. Time variation is considered slow if two criteria are satisfied:

(a) The variation is such that the displacement current density $|\dot{\mathbf{D}}|$ is negligible compared with the conduction current density $|\mathbf{J}|$ in the conductors where the currents flow.

(b) The period of variation is large compared with the time taken for electromagnetic effects to propagate through the region of interest, e.g. from one end of the apparatus of an experiment to the other or from a source point to the field point, so that the velocity of propagation in the medium, in which the conductors are imbedded, may be treated as infinite.

To put these criteria in a more quantitative way, consider a field which varies sinusoidally with time with an angular frequency ω:

$$\mathbf{E} = \mathbf{E}_0 e^{-i\omega t} \quad ,$$

where, as in what follows, the real part is the part understood to be physically significant. The criterion (a) requires that

$$\sigma E >> \omega \varepsilon E \quad , \quad \text{or} \quad \omega << \frac{\sigma}{\varepsilon} \quad ,$$

i.e. the frequency of field oscillation is very much smaller than the reciprocal of the relaxation time of the conductors. For metals in general: $\varepsilon \simeq \varepsilon_0$, $\sigma \simeq 10^7$ $\text{ohm}^{-1}\text{m}^{-1}$, we require $\omega << 10^{18}$ s^{-1}. This is satisfied for frequencies up to the optical region.

Criterion (b) requires that the linear dimention ℓ of the region of interest and the period T of field variation should satisfy

$$\frac{\ell}{v} << T = \frac{2\pi}{\omega} \quad , \quad \text{or} \quad \omega << \frac{2\pi v}{\ell} \quad ,$$

where v is the velocity of the electromagnetic waves in the medium. In air or vacuum $v = 3 \times 10^8 \text{m s}^{-1}$, ℓ is usually of the order of meters, so that this condition requires the frequency to be smaller than about 10^8 Hz, much more stringent than that required by criterion (a). Occasionally the term quasi-stationary also applies to cases where criterion (b) is not satisfied, such as in the theory of transmission lines.

If criterion (a) is satisfied Maxwell's equations can be

written as

$$\nabla \times \mathbf{H} = \mathbf{J} \quad , \tag{4.1}$$

$$\nabla \times \mathbf{E} = -\dot{\mathbf{B}} \quad , \tag{4.2}$$

$$\nabla \cdot \mathbf{B} = 0 \tag{4.3}$$

and

$$\nabla \cdot \mathbf{D} = \rho \quad . \tag{4.4}$$

These equations differ from those for the static fields by the presence of $\dot{\mathbf{B}}$ in Eq. (4.2). This difference arises from the interaction between the electric and magnetic fields and prevents their separate analysis.

If criterion (b) is satisfied, we can take v to be infinitely large and write the differential equations for the potentials in the Lorentz gauge as

$$\nabla^2 \Phi = -\frac{\rho}{\varepsilon} \tag{4.5}$$

and

$$\nabla^2 \mathbf{A} = -\mu \mathbf{J} \quad , \tag{4.6}$$

which are the same as for static fields. In terms of the potentials the electric and magnetic fields are given respectively by

$$\mathbf{E} = -\nabla\Phi - \dot{\mathbf{A}} \tag{4.7}$$

and

$$\mathbf{B} = \nabla \times \mathbf{A} \quad . \tag{4.8}$$

A complication arises from the fact that curl $\mathbf{E}$ no longer vanishes. The electric intensity $\mathbf{E}$ in general will not be conservative as in the electrostatic field. This is due of course to the interconversion of energy between electric and magnetic fields. Furthermore, in addition to the electric field $\mathbf{E}$ which is part of an electromagnetic field, we may often have to deal with the field $\mathbf{E}'$ set up by applied electromative forces. The current density $\mathbf{J}$ in a conductor of conductivity σ will then be given by

$$\mathbf{J} = \sigma(\mathbf{E} + \mathbf{E}') \quad . \tag{4.9}$$

Under the criteria for quasi-stationary fields, we include most of the electromagnetic fields encountered in power engineering as well as in laboratory circuitry. These form an important part of the study of electromagnetism and are thoroughly treated elsewhere. Only some general principles derivable from Maxwell's equations will be considered here. Other important topics, the skin effect and transmission lines, will however be left to a later chapter.

4.2 Kirchhoff's Equations

Consider a system of linear conductors which form part of a linear circuit and integrate Eq. (4.9) along a closed conducting path C:

$$\oint_C \frac{\mathbf{J}\cdot d\mathbf{r}}{\sigma} = \oint_C \mathbf{E}\cdot d\mathbf{r} + \oint_C \mathbf{E}'\cdot d\mathbf{r} \quad . \tag{4.10}$$

The first integral on the right-hand side can be transformed with the help of Stokes' theorem:

$$\oint_C \mathbf{E}\cdot d\mathbf{r} = -\oint_C (\nabla\Phi + \dot{\mathbf{A}})\cdot d\mathbf{r}$$

$$= -\frac{d}{dt}\oint_C \mathbf{A}\cdot d\mathbf{r} = -\frac{d}{dt}\int_S \nabla\times\mathbf{A}\cdot d\mathbf{S}$$

$$= -\frac{d}{dt}\int_S \mathbf{B}\cdot d\mathbf{S} = -\frac{d}{dt}\mathscr{F} \quad ,$$

where S is an open, two-sided surface bounded by C and $\mathscr{F}$ the total magnetic flux crossing this surface. The second integral on the right-hand side of Eq. (4.10) is by definition the applied electromotive force ε. If the path C should cross several sites of applied electromotive forces, then

$$\varepsilon = \sum_s \varepsilon_s \quad .$$

For a linear conductor, $\mathbf{J}\cdot d\mathbf{r} = J d\ell$, where $d\ell$ is a length element. The integral on the left-hand side of Eq. (4.10) is then

$$\oint_C \frac{J d\ell}{\sigma} = \int_{C_1 + C_2 + \ldots} \frac{J d\ell}{\sigma} = \sum_m I_m \int_{C_m} \frac{d\ell}{a_m \sigma_m} = \sum_m I_m R_m \tag{4.11}$$

where C_m represents the m-th segment of C, which has a cross section of area a_m and carries current I_m. Note that if the current density is uniform over a_m, then $I_m = J_m a_m$. For high frequencies the current density will not be uniform. However, in this case we may regard the segment as comprising a bundle of N parallel tubes of equal small cross sectional area α, in each of which the current density may be considered uniform. The mean current density is then

$$J_m = \frac{1}{N} \sum_{s=1}^{N} \frac{I_s}{\alpha} = \frac{I_m}{N\alpha} = \frac{I_m}{a_m} ,$$

where I_s is the current in the s-th tube. The relation $I_m = J_m a_m$ again applies with J_m interpreted as the mean current density. The integral $\int (a\sigma)^{-1} d\ell$, which depends only on the geometry and conductivity of the conductor, and possibly on the frequency of the current flowing, but not on its magnitude, is by definition the *resistance* R of the conductor.

Equation (4.10) can then be written as

$$\sum_m I_m R_m = \sum_s \varepsilon_s - \frac{d}{dt} \mathscr{F} \tag{4.12}$$

This equation is one of Kirchhoff's equations.

The other Kirchhoff's equation is obtained by considering a junction in a circuit. If there is no accumulation of charge at that point the continuity equation requires that

$$\nabla \cdot \mathbf{J} = - \frac{\partial \rho}{\partial t} = 0 \quad .$$

Integrating over a small volume V of boundary surface S enclosing the junction, as $\rho = 0$ everywhere in V we find that

$$\int_V \nabla \cdot \mathbf{J} dV = \oint_S \mathbf{J} \cdot d\mathbf{S} = 0 \quad ,$$

using the divergence theorem. The surface integral represents the net current leaving V. As S is arbitrary as long as it encloses the junction, this can be written as the second Kirchhoff's equation

$$\sum_s I_s = 0 \quad ,$$

where I_s is the s-th current leaving the junction.

4.3 Magnetic Field Energy

Consider a process in which magnetic field is set up by the application of an electromotive force $\oint \mathbf{E}' \cdot d\mathbf{r}$ to a closed circuit C. Scalar-multiply both sides of Eq. (4.9) with $\mathbf{J}$ and integrate over all space to give

$$\int \mathbf{E}' \cdot \mathbf{J}\, dV = \int \frac{J^2}{\sigma} dV - \int \mathbf{E} \cdot \mathbf{J}\, dV \quad . \qquad (4.13)$$

The last integral on the right-hand side can be transformed using Maxwell's equations and Gauss' theorem:

$$\begin{aligned} \int \mathbf{E} \cdot \mathbf{J}\, dV &= \int \mathbf{E} \cdot \nabla \times \mathbf{H}\, dV \\ &= \int \mathbf{H} \cdot \nabla \times \mathbf{E}\, dV - \int \nabla \cdot (\mathbf{E} \times \mathbf{H}) dV \\ &= -\int \mathbf{H} \cdot \dot{\mathbf{B}}\, dV - \oint \mathbf{E} \times \mathbf{H} \cdot d\mathbf{S} \quad . \end{aligned}$$

The surface integral involves an infinite surface. As the distance r of a surface element from the region of interest becomes arbitrarily large, $E \sim r^{-2}$, $H \sim r^{-2}$, $dS \sim r^2$, and the integral $\sim r^{-2} \to 0$ as $r \to \infty$. Equation (4.13) can therefore be written as

$$\int \mathbf{E}' \cdot \mathbf{J}\, dV = \int \frac{J^2}{\sigma}\, dV + \int \mathbf{H} \cdot \dot{\mathbf{B}}\, dV$$

$$= \int \frac{J^2}{\sigma}\, dV + \frac{d}{dt}\left\{\frac{1}{2}\int \mathbf{H} \cdot \mathbf{B}\, dV\right\} \quad , \qquad (4.14)$$

where we have made use of the linearity of the medium.

To interpret this equation we consider a simple linear circuit C of uniform cross section a carrying a low-frequency current I. The integrand $\mathbf{E}' \cdot \mathbf{J}$ of the left-hand side represents the work done per unit volume per unit time by the applied field $\mathbf{E}'$ on the charges whose motion gives rise to the current I (Sec. 2.6). The integral therefore represents the rate at which the applied electromotive force is doing work. As $\mathbf{J}$ vanishes outside C and is uniform over a, the first integral on the right-hand side can be written

$$\int \frac{\mathbf{J} \cdot \mathbf{J}}{\sigma}\, dV = I \oint_C \frac{\mathbf{J} \cdot d\mathbf{r}}{\sigma} = I^2 \oint_C \frac{d\ell}{a\sigma} = I^2 R \quad ,$$

where R is the resistance of the circuit. By Joule's law I^2R is the rate at which energy is converted into heat in the circuit.

The second term on the right-hand side of Eq. (4.14), which involves the magnetic field vectors only, must then be interpreted as the rate at which energy is stored up in the magnetic field. Since initially no magnetic field is present, the integral

$$W_m = \frac{1}{2}\int \mathbf{H} \cdot \mathbf{B}\, dV \quad , \qquad (4.15)$$

which is to be taken over all space, must be considered as representing the energy of the magnetic field set up by the process.

In the expression for W_m, only the final magnetic field strength appears but not the nature of the process setting up the field.

It may thus be considered as representing the magnetic field energy in general, irrespective of origin. If, as in the case of the electric field energy, the magnetic energy is assumed to be localized at every point of the field, we may define the *magnetic energy density* by

$$U_m = \frac{1}{2}\mathbf{H}\cdot\mathbf{B} \quad . \tag{4.16}$$

It was further suggested by Maxwell that this may be used to represent the energy density of magnetic fields in general, stationary or time-varying.

In this section and in Sec. 3.6 we have demonstrated for electrostatic and quasi-stationary fields how expressions for the electric and magnetic field energies may be obtained. If localization of the field energies is assumed we can also obtain expressions for the energy densities. Maxwell furthermore suggested the extension of the validity of these formulae to electric and magnetic fields in general. On account of the hypothetical nature of these formulae they should rank as additional assumptions, and not as derived laws. We must also bear in mind the inadequate knowledge we have of the role played by self-energy. Nevertheless, the concept of localized energies and the use of Eqs. (3.40) and (4.16) to represent the field energy densities have been found to be both plausible and useful. Accepting these assumptions the energy density of an electromagnetic field is

$$U = U_e + U_m = \frac{1}{2}\mathbf{E}\cdot\mathbf{D} + \frac{1}{2}\mathbf{H}\cdot\mathbf{B} \quad . \tag{4.17}$$

Equation (4.15) can also be written in terms of the vector potential :

$$\begin{aligned} W_m &= \frac{1}{2}\int \mathbf{H}\cdot\nabla\times\mathbf{A}\,dV \\ &= \frac{1}{2}\int \nabla\cdot(\mathbf{H}\times\mathbf{A})dV + \frac{1}{2}\int \mathbf{A}\cdot\nabla\times\mathbf{H}\,dV \\ &= \frac{1}{2}\oint \mathbf{H}\times\mathbf{A}\cdot d\mathbf{S} + \frac{1}{2}\int \mathbf{A}\cdot\mathbf{J}\,dV = \frac{1}{2}\int \mathbf{A}\cdot\mathbf{J}\,dV \quad . \end{aligned} \tag{4.18}$$

The surface integral vanishes as the volume integral is to be taken over all space. In practice of course the volume integral vanishes outside the current distribution.

Equation (4.18) enables us to find the energy of the magnetic field of a system of n linear closed circuits $C_1, C_2, \ldots, C_n$ in a medium of permeability μ carrying currents $I_1, I_2, \ldots, I_n$ of densities $\mathbf{J}_1, \mathbf{J}_2, \ldots, \mathbf{J}_n$ respectively. It gives

$$W_m = \frac{\mu}{8\pi}\int_{\Sigma C_s}\int_{\Sigma C_t}\frac{\mathbf{J}_s\cdot\mathbf{J}_t}{\bar{r}_{st}}\,dV_s dV_t' = \frac{\mu}{8\pi}\sum_{s=1}^{n}\sum_{t=1}^{n}\int_{C_s}\int_{C_t}\frac{\mathbf{J}_s\cdot\mathbf{J}_t}{\bar{r}_{st}}\,dV_s dV_t' ,$$

where $\bar{r}_{st} = |\mathbf{r}_s - \mathbf{r}_t'|$ is the distance between a "field" point and a "source" point which may be on the same or on different circuits. As $\mathbf{J}\,dV = I d\mathbf{r}$ for linear conductors, the above may be written as

$$W_m = \frac{\mu}{8\pi}\sum_s\sum_t I_s I_t \oint_{C_s}\oint_{C_t}\frac{d\mathbf{r}_s\cdot d\mathbf{r}_t'}{\bar{r}_{st}} = \frac{1}{2}\sum_s\sum_t L_{st} I_s I_t , \tag{4.19}$$

where

$$L_{st} = \frac{\mu}{4\pi}\oint_{C_s}\oint_{C_t}\frac{d\mathbf{r}_s\cdot d\mathbf{r}_t'}{\bar{r}_{st}} ,$$

which depend only on the geometrical configuration of the conductors and the permeability of the medium but are independent of the currents, are called *coefficients of mutual induction* if $s \neq t$, or *coefficients of self-induction* if $s = t$. It is obvious from the last formula that $L_{st} = L_{ts}$.

In the special case where one conductor only is present, Eq. (4.19) reduces to

$$W_m = \frac{1}{2} L I^2 . \tag{4.20}$$

L being the coefficient of self-induction of the conductor.

The circuit equation, (Eq. 4.12), can also be obtained from a consideration of energy conservation, which states that

$$I\varepsilon = \frac{dW_e}{dt} + \frac{dW_m}{dt} + \int_V \frac{J^2}{\sigma} dV \quad . \tag{4.21}$$

Here $I\varepsilon$ is the rate at which energy is supplied by the source of the electromotive force, since by definition ε is the work done in moving a unit charge once round the circuit. The integral over the volume V of the conductors of the circuit represents the rate of Joule heating. Consider for example a simple series circuit comprising a capacitor C, a self-inductor L and a resistance R. As

$$W_e = \frac{1}{2}\frac{q^2}{C} \quad , \quad W_m = \frac{1}{2} LI^2 \quad , \quad \int \frac{J^2}{\sigma} dV = I^2 R \quad ,$$

Equation (4.21) can be written as

$$\varepsilon = \frac{q}{C} + L\dot{I} + IR \quad .$$

That this equation is identical with Eq. (4.12) as applied to the present example may be seen from the following. Equation (3.45) for a capacitor gives $\frac{q}{C} = \Phi_2 - \Phi_1$, which is the potential difference between the two conductors of the capacitor and may be considered as a back electromotive force. The total magnetic flux crossing the circuit is by definition

$$\mathscr{F} = \int_S \mathbf{B} \cdot d\mathbf{S} = \int_S \nabla \times \mathbf{A} \cdot d\mathbf{S} = \oint \mathbf{A} \cdot d\mathbf{r} = \frac{\mu I}{4\pi} \oint\oint \frac{d\mathbf{r}' \cdot d\mathbf{r}}{\bar{r}} = LI$$

where the surface integrals are over a surface S bounded by the circuit and the closed line integrals are over the circuit. Differentiating we have

$$\frac{d\mathscr{F}}{dt} = L\dot{I} \quad .$$

The application of the circuit equation to a series circuit

containing a resistance R, inductance L, capacitance C and an alternating source of electromotive force $\varepsilon = \varepsilon_0 \exp(-i\omega t)$ yields the familiar result $\varepsilon = ZI$, where $Z = R + i\left(\frac{1}{\omega C} - \omega L\right)$ is the *impedance* of the circuit.

4.4 Potentials of a Rapidly Varying Field

For a rapidly varying field, full Maxwell's equations must be used and, as in the case of a quasi-stationary field, it will not be possible to solve for the electric and the magnetic field separately. The electromagnetic field is specified by vector and scalar potentials **A** and Φ, which in a linear, isotropic and homogeneous medium of permittivity ε and permeability μ are given by the differential equations

$$\nabla^2 \mathbf{A} - \frac{1}{v^2}\ddot{\mathbf{A}} = -\mu \mathbf{J} \tag{4.22}$$

and

$$\nabla^2 \Phi - \frac{1}{v^2}\ddot{\Phi} = -\frac{\rho}{\varepsilon} \quad , \tag{4.23}$$

where

$$v = (\mu\varepsilon)^{-\frac{1}{2}} \quad .$$

These equations differ from the corresponding equations under static or quasi-static conditions by the time derivative term $\frac{1}{v^2}\frac{\partial^2}{\partial t^2}$. On the other hand, in a source-free region the equations are reduced to wave equations, showing that the electromagnetic field propagates through the medium as waves with a phase velocity v. The time derivative terms in Eqs. (4.22) and (4.23) take explicit account of the finite velocity v of propagation of electromagnetic effects, for if the velocity of propagation were infinite these equations would not be different from the static equations. Then, if conditions at the source are rapidly varying, the finite time taken for these effects to go from a source point to the field point cannot be ignored. The observed field would then be expected to respond to conditions at the source at an earlier time, the interval being equal to the time taken for the effects to propagate from the source to the observer. We would therefore expect the solutions of these equations to be similar to those for the static case, but with due allowance made for time retardation. The solutions turn out as expected to be

$$\mathbf{A}(\mathbf{r}, t) = \frac{\mu}{4\pi} \int \frac{\mathbf{J}(\mathbf{r}', t')dV'}{\bar{r}} \tag{4.24}$$

and

$$\Phi(\mathbf{r}, t) = \frac{1}{4\pi\varepsilon} \int \frac{\rho(\mathbf{r}', t')dV'}{\bar{r}} \quad , \tag{4.25}$$

where $\bar{r} = |\mathbf{r} - \mathbf{r}'|$, $t' = t - \bar{r}/v$, and the integrals are to be taken over all space.

The implicit dependence of $\mathbf{J}$ and ρ on the time t' means that, in evaluating the potentials at the field point at time t, the values of these quantities to be used are those appropriate to a time $\bar{r}/v$ earlier. These quantities are therefore said to be *retarded* and the potentials are known as *retarded potentials*. It is seen that if the distribution of these quantities should remain stationary or if the velocity v should become infinitely large, the potentials would reduce to the static potentials as required.

The solutions (4.24) and (4.25) may be obtained by *Kirchhoff's method* of integration, similar to the technique used in the integration of *Poisson's equation*. The procedure is however rather involved and will not be given here. It suffices to show that these expressions do satisfy the respective differential equations. It should perhaps be remarked that mathematically solutions with $t' = t + \dfrac{\bar{r}}{v}$ are equally acceptable. Physically, these additional solutions would imply that effect preceeds cause and are for this reason discarded.

Consider as example the scalar potential Φ. We can divide the entire space into two parts V_1 and V_2; V_1 is a very small volume containing the field point and V_2 the remaining space. Equation (4.25) can then be written as

$$\Phi = \frac{1}{4\pi\varepsilon}\int_{V_1} \frac{[\rho]dV'}{\bar{r}} + \frac{1}{4\pi\varepsilon}\int_{V_2} \frac{[\rho]dV'}{\bar{r}} = \Phi_1 + \Phi_2 \quad ,$$

say, *enclosing the retarded quantities with square brackets.*

We can make V_1 so small that the time of propagation of electromagnetic effects within it is negligible. Hence

$$\Phi_1(\mathbf{r}, t) = \frac{1}{4\pi\varepsilon}\int_{V_1} \frac{\rho(\mathbf{r}', t)dV'}{\bar{r}} \quad .$$

It follows that

$$\nabla^2\Phi_1(\mathbf{r}, t) = \frac{1}{4\pi\varepsilon}\int_{V_1} \rho(\mathbf{r}', t)\nabla^2 \left(\frac{1}{\bar{r}}\right)dV'$$

$$= \frac{1}{4\pi\varepsilon}\int_{V_1} \rho(\mathbf{r}', t)\nabla'^2 \left(\frac{1}{\bar{r}}\right)dV'$$

$$= -\frac{1}{\varepsilon}\int_{V_1} \rho(\mathbf{r}', t)\delta(\mathbf{r}' - \mathbf{r})dV'$$

$$= -\frac{\rho(\mathbf{r}, t)}{\varepsilon} \quad , \tag{4.26}$$

implying that

$$\frac{1}{v^2}\frac{\partial^2}{\partial t^2}\Phi_1 = 0 \tag{4.27}$$

when it is compared with Eq. (4.23). Note that as $\nabla'^2\left(\frac{1}{\bar{r}}\right) = 0$ everywhere except at $\bar{r} = 0$, V_1 can be made arbitrarily small as long as it contains the field point $\mathbf{r}$.

Next consider

$$\nabla^2\Phi_2 = \frac{1}{4\pi\varepsilon}\int_{V_2} \nabla^2\left\{\frac{[\rho]}{\bar{r}}\right\}dV' = \frac{1}{4\pi\varepsilon}\int_{V_2} \bar{\nabla}^2\left\{\frac{[\rho]}{\bar{r}}\right\}dV' \quad .$$

As $[\rho]$ is a function of x_i' and $t-\frac{\bar{r}}{v}$, it is a function of x_i through $\bar{r}$ only. Then as far as partial differentiation with respect to the field coordinates x_i is concerned, $[\rho]/\bar{r}$ is a function of $\bar{r}$. Relation (1.5) is therefore applicable. Furthermore, since for a function $f(r)$ we have

$$\nabla^2 f(r) = \frac{1}{r}\frac{d^2}{dr^2}\{rf(r)\} \quad ,$$

the above becomes

$$\nabla^2\Phi_2 = \frac{1}{4\pi\varepsilon}\int_{V_2}\frac{1}{\bar{r}}\frac{\partial^2}{\partial\bar{r}^2}[\rho]dV' \quad .$$

With respect to the variables $\bar{r}$ and t, $[\rho]$ is a function of $(t-\bar{r}/v)$ and is thus a solution of the wave equation (Sec. 2.8)

$$\frac{\partial^2}{\partial\bar{r}^2}[\rho] = \frac{1}{v^2}\frac{\partial^2}{\partial t^2}[\rho] \quad .$$

It follows that

$$\begin{aligned}\nabla^2\Phi_2 &= \frac{1}{4\pi\varepsilon}\int_{V_2}\frac{1}{v^2}\frac{1}{\bar{r}}\frac{\partial^2}{\partial t^2}[\rho]dV' \\ &= \frac{1}{v^2}\frac{\partial^2}{\partial t^2}\left\{\frac{1}{4\pi\varepsilon}\int_{V_2}\frac{[\rho]}{\bar{r}}dV'\right\} \\ &= \frac{1}{v^2}\frac{\partial^2\Phi_2}{\partial t^2} \quad . \end{aligned} \tag{4.28}$$

Combining the Eqs. (4.26) to (4.28), we find

$$\nabla^2\Phi = \nabla^2\Phi_1 + \nabla^2\Phi_2 = -\frac{\rho}{\varepsilon}+\frac{1}{v^2}\frac{\partial^2}{\partial t^2}(\Phi_1+\Phi_2) = -\frac{\rho}{\varepsilon}+\frac{1}{v^2}\frac{\partial^2\Phi}{\partial t^2} \quad ,$$

or $$\nabla^2\Phi - \frac{1}{v^2}\frac{\partial^2\Phi}{\partial t^2} = -\frac{\rho}{\varepsilon}$$

as required.

We shall use the retarded potentials to obtain the electromagnetic fields of a moving charge and some simple radiating systems. To avoid any complication that may arise from the variation of the properties of material media with the frequency of the electromagnetic field, we shall mainly consider fields in free space only.

4.5 The Liénard-Wiechert Potentials

For the application of Eqs. (4.24) and (4.25) we may imagine the space as being divided into regions by concentric spheres with the centre at the field point P(r). For the region bounded by spheres of radii $\bar{r}$ and $\bar{r}+d\bar{r}$, the densities to be used in the integrals are those appropriate to the retarded time $t' = t - \bar{r}/c$. Alternatively, we may imagine the information about the space surrounding P as being collected by an *information sphere* centred at P, whose radius $\bar{r}$ contracts at the uniform rate of $\frac{d\bar{r}}{dt'} = -c$ from infinity to become zero at the instant t, at which time the potentials at P are to be evaluated. The information on an element of charge or current at a distance $\bar{r}$ from P is picked up by the sphere as it sweeps across it at the instant $t - \bar{r}/c$. We shall make use of this picture to obtain the potentials of a point charge q moving in free space with a velocity u comparable to but smaller than the velocity of light c.

The point charge is assumed to have some unspecified structure in a small volume, all elements of which are rigidly connected to one another and have at a given time t' the same velocity $\mathbf{u}(t')$. Considering the integral for Φ, we see that each element $[\rho]dV'$ is to be evaluated at a different retarded time. Consequently the integral $\int [\rho]dV'$ does not represent the total charge. It will be necessary, therefore, before going into the limit of a point charge, first to transform the integral into one over the charge element dq which is encountered by the information sphere once and once only.

If the system of charge were stationary, the amount that would be encountered by an area element dS of the information sphere in the time interval dt' during which the radius decreases by $d\bar{r}$, is $[\rho]d\bar{r}dS = [\rho]dV'$. However, as the charge has a radial velocity u_r toward P, an amount of charge $[\rho u_r]dt'dS = [\rho u_r]dV'/c$ will have left the volume dV' and not be encountered by the sphere during dt' (see Fig. 4.1). The amount of charge actually encountered is therefore

$$dq = [\rho]\left[1 - \frac{u_r}{c}\right]dV' = [\rho]\left[1 - \frac{\mathbf{u}\cdot\bar{\mathbf{r}}}{\bar{r}c}\right]dV' \quad ,$$

giving

$$[\rho]dV' = \frac{dq}{\left[1 - \frac{\mathbf{u}\cdot\bar{\mathbf{r}}}{\bar{r}c}\right]} \quad .$$

Similarly we have $[\mathbf{J}]dV' = [\rho\mathbf{u}]dV' = \dfrac{[\mathbf{u}]dq}{\left[1 - \frac{\mathbf{u}\cdot\bar{\mathbf{r}}}{\bar{r}c}\right]}$.

Substitution in the expressions for the retarded potentials gives

$$\mathbf{A} = \frac{\mu_0}{4\pi}\frac{[\mathbf{u}]}{\left[\bar{r} - \frac{\mathbf{u}\cdot\bar{\mathbf{r}}}{c}\right]}\int dq = \frac{\mu_0}{4\pi}\frac{[\mathbf{u}]q}{\left[\bar{r} - \frac{\mathbf{u}\cdot\bar{\mathbf{r}}}{c}\right]} \tag{4.29}$$

and

$$\Phi = \frac{1}{4\pi\varepsilon_0}\frac{1}{\left[\bar{r} - \frac{\mathbf{u}\cdot\bar{\mathbf{r}}}{c}\right]}\int dq = \frac{1}{4\pi\varepsilon_0}\frac{q}{\left[\bar{r} - \frac{\mathbf{u}\cdot\bar{\mathbf{r}}}{c}\right]} \quad , \tag{4.30}$$

in the limit of a point charge, for as every part of q is encountered once and once only by the information sphere the integral of dq gives q. These potentials are known as the *Liénard-Wiechert potentials*. Note that for $u/c \to 0$, these become the *Biot-Savart* and the *Coulomb potentials*.

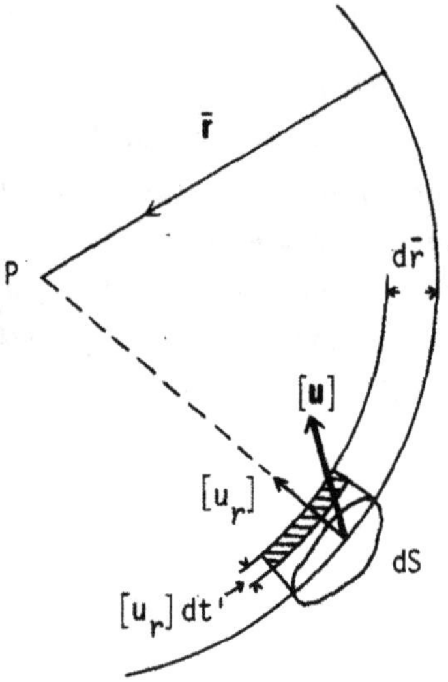

Fig. 4.1 In the time interval dt' a surface element dS of the information sphere sweeps across a volume $dV' = d\bar{r}\,dS$. In the same interval, an amount of charge $[\rho u_r]dt'\,dS = [\rho u_r]d\bar{r}dS/c$ will have left the volume dV'.

4.6 Field of a Uniformly Moving Charge

The Liénard-Wiechert potentials are expressed in terms of the retarded position and velocity of a point charge. These can be expressed as functions of the "present" quantities in the special case of uniform motion. This is possible because, as we shall see later, uniformly moving charge does not radiate energy.

Let the retarded position of a charge q be $Q'(\mathbf{r}_0')$, its "present" position be $Q(\mathbf{r}')$, the radius vectors of the field point $P(\mathbf{r})$ relative to Q' and Q be $[\mathbf{r}]$ and $\mathbf{r}$ respectively as shown in Fig. 4.2. The displacement of the charge during the retardation time is

$$\overrightarrow{Q'Q} = \mathbf{u}(t - t') = \mathbf{u}\frac{[\bar{r}]}{c} = [\bar{\mathbf{r}}] - \bar{\mathbf{r}} \qquad . \qquad (4.31)$$

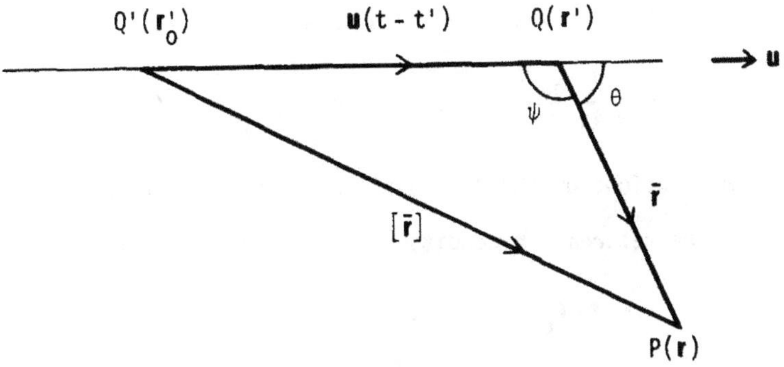

Fig. 4.2 A charge q moves with uniform velocity $\mathbf{u}$ from Q' to Q in the time interval t - t', during which the electromagnetic effects originated from the charge when it was at Q' reaches the field point P.

If we denote

$$[\bar{r}] - \frac{\mathbf{u}\cdot[\bar{\mathbf{r}}]}{c} = s, \tag{4.32}$$

then as $(\mathbf{u}\times\bar{\mathbf{r}})^2 = (\mathbf{u}\times[\bar{\mathbf{r}}])^2 = u^2[\bar{\mathbf{r}}]^2 - (\mathbf{u}\cdot[\bar{\mathbf{r}}])^2$ we have, making use of Eq. (4.31),

$$s^2 = [\bar{r}]^2 - \frac{2\mathbf{u}\cdot[\bar{\mathbf{r}}]}{c}[\bar{r}] + \frac{1}{c^2}(\mathbf{u}\cdot[\bar{\mathbf{r}}])^2$$

$$= ([\bar{\mathbf{r}}] - \frac{\mathbf{u}[\bar{r}]}{c})^2 - \frac{1}{c^2}(\mathbf{u}\times\bar{\mathbf{r}})^2$$

$$= \bar{r}^2 - \frac{1}{c^2}(\mathbf{u}\times\bar{\mathbf{r}})^2 = \bar{r}^2 - \frac{1}{c^2}u^2\bar{r}^2 + \frac{1}{c^2}(\mathbf{u}\cdot\bar{\mathbf{r}})^2$$

$$= \alpha\bar{r}^2 + \frac{1}{c^2}(\mathbf{u}\cdot\bar{\mathbf{r}})^2 \quad , \tag{4.33}$$

where

$$\alpha = 1 - (\frac{u}{c})^2 \quad . \tag{4.34}$$

The Liénard-Wiechert potentials

$$\mathbf{A} = \frac{\mu_o}{4\pi}\frac{q\mathbf{u}}{s} \qquad , \qquad \Phi = \frac{1}{4\pi\varepsilon_o}\frac{q}{s}$$

are now functions of the "present" position through s.

The retardation condition Eq. (4.31) may be written as

$$\bar{\mathbf{r}} = \mathbf{r} - \mathbf{r}_o' - \mathbf{u}(t - t') \qquad , \tag{4.35}$$

and the differentiation of s with respect to the field quantities $\mathbf{r}$ and t is to be effected through differentiation with respect to $\bar{\mathbf{r}}$. As

$$\nabla = \bar{\nabla} \qquad , \qquad \frac{\partial}{\partial t} = \frac{\partial \bar{x}_j}{\partial t}\frac{\partial}{\partial \bar{x}_j} = -(\mathbf{u}\cdot\bar{\nabla}) \qquad ,$$

the field vectors are given by

$$\mathbf{E} = -\bar{\nabla}\Phi + (\mathbf{u}\cdot\bar{\nabla})\mathbf{A} \qquad , \qquad \mathbf{B} = \bar{\nabla}\times\mathbf{A} \qquad .$$

Differentiating Eq. (4.33), we find

$$\frac{\partial s}{\partial \bar{x}_i} = \frac{1}{s}\left(\alpha\bar{x}_i + \frac{\mathbf{u}\cdot\bar{\mathbf{r}}}{c^2}u_i\right) \qquad ,$$

or

$$\bar{\nabla}s = \frac{1}{s}\left(\alpha\bar{\mathbf{r}} + \frac{\mathbf{u}\cdot\bar{\mathbf{r}}}{c^2}\mathbf{u}\right) \qquad .$$

Hence

$$\mathbf{E} = \frac{q}{4\pi\varepsilon_o s^3}\left\{\alpha\bar{\mathbf{r}} + \frac{(\mathbf{u}\cdot\bar{\mathbf{r}})}{c^2}\mathbf{u} - \frac{\alpha(\mathbf{u}\cdot\bar{\mathbf{r}})}{c^2}\mathbf{u} - \frac{u^2}{c^4}(\mathbf{u}\cdot\bar{\mathbf{r}})\mathbf{u}\right\}$$

$$= \frac{q}{4\pi\varepsilon_o}\frac{\alpha\bar{\mathbf{r}}}{s^3} \qquad , \tag{4.36}$$

$$\mathbf{B} = -\frac{\mu_o}{4\pi}\frac{q}{s^3}\left\{\alpha\bar{\mathbf{r}}\times\mathbf{u} + \frac{(\mathbf{u}\cdot\bar{\mathbf{r}})}{c^2}\,\mathbf{u}\times\mathbf{u}\right\} = \frac{\mu_o}{4\pi}\frac{q}{s^3}\,\alpha\,\mathbf{u}\times\bar{\mathbf{r}}$$

$$= \frac{1}{c^2}\,\mathbf{u}\times\mathbf{E} \qquad . \qquad (4.37)$$

It may be noted that **E** is radial to the present position of the charge, while **H** is perpendicular to both the radial direction and the direction of uniform motion.

For low velocities $(u \ll c)$, $\alpha \simeq 1$, $s \simeq \bar{r}$, the above expressions reduce to those for the Coulomb and the Biot-Savart field. For high velocities both the electric and magnetic fields are increased from their low velocity values by an additional factor

$$(1-\frac{u^2}{c^2})\left\{1-(\frac{\mathbf{u}\times\bar{\mathbf{r}}}{\bar{r}c})^2\right\}^{-3/2} \qquad .$$

This means that in a direction transverse to the motion both fields are enhanced by a factor $(1-\frac{u^2}{c^2})^{-\frac{1}{2}}$. In the longitudinal direction the magnetic field remains zero while the electric field is reduced by a factor $(1-\frac{u^2}{c^2})$. Thus as the velocity approaches the velocity of light, the electromagnetic field of a moving charge resembles more and more that of a plane electromagnetic wave, i.e. **E** and **B** are both transverse and mutually perpendicular. The relation between the magnitudes of the electric and the magnetic vector,

$$\sqrt{\varepsilon_o}\,|\mathbf{E}| = \sqrt{\mu_o}\,|\mathbf{H}| \qquad ,$$

which is characteristic of plane waves, also applies for velocities approaching the velocity of light.

A charge moving uniformly in vacuum does not radiate energy. In general, only the $\bar{r}^{-1}$-dependent field due to a point source can radiate energy. This can be readily seen when we realize that $\oint_S \mathbf{E}\times\mathbf{H}\cdot d\mathbf{S}$ must be the same for all closed surfaces S enclosing the point source, which in the present case is the retarded position of the charge. As $dS \sim \bar{r}^2$ and $|\mathbf{E}| \propto |\mathbf{H}|$, we require that $|\mathbf{E}| \sim \bar{r}^{-1}$ and $|\mathbf{H}| \sim \bar{r}^{-1}$ for the radiation field of a point source. As the fields of a uniformly moving charge have $\bar{r}^{-2}$-dependence, no energy can be radiated. For a more detailed consideration, take the z-axis along the trajectory and the origin O at its present position. Consider a cylinder of radius ρ, which is finite but arbitrary, with its axis along the z-axis, and extending from $-\infty$ to $+\infty$. All radiation from the charge must cross the surface of the cylinder. Then since the motion of the charge is uniform its power loss is equal to the radiated energy crossing the surface of the cylinder per unit time:

$$
\begin{aligned}
-\frac{dW}{dt'} &= \int \mathbf{E} \times \mathbf{H} \cdot d\mathbf{S} \\
&= \varepsilon_0 \int \mathbf{E} \times (\mathbf{u} \times \mathbf{E}) \cdot d\mathbf{S} \\
&= \varepsilon_0 \int E^2 \mathbf{u} \cdot d\mathbf{S} - \varepsilon_0 \int (\mathbf{u} \cdot \mathbf{E}) \mathbf{E} \cdot d\mathbf{S} \\
&= -\varepsilon_0 \int (\mathbf{u} \cdot \mathbf{E}) \mathbf{E} \cdot d\mathbf{S} \quad ,
\end{aligned}
$$

where as $\mathbf{u}$ and $d\mathbf{S}$ are mutually perpendicular the first integral vanishes.

Let the angle between $\mathbf{u}$ and $\bar{\mathbf{r}}$ be θ. As $\mathbf{E}$ is radial to the present position it makes an angle $\frac{\pi}{2} - \theta$ with $d\mathbf{S}$. If we now take as surface element a ring of width dz at z, we have

$$
\mathbf{E} \cdot d\mathbf{S} = E \sin\theta \, 2\pi\bar{r} \sin\theta \, dz \quad .
$$

Using also

$$
\bar{r}\cos\theta = z \qquad \text{and} \qquad \bar{r}^2 = z^2 + \rho^2 \quad ,
$$

we can write

$$
-\frac{dW}{dt'} = -\frac{q^2\alpha^2\rho^2 u}{8\pi\varepsilon_0} \int_{-\infty}^{\infty} \frac{z\,dz}{(z^2 + \alpha\rho^2)^3} \quad .
$$

The integral is zero since the integrand is an odd function of z. In other words, contributions from the $z<0$ and $z>0$ regions cancel out.

It may be remarked here that since uniformly moving charges radiate no energy so does a steady current.

In the derivation of the Liénard-Wiechert potentials it was tacitly assumed that the velocity u of the charge is smaller than the velocity of light c. In a material medium of permittivity ε and permeability μ, the velocity of a charge, which is limited only by special relativity to be smaller than c, may be greater than the velocity of propagation of electromagnetic effects in the medium,

$$v = (\mu\varepsilon)^{-\frac{1}{2}} = \frac{c}{n} \quad ,$$

where $n = (\mu\varepsilon/\mu_0\varepsilon_0)^{\frac{1}{2}}$ is the refractive index of the medium. If this is the case, the charge may be encountered by the information sphere twice, and it is readily seen that in the first encounter, in which the charge impinges on the sphere, the relation between dq and $[\rho]dV'$ is opposite in sign to that given in Sec. 4.5. With respect to the present position $\mathbf{r}$, the two retarded positions will both make $\cos\theta$ negative. In fact, for the potentials to be nonzero, the field point P must be situated such that

$$\psi < \psi_0 = \arccos\sqrt{1 - \frac{v^2}{u^2}} = \arcsin\left(\frac{v}{u}\right) \quad ,$$

where $\psi = \pi - \theta$, [Prob. (4.4)].

Thus for a charge of velocity $u > v$, the potentials are zero outside the backward cone of half-angle ψ_0 with vertex at the present position. This is known as the *Čerenkov cone*. Inside this cone, for every field point there are two retarded positions. As both satisfy the retardation condition Eq. (4.31), the two values of s are solutions of Eq. (4.33), i.e.

$$s = \pm\sqrt{\alpha\bar{r}^2 + \left(\frac{\mathbf{u}\cdot\bar{\mathbf{r}}}{v}\right)^2} = \pm\,\bar{r}\sqrt{1 - \left(\frac{u}{v}\sin\psi\right)^2} \quad .$$

The potentials for such a field point are then

$$\Phi = \frac{1}{4\pi\varepsilon}\frac{q}{\bar{r}\{1-(\frac{u}{v}\sin\psi)^2\}^{\frac{1}{2}}} + \frac{1}{4\pi\varepsilon}\frac{q}{-\left[-\bar{r}\{1-(\frac{u}{v}\sin\psi)^2\}^{\frac{1}{2}}\right]}$$

$$= \frac{1}{4\pi\varepsilon}\frac{2q}{\bar{r}\{1-(\frac{u}{v}\sin\psi)^2\}^{\frac{1}{2}}}$$

and

$$\mathbf{A} = \frac{\mathbf{u}}{v^2}\Phi \quad .$$

The field vectors are the same as those given by Eqs. (4.36) and (4.37) but for an additional factor of 2 and the replacements $\varepsilon_0 \to \varepsilon$ and $c \to v$.

The radiation loss of the charge per unit time is given by the integral of the Poynting vector over the part of the cylindrical surface considered above which is inside the Čerenkov cone:

$$-\frac{dW}{dt'} = -\frac{q^2\alpha^2\rho^2 u}{2\pi\varepsilon}\int_{-\infty}^{-\gamma\rho}\frac{z\,dz}{(z^2-\gamma^2\rho^2)^3} \quad ,$$

where for convenience we have replaced α, which is now negative, by $-\gamma^2$. Without the cancellation due to the opposite contribution of the region $z > 0$, the integral no longer vanishes. Radiation is therefore expected from a charge moving uniformly in a medium for which $v < u$. Such radiation was first discovered by Čerenkov in 1934 and is known as the *Čerenkov radiation*.

The Poynting vector becomes infinite at the surface of the Čerenkov cone, which of course cannot be correct physically. The singularity arises from our assumption that the phase velocity v of the radiation is independent of frequency. It is removed when dispersion is taken into account. In any case it can be seen that the major contributions to the radiation come from regions forming the surface of the Čerenkov cone, where the fields have a sharp maximum. To get a meaningful expression for the radiation per unit time, we replace the cone by a thin absorption layer and use as the upper limit of integration $-\gamma\rho-\delta$, where δ is a small number. This gives

$$-\frac{dW}{dt'} = -\frac{q^2\alpha u}{32\pi\varepsilon\delta^2} = \frac{q^2u^3}{32\pi\varepsilon v^2\delta^2}\left(1-\frac{c^2}{n^2u^2}\right)$$

as the Čerenkov radiation emitted by the charge per unit time. It can be seen that the threshold occurs at

$$u > \frac{c}{n} \qquad ,$$

below which no radiation is emitted.

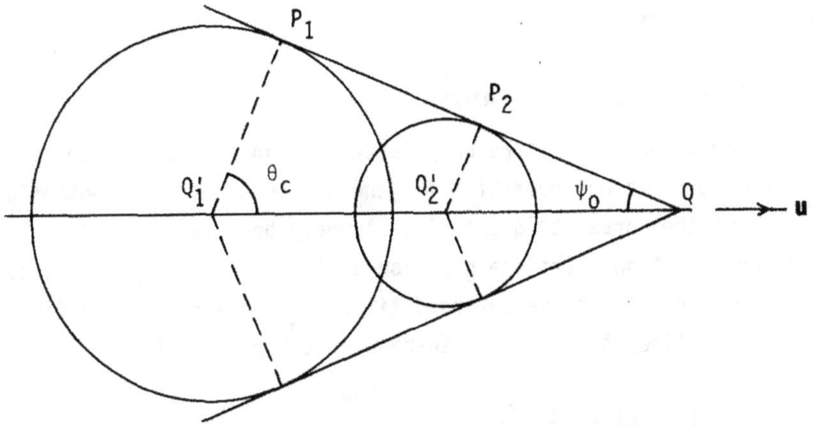

Fig. 4.3 Radiation emitted by the charge at time t_1', when it was at Q_1' arrives at P_1 at time t, when the charge arrives at Q, i.e., $Q_1'Q = (t-t_1')u$, $Q_1'P_1 = (t-t_1')v$. Radiation emitted when the charge was at some intermediate point Q_2' at time t_2' arrives at P_2 at time t. If $t-t_2' = h(t-t_1')$, we have $Q_2'Q = (t-t_2')u = hQ_1'Q$, $Q_2'P_2 = (t-t_2')v = hQ_1'P_1$, hence $Q_2'P_2 : Q_1'P_1 = Q_2'Q : Q_1'Q$ and P_2 falls on the straight line QP_1. Thus QP_1 forms the wavefront of all the radiation emitted prior to t. The direction of propagation of the radiation is given by $\cos\theta_c = \sin\psi_o = v/u$.

The radiation is confined mainly to the surface of the Čerenkov cone. In fact, the cone forms the wavefront of the radiation. For large ρ, i.e. at large distances from the trajectory, the wavefront can be considered plane. The direction of propagation of the Čerenkov radiation is normal to the wavefront and makes an angle θ_c to the trajectory given by

$$\cos\theta_c = \sin\psi_o = \frac{c}{nu} \quad ,$$

as can be seen from Fig. 4.3. The angle θ_c is known as the *Čerenkov angle*. For a given medium it is a function of the charge velocity only and is used as the basis for velocity determination for high energy charged particles.

4.7 Field of an Accelerated Charge

The situation is more complicated for an accelerated charge for in this case it is not possible, in general, to express the potentials in terms of the "present" quantities, largely because of the loss of energy by radiation. Let the position of the charge be specified by $\mathbf{r}'(t')$, where t' is the retarded time. Its potentials at a field point $\mathbf{r}$ at time t are the Liénard-Wiechert potentials

$$\Phi(\mathbf{r}, t) = \frac{1}{4\pi\varepsilon_o}\frac{q}{s}$$

and

$$\mathbf{A}(\mathbf{r}, t) = \frac{\mu_o}{4\pi}\frac{q[\mathbf{u}]}{s} \quad ,$$

where

$$s = \left[\bar{r} - \frac{\mathbf{u}\cdot\bar{\mathbf{r}}}{c}\right]$$

with

$$\bar{r} = |\mathbf{r} - \mathbf{r}'| = c(t - t') \quad ,$$

or

$$\bar{r}^2 = (x_j - x_j')^2 = c^2(t - t')^2 \quad . \tag{4.38}$$

Equation (4.38) is the retardation condition.

The electric and magnetic fields are given by

$$\mathbf{E} = -\nabla\Phi - \frac{\partial \mathbf{A}}{\partial t} \quad , \qquad \mathbf{B} = \nabla \times \mathbf{A}$$

o that the differentials required are $(\frac{\partial}{\partial t})_x$ and[a] $(\frac{\partial}{\partial x_i})_t$.

As Φ and **A** are given in terms of the retarded quantities $\bar{r}$ nd **u** which have the following functional dependences

$$\bar{r} = \bar{r}(x_i, x_i'(t')) = \bar{r}(x_i, t') \quad ,$$

$$\mathbf{u} = \mathbf{u}(t') \quad ,$$

hey are functions of x_i and t'. They are functions of t through he retardation condition Eq. (4.38) only, which may be written as

$$t' = t'(x_i, t) \quad .$$

t follows that for the differentiation of **A** and Φ, we have

$$(\frac{\partial}{\partial t})_x = (\frac{\partial t'}{\partial t})_x \, (\frac{\partial}{\partial t'})_x \quad . \tag{4.39}$$

n the other hand, the potentials are functions of x_i through their xplicit dependence on x_i as well as through t' via Eq. (4.38). ence

$$(\frac{\partial}{\partial x_i})_t = (\frac{\partial}{\partial x_i})_{t'} + (\frac{\partial t'}{\partial x_i})_t \, (\frac{\partial}{\partial t'})_x \quad . \tag{4.40}$$

The factors $(\frac{\partial t'}{\partial t})_x$ and $(\frac{\partial t'}{\partial x_i})_t$ are to be evaluated from the etardation condition Eq. (4.38). First we differentiate it with

---

[a] It should properly be written as $(\frac{\partial}{\partial x_i})_{t,x_j}$ where $j \neq i$. For convenience the subscript x_j is omitted but understood. (See also the footnote on p.3).

respect to t keeping x fixed:

$$-(x_j - x_j')\frac{dx_j'}{dt'}\frac{\partial t'}{\partial t} = c^2(t-t')\left(1-\frac{\partial t'}{\partial t}\right) ,$$

i.e.

$$-\mathbf{u}\cdot\bar{\mathbf{r}}\frac{\partial t'}{\partial t} = c\bar{r}\left(1-\frac{\partial t'}{\partial t}\right) ,$$

hence

$$\left(\frac{\partial t'}{\partial t}\right)_x = \frac{\bar{r}}{\bar{r}-\frac{\mathbf{u}\cdot\bar{\mathbf{r}}}{c}} = \frac{\bar{r}}{s} .$$

Next we differentiate Eq. (4.38) with respect to x_i keeping t fixed:

$$(x_j - x_j')\left(\delta_{ij} - \frac{dx_j'}{dt'}\frac{\partial t'}{\partial x_i}\right) = -c^2(t-t')\frac{\partial t'}{\partial x_i} ,$$

i.e.

$$(x_i - x_i') - \mathbf{u}\cdot\bar{\mathbf{r}}\frac{\partial t'}{\partial x_i} = -c\bar{r}\frac{\partial t'}{\partial x_i}$$

hence

$$\left(\frac{\partial t'}{\partial x_i}\right)_t = -\frac{x_i - x_i'}{sc} ,$$

or

$$(\nabla t')_t = -\frac{\bar{\mathbf{r}}}{sc} .$$

Equations (4.39) and (4.40) can now be written as operator equations:

$$\left(\frac{\partial}{\partial t}\right)_x = \frac{\bar{r}}{s}\left(\frac{\partial}{\partial t'}\right)_x , \tag{4.41}$$

and

$$(\nabla)_t = (\nabla)_{t'} - \frac{\bar{\mathbf{r}}}{sc}\left(\frac{\partial}{\partial t'}\right)_x . \tag{4.42}$$

To apply these operator equations to Φ and $\mathbf{A}$, we first note that the relation $\bar{r}^2 = (x_j - x_j')^2$ gives

$$\left(\frac{\partial \bar{r}}{\partial t'}\right)_x = -\frac{\mathbf{u}\cdot\bar{\mathbf{r}}}{\bar{r}} , \qquad (\nabla\bar{r})_{t'} = \frac{\bar{\mathbf{r}}}{\bar{r}} .$$

We also have

$$\nabla_{t'}(\mathbf{u}\cdot\bar{\mathbf{r}}) = \mathbf{u} \qquad , \qquad \left\{\frac{\partial}{\partial t'}(\mathbf{u}\cdot\bar{\mathbf{r}})\right\}_x = \dot{\mathbf{u}}\cdot\bar{\mathbf{r}} - u^2 \qquad .$$

Here it should be noted that as $x_i' = x_i'(t)$, x_i' and $\mathbf{u}$ do not change if t' is held fixed. It is now only one step further to obtain the differentials of s and $\mathbf{u}$.

Using these relations and Eqs. (4.41) and (4.42) we find

$$\left(\frac{\partial s}{\partial t}\right)_x = \frac{\bar{r}}{s}\left(\frac{\partial s}{\partial t'}\right)_x \qquad , \qquad \left(\frac{\partial \mathbf{u}}{\partial t}\right)_x = \frac{\bar{r}}{s}\dot{\mathbf{u}} \qquad ,$$

$$(\nabla s)_t = \frac{\bar{\mathbf{r}}}{\bar{r}} - \frac{\mathbf{u}}{c} - \frac{\bar{\mathbf{r}}}{sc}\left(\frac{\partial s}{\partial t'}\right)_x \qquad ,$$

where

$$\left(\frac{\partial s}{\partial t'}\right)_x = \frac{u^2}{c} - \frac{\mathbf{u}\cdot\bar{\mathbf{r}}}{\bar{r}} - \frac{\dot{\mathbf{u}}\cdot\bar{\mathbf{r}}}{c} \qquad .$$

Hence

$$\mathbf{E} = \frac{q}{4\pi\varepsilon_0 s^3}\left\{\left(\frac{\bar{\mathbf{r}}}{\bar{r}} - \frac{\mathbf{u}}{c}\right)s - \frac{\bar{r}s\dot{\mathbf{u}}}{c^2} + \frac{1}{c^2}(\bar{r}\mathbf{u} - c\bar{\mathbf{r}})\left(\frac{\partial s}{\partial t'}\right)_x\right\} \qquad .$$

If we denote

$$\alpha = 1 - \frac{u^2}{c^2} \qquad , \tag{4.43}$$

$$\mathbf{r}_0 = \bar{\mathbf{r}} - \frac{\bar{r}}{c}\mathbf{u} \qquad , \tag{4.44}$$

then as

$$\left(\frac{\bar{\mathbf{r}}}{\bar{r}} - \frac{\mathbf{u}}{c}\right)s = \mathbf{r}_0 - \frac{\mathbf{u}\cdot\bar{\mathbf{r}}}{\bar{r}c}\bar{\mathbf{r}} + \frac{\mathbf{u}\cdot\bar{\mathbf{r}}}{c^2}\mathbf{u}$$

and

$$\bar{r}s = \mathbf{r}_0 \cdot \bar{\mathbf{r}} \qquad ,$$

the expression for $\mathbf{E}$ can be simplified to

$$\mathbf{E} = \frac{q}{4\pi\varepsilon_0 s^3}\left\{\alpha\mathbf{r}_0 + \frac{1}{c^2}\bar{\mathbf{r}}\times(\mathbf{r}_0\times\dot{\mathbf{u}})\right\} \qquad . \tag{4.45}$$

To obtain the magnetic vector **B** we have to calculate $\nabla\times\mathbf{A}$. Consider its i-th component:

$$(\nabla\times\mathbf{A})_i = \varepsilon_{ijk}\left(\frac{\partial A_k}{\partial x_j}\right)_t = \frac{\mu_0 q}{4\pi}\varepsilon_{ijk}\left(\frac{\partial}{\partial x_j}\frac{u_k}{s}\right)_t$$

$$= \frac{\mu_0 q}{4\pi s^2}\varepsilon_{ijk}\left\{ s\left(\frac{\partial u_k}{\partial x_j}\right)_t - u_k\left(\frac{\partial s}{\partial x_j}\right)_t \right\} .$$

Using Eq. (4.42) and bearing in mind $\mathbf{u}=\mathbf{u}(t')$, we find

$$\left(\frac{\partial u_k}{\partial x_j}\right)_t = \left(\frac{\partial u_k}{\partial x_j}\right)_{t'} - \frac{x_j - x_j'}{sc}\frac{du_k}{dt'} = -\frac{(x_j - x_j')\dot{u}_k}{sc} .$$

This expression together with that for $(\nabla s)_t$ gives

$$\nabla\times\mathbf{A} = \frac{\mu_0 q}{4\pi s^3}\left\{ \left(\frac{\dot{\mathbf{u}}\times\bar{\mathbf{r}}}{c}+\frac{\mathbf{u}\times\bar{\mathbf{r}}}{\bar{r}}\right)\left(\bar{r}-\frac{\bar{\mathbf{r}}\cdot\mathbf{u}}{c}\right)+\frac{\bar{\mathbf{r}}\times\mathbf{u}}{c}\left(\frac{\partial s}{\partial t'}\right)_x + \frac{\mathbf{u}\times\mathbf{u}s}{c}\right\}$$

$$= \frac{\mu_0 q}{4\pi s^3}\left\{ \left(1-\frac{u^2}{c^2}\right)\mathbf{u}\times\bar{\mathbf{r}} - (\bar{\mathbf{r}}\cdot\dot{\mathbf{u}})\frac{\bar{\mathbf{r}}\times\mathbf{u}}{c^2} \right.$$

$$\left. + (\bar{\mathbf{r}}\cdot\mathbf{u})\frac{\bar{\mathbf{r}}\times\dot{\mathbf{u}}}{c^2} - \frac{\bar{r}}{c}(\bar{\mathbf{r}}\times\dot{\mathbf{u}}) \right\}$$

$$= \frac{\mu_0 q}{4\pi s^3}\left\{ \left(1-\frac{u^2}{c^2}\right)\mathbf{u}\times\bar{\mathbf{r}} + \frac{\bar{\mathbf{r}}}{\bar{r}c}\times\left[\bar{\mathbf{r}}\times\left\{\left(\bar{\mathbf{r}}-\frac{\bar{r}}{c}\mathbf{u}\right)\times\dot{\mathbf{u}}\right\}\right]\right\}$$

i.e.

$$\mathbf{B} = \frac{\mu_0}{4\pi}\frac{q}{s^3}\left\{\alpha\,\mathbf{u}\times\bar{\mathbf{r}} + \frac{\bar{\mathbf{r}}}{\bar{r}c}\times\left[\bar{\mathbf{r}}\times(\mathbf{r}_0\times\dot{\mathbf{u}})\right]\right\} . \qquad (4.46)$$

As $\bar{\mathbf{r}}\times\mathbf{r}_0 = \frac{\bar{r}}{c}(\mathbf{u}\times\bar{\mathbf{r}})$ and $\mu_0\varepsilon_0 = c^{-2}$, Eq. (4.46) can also be written as

$$\mathbf{B} = \frac{q}{4\pi\varepsilon_0 c^2 s^3}\left\{\frac{\alpha c}{\bar{r}}(\bar{\mathbf{r}}\times\mathbf{r}_0) + \frac{\bar{\mathbf{r}}}{\bar{r}c}\times[\bar{\mathbf{r}}\times(\mathbf{r}_0\times\dot{\mathbf{u}})]\right\} = \frac{\bar{\mathbf{r}}\times\mathbf{E}}{\bar{r}c} .$$

or

$$\mathbf{H} = \sqrt{\frac{\varepsilon_0}{\mu_0}}\,\frac{\bar{\mathbf{r}} \times \mathbf{E}}{\bar{r}} \quad . \tag{4.47}$$

Expressions (4.45) and (4.47) give the electric and magnetic fields of an accelerated charge. It is seen that $\mathbf{H}$ is normal to $\mathbf{E}$ and is always transverse to the direction of observation from the retarded position of the charge. $\mathbf{E}$ however has a longitudinal component. It should also be noted that the quantities $\bar{\mathbf{r}}$, $\mathbf{u}$ and $\dot{\mathbf{u}}$ in these expressions are retarded quantities. In general it is not possible in the case of an accelerated charge to express the fields in terms of the "present" quantities as in the case of uniform motion.

It is convenient to decompose the electromagnetic field of an accelerated charge into components on the basis of acceleration dependence. These are known as the acceleration and induction fields and vary respectively as $\bar{r}^{-1}$ and $\bar{r}^{-2}$ for large distances from the charge. The *induction field* which is obtained by putting $\dot{\mathbf{u}} = 0$ in Eqs. (4.45) and (4.46) has the vectors

$$\mathbf{E}_{ind} = \frac{q}{4\pi\varepsilon_0 s^3}\left(1 - \frac{u^2}{c^2}\right)\mathbf{r}_0 \tag{4.48}$$

and

$$\mathbf{B}_{ind} = \frac{\mu_0 q}{4\pi s^3}\left(1 - \frac{u^2}{c^2}\right)\mathbf{u} \times \mathbf{r}_0 \quad . \tag{4.49}$$

These are formally identical with the expressions (4.36) and (4.37) for the electric and magnetic fields of a charge in uniform motion. In the present case, however, $\mathbf{r}_0$ is a retarded quantity $[\bar{\mathbf{r}} - \bar{r}\mathbf{u}/c]$. Only when $\mathbf{u}$ is constant will $\mathbf{r}_0$ be identical with the "present" quantity $\mathbf{r}$ appearing in the latter case.

$\mathbf{E}_{ind}$ can be written as the sum of two parts:

$$\mathbf{E}_{ind} = \frac{q\alpha\bar{\mathbf{r}}}{4\pi\varepsilon_0 s^3} - \frac{q}{4\pi\varepsilon_0}\,\frac{\alpha\bar{r}\mathbf{u}}{s^3 c} \quad .$$

The first part represents a radial field and varies approximately as $\bar{r}^{-2}$. It may be considered an extension of the Coulomb field. The second part represents a field in the direction of the velocity. Together with $\mathbf{B}_{ind}$ which is purely velocity-dependent, it constitutes the *velocity field.* $\mathbf{B}_{ind}$ itself may be considered as an extension of the Biot-Savart field. Note that the Coulomb field has no magnetic counterpart.

The acceleration field, which is $\bar{r}^{-1}$-dependent, is also known as the *radiation field* since it is the component that is solely responsible for radiation of energy from an accelerated charge. The radiation field of an accelerated charge is represented by the following vectors:

$$\mathbf{E}_{rad} = \frac{q}{4\pi\epsilon_0 c^2 s^3}\, \bar{\mathbf{r}} \times (\mathbf{r}_0 \times \dot{\mathbf{u}}) \quad , \tag{4.50}$$

$$\mathbf{B}_{rad} = \frac{\mu_0}{4\pi c}\, \frac{q}{s^3 \bar{r}}\, \bar{\mathbf{r}} \times \{ \bar{\mathbf{r}} \times (\mathbf{r}_0 \times \dot{\mathbf{u}}) \} \quad . \tag{4.51}$$

Here the electric and magnetic fields are both transverse to $\mathbf{r}$, which is also in the direction of $\mathbf{E} \times \mathbf{H}$, i.e. the direction of the Poynting vector, and are mutually perpendicular. Furthermore,

$$\sqrt{\epsilon_0}\, |\mathbf{E}| = \sqrt{\mu_0}\, |\mathbf{H}| \quad ,$$

so that the structure of the radiation field is identical with that of the field of a plane electromagnetic wave. This means that the radiation from an accelerated charge propagates through space in the form of plane electromagnetic waves.

4.8 Radiation from an Accelerated Charge

An accelerated charge radiates energy in the form of plane electromagnetic waves. The amount of radiated energy per unit time per unit area normal to the direction of flow is given by the Poynting vector $\mathbf{N} = \mathbf{E}_{rad} \times \mathbf{H}_{rad}$. It should be noted however that, while the field vectors are given in terms of the quantities appropriate to the charge at the retarded time t', they give the Poynting vector appropriate to the field time t. The fine distinction between t and t' in the calculation of energy loss may be neglected if the velocity of the charge is small compared with the velocity of light as is obvious from Eq. (4.41).

At low velocities $(u \ll c)$, we may make approximations $\mathbf{r}_0 \simeq \bar{\mathbf{r}}$, $s \simeq \bar{r}$ and $\alpha \simeq 1$, so that

$$\mathbf{E}_{rad} = \frac{q}{4\pi\varepsilon_0 c^2 \bar{r}^3}\, \bar{\mathbf{r}} \times (\bar{\mathbf{r}} \times \dot{\mathbf{u}}) \quad , \tag{4.52}$$

$$\mathbf{B}_{rad} = \frac{\mu_0}{4\pi c}\frac{q}{\bar{r}^2}(\dot{\mathbf{u}} \times \bar{\mathbf{r}}) \quad , \tag{4.53}$$

and

$$\mathbf{N} = \frac{q^2(\dot{\mathbf{u}} \times \bar{\mathbf{r}})^2}{16\pi^2\varepsilon_0 c^3 \bar{r}^5}\,\bar{\mathbf{r}} = \frac{q^2\dot{u}^2 \sin^2\theta}{16\pi^2\varepsilon_0 c^3 \bar{r}^3}\,\bar{\mathbf{r}} \quad , \tag{4.54}$$

where θ is the angle between the directions of propagation and acceleration. As the Poynting vector is directed radially from the retarded position of the charge, the power radiated per unit solid angle is

$$-\frac{dw(\theta, \phi)}{dt'} = -\frac{dw(\theta, \phi)}{dt} = \frac{N}{\bar{r}^{-2}} = \frac{q^2\dot{u}^2 \sin^2\theta}{16\pi^2\varepsilon_0 c^3} \quad , \tag{4.55}$$

which is independent of $\bar{r}$, a condition necessary for energy dissipation from a point source. It is seen that the energy radiated per unit time per unit solid angle, $-\frac{dw}{dt'}$ varies as $\sin^2\theta$. The radiation is therefore strongest in a direction normal to the acceleration but vanishes along the direction of the acceleration. The total radiation emitted by the charge per unit time is obtained by integrating over a sphere S of radius $\bar{r}$ centred at the charge:

$$-\frac{dW}{dt'} = \oint_S \mathbf{N} \cdot d\mathbf{S} = \int_0^\pi 2\pi\bar{r}^2 \sin\theta\, N d\theta = \frac{q^2\dot{u}^2}{6\pi\varepsilon_0 c^3} \quad . \tag{4.56}$$

The frequently quoted formula in Gaussian units is

$$-\frac{dW}{dt'} = \frac{2q^2\dot{u}^2}{3c^3} \qquad \text{(c.g.s.)} \quad .$$

For high velocities the Poynting vector of the radiation is

$$\mathbf{N} = \frac{q^2}{16\pi^2\varepsilon_0 c^3}\frac{\{\bar{\mathbf{r}}\times(\mathbf{r}_0\times\dot{\mathbf{u}})\}^2}{s^6}\frac{\bar{\mathbf{r}}}{\bar{r}} \quad . \tag{4.57}$$

The integration of this expression for the rate of total radiation loss is rather involved. Only the result[b] will be quoted:

$$-\frac{dW}{dt'} = \frac{q^2}{6\pi\varepsilon_0 c^3}\left(1-\frac{u^2}{c^2}\right)^{-3}\left\{\dot{u}^2-\left(\frac{\mathbf{u}\times\dot{\mathbf{u}}}{c}\right)^2\right\} \quad . \tag{4.58}$$

An important case which is also simple mathematically is when the acceleration $\dot{\mathbf{u}}$ is parallel to the velocity $\mathbf{u}$. In this case

$$\mathbf{r}_0\times\dot{\mathbf{u}} = \left(\bar{\mathbf{r}}-\frac{\bar{r}\mathbf{u}}{c}\right)\times\dot{\mathbf{u}} = \bar{\mathbf{r}}\times\dot{\mathbf{u}}$$

so that the expression (4.57) for the Poynting vector is the same as for small velocities but for the factor s^6 instead of $\bar{r}^6$ in the denominator. The radiation emitted per unit time per unit solid angle as observed at the field point is

$$-\frac{dw(\theta,\phi)}{dt} = N\bar{r}^2 = \frac{q^2\dot{u}^2\bar{r}^6\sin^2\theta}{16\pi^2\varepsilon_0 c^3 s^6} \quad .$$

The rate of energy loss by the charge per unit solid angle, on the

---

[b] See Panofsky and Phillips, *Classical Electricity and Magnetism* 2nd ed., Addison-Wesley, Reading, Mass. (1962), p. 370.

other hand, is

$$-\frac{dw}{dt'} = -\frac{dw}{dt}\frac{\partial t}{\partial t'} = -\frac{s}{\bar{r}}\frac{dw}{dt} = \frac{q^2\dot{u}^2}{16\pi^2\varepsilon_0 c^3}\frac{\sin^2\theta}{(1-\frac{u}{c}\cos\theta)^5} \quad .$$

The additional factor $(1-\frac{u}{c}\cos\theta)^5$ in the denominator of the formula for high velocities has the effect of peaking the radiation in the forward hemisphere of motion as can be seen from the angular distributions plotted in Fig. 4.4 for a given acceleration $\dot{\mathbf{u}}$ but different values of $\frac{u}{c}$. The length of the vector from the charge q is proportional to the rate of radiation per unit solid angle in its direction. The actual distributions are given by the surfaces of revolution obtained by rotating the curves about the $\mathbf{u}$, $\dot{\mathbf{u}}$-axis.

We have seen that energy is radiated when the velocity $\mathbf{u}$ of a charge changes, be it arising from acceleration, deceleration or mere deflection without change of speed. The radiation loss will cause a reduction in the kinetic energy which the charge would otherwise possess. To include this additional loss of energy in a mechanical description it would be necessary to add to the external forces acting on the charge a force analogous to friction in mechanics, called the *radiation reaction*. For example, for a charge q moving with a velocity $\mathbf{u}$ in a magnetic field of induction $\mathbf{B}$, the force acting is the magnetic force $q\mathbf{u}\times\mathbf{B}$. It has the effect of accelerating the charge in a direction normal to $\mathbf{u}$ without changing its speed since its rate of doing work $q\mathbf{u}\times\mathbf{B}\cdot\mathbf{u}$ vanishes. Nevertheless, energy is dissipated through radiation as the charge undergoes deflection, which will of course entail a change of speed. To account for this loss it will be necessary to add to the magnetic force a radiation reaction. Fortunately, for low velocities $(u \ll c)$ the radiation reaction is exceedingly small so that a satisfactory account of the motion is still achieved even if the radiation reaction is ignored completely. For high velocities the radiation reaction becomes important and must be fully taken into account.

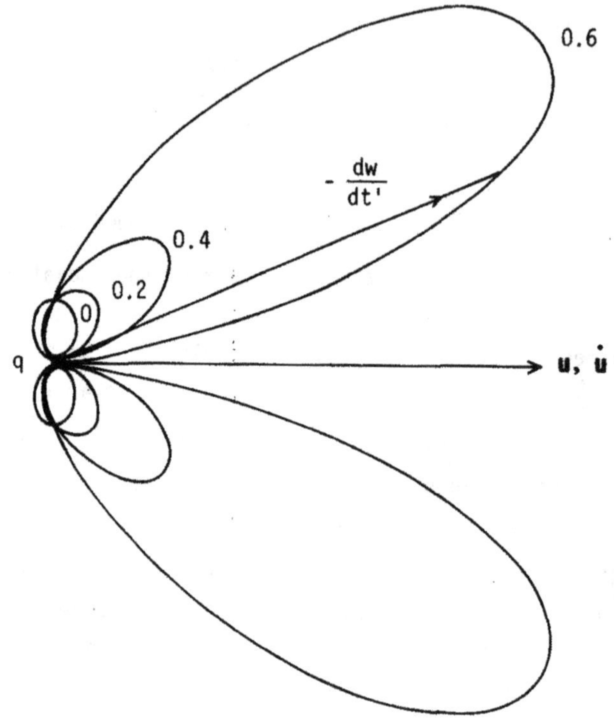

Fig. 4.4 The angular distributions of the radiation of a charge q undergoing acceleration in the direction of motion for values of $u/c = 0$, 0.2, 0.4 and 0.6. The length of a vector from the charge is proportional to the rate of emission of radiation per unit solid angle in the direction of the vector. The actual distributions are given by the surfaces of revolution obtained by rotating the curves about the **u**, **u̇**-axis. It is seen that while the radiation in the longitudinal distribution remains zero there is increasing forward peaking as the velocity increases towards the velocity of light c.

A charged particle in traversing matter suffers acceleration and deceleration as it interacts electromagnetically with the electrons and nuclei of the medium. As a result radiation is emitted, which is commonly known as *Bremsstrahlung*.

4.9 Field of a Hertzian Dipole

A *Hertzian dipole* is an electric dipole whose moment $\mathbf{p}$ oscillates sinusoidally:

$$\mathbf{p} = \mathbf{p}_0 e^{-i\omega t'} \quad , \tag{4.60}$$

where ω is the angular frequency of oscillation. The dipole may be considered as consisting of a stationary negative charge $-q$ and a mobile positive charge $+q$ at displacement $\boldsymbol{\ell}$ from the former. The oscillation of $\mathbf{p}$ then arises from the periodic variation of $\boldsymbol{\ell}$:

$$\dot{\mathbf{p}} = q\dot{\boldsymbol{\ell}} \quad .$$

The dipole may also be thought of as an oscillating current I flowing in a short conductor $\boldsymbol{\ell}$ so that

$$\dot{\mathbf{p}} = \dot{q}\boldsymbol{\ell} = I\boldsymbol{\ell} \quad .$$

It will be assumed that the dipole is so short that (a) the retarded time t' may be taken to be the same for the entire dipole and (b) at any instant t', either all the elements of the mobile charge have the same velocity or the current is uniform throughout the length ℓ as the case may be. For these assumptions to be valid we require the distance of observation r to be large compared with the length of the dipole, i.e. $\ell \ll r$, and, in addition, the period of variation of the dipole to be large compared with the time taken for electromagnetic effects to travel across its length i.e. $\frac{\ell}{c} \ll \frac{2\pi}{\omega}$ or $\ell \ll \lambda$, where $\lambda = \frac{2\pi c}{\omega}$ is the wavelength of electromagnetic waves of the same frequency. Thus the conditions to be assumed are such that

$$\ell \ll \lambda \ll r \quad ,$$

where for simplicity of calculation we assume $\lambda \ll r$ also and limit the validity of the results to large distances. The results obtained for the near regions are approximate only (cf. Sec. 4.12).

The retarded vector potential $\mathbf{A}(\mathbf{r})$ of a Hertzian dipole is

$$\mathbf{A} = \frac{\mu_0}{4\pi}\int \frac{[\mathbf{J}]dV'}{\bar{r}} \simeq \frac{\mu_0}{4\pi r}\int [\rho\dot{\ell}]dV' \simeq \frac{\mu_0}{4\pi}\frac{[\dot{\ell}]}{r}\int [\rho]dV'$$

$$\simeq \frac{\mu_0}{4\pi} q \frac{[\dot{\ell}]}{r} = \frac{\mu_0}{4\pi r}\frac{d}{dt'}[\mathbf{p}] \quad ,$$

or

$$\mathbf{A} = \frac{\mu_0}{4\pi}\int \frac{[I]d\ell}{\bar{r}} \simeq \frac{\mu_0}{4\pi r}[I]\int d\ell = \frac{\mu_0}{4\pi}\frac{[I]\ell}{r} = \frac{\mu_0}{4\pi r}\frac{d}{dt'}[\mathbf{p}] \quad ,$$

where in the expansion of $\bar{r}^{-1} = |\mathbf{r} - \mathbf{r}'|^{-1}$ only the first term has been retained. Higher order terms would give rise to quadrupole and higher multipole potentials. We have also neglected the variation of the retarded time t' across the length of the dipole. Both approximations are allowed under the assumption $\ell \ll \lambda \ll r$.

Choose a coordinate system with the origin at the dipole and the z-axis along the dipole axis as shown in Fig. 4.5. In spherical coordinates the vector potential may be expressed as

$$\mathbf{A}(\mathbf{r}, t) = -\frac{i\omega\mu_0}{4\pi r}\mathbf{p}(0, t') = -\frac{i\omega\mu_0}{4\pi r} p_0 e^{-i\omega t'}(\cos\theta, -\sin\theta, 0) \quad , \tag{4.61}$$

where $t' = t - \frac{r}{c}$ is the retarded time.

The magnetic intensity $\mathbf{H}$ is given by

$$\mathbf{H} = \frac{1}{\mu_0}\nabla \times \mathbf{A} = \frac{1}{\mu_0}\begin{vmatrix} \dfrac{\mathbf{i}_r}{r^2\sin\theta} & \dfrac{\mathbf{i}_\theta}{r\sin\theta} & \dfrac{\mathbf{i}_\phi}{r} \\ \dfrac{\partial}{\partial r} & \dfrac{\partial}{\partial\theta} & \dfrac{\partial}{\partial\phi} \\ A_r & rA_\theta & r\sin\theta A_\phi \end{vmatrix}$$

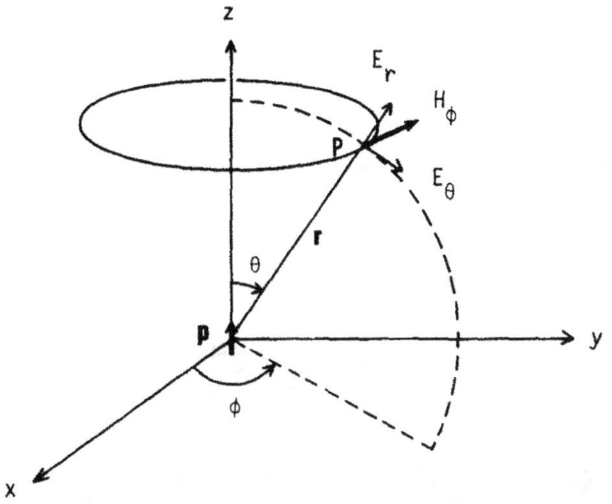

Fig. 4.5 A Hertzian dipole **p** lies at the origin along the z-axis. At a field point P, the electric vector lies in the meridian plane containing P and the magnetic vector is perpendicular to this plane, its lines of force forming coaxial circles about the z-axis.

Its components are

$$H_r = H_\theta = 0 \quad ,$$

$$H_\phi = \frac{1}{4\pi}\left(-\frac{\omega^2}{rc} - \frac{i\omega}{r^2}\right)\sin\theta\, p_0 e^{-i\omega t'} \quad . \tag{4.62}$$

The evaluation of the scalar potential Φ can be avoided if we calculate **E** by means of the relation

$$\nabla \times \mathbf{H} = \dot{\mathbf{D}} + \mathbf{J} = \dot{\mathbf{D}} \quad ,$$

J being zero outside the dipole. As the field vector **H** varies sinusoidally **D** must likewise vary sinusoidally. Then as the dipole location remains fixed, we have

$$\frac{\partial \mathbf{D}}{\partial t} = \frac{\partial \mathbf{D}}{\partial t'} = -\, i\omega\varepsilon_0 \mathbf{E} \quad ,$$

giving

$$\mathbf{E} = \frac{i}{\omega\varepsilon_0} \nabla \times \mathbf{H} \quad . \tag{4.63}$$

From the expressions for the components of **H** we obtain

$$E_r = \frac{i}{\omega\varepsilon_0} \frac{1}{r^2 \sin\theta} \frac{\partial}{\partial\theta} (r \sin\theta H_\phi)$$

$$= \frac{1}{2\pi\varepsilon_0} \left(-\frac{i\omega}{r^2 c} + \frac{1}{r^3}\right) \cos\theta \, p_0 e^{-i\omega t'} \quad ,$$

$$E_\theta = -\frac{i}{\omega\varepsilon_0} \frac{1}{r \sin\theta} \frac{\partial}{\partial r} (r \sin\theta \, H_\phi)$$

$$= \frac{1}{4\pi\varepsilon_0} \left(-\frac{\omega^2}{rc^2} - \frac{i\omega}{r^2 c} + \frac{1}{r^3}\right) \sin\theta \, p_0 e^{-i\omega t'} \quad ,$$

$$E_\phi = 0 \quad . \tag{4.64}$$

It can be seen that as $E_\phi = H_r = H_\theta = 0$, **E** lies in the meridian plane of the dipole containing the field point and **H** is normal to this plane, the magnetic lines of force forming coaxial circles about the dipole axis (see Fig. 4.5). If we let the angular frequency ω tend to zero, the only components which remain are

$$E_r = \frac{2p_0 \cos\theta}{4\pi\varepsilon_0 r^3} \quad , \qquad E_\theta = \frac{p_0 \sin\theta}{4\pi\varepsilon_0 r^3} \quad ,$$

as would be expected for a static electric dipole.

In describing the electromagnetic field of a Hertzian dipole, it is convenient to divide the surrounding space into zones. In the immediate zone, $\ell \ll r \ll \lambda$ or $\omega r \ll c$, only the r^{-3}-dependent component of the field is important. This component consists of electric field alone and, as it is similar to the field of a static dipole, is called the *static field*. In the far zone, $\lambda \ll r$, or $c \ll \omega r$, the

r^{-1}-dependent component predominates. As this is the part of the field which dissipates energy by radiation (Sec. 4.6) it is called the *radiation field*. In the intermediate region, $\lambda \leq r$ or $c \leq \omega r$, the r^{-2}-dependent component is important. This part of the field is called the *induction* or *transition field*. Note that the induction magnetic field is similar to the magnetostatic field of a current element $Id\ell = \dot{\mathbf{p}}$ and may be regarded as an extension of the Biot-Savart field. The magnitudes of static, induction and radiation fields are in the ratio $(\lambda/2\pi r)^2 : \lambda/2\pi r : 1$.

The radiation field of a Hertzian dipole has components

$$E_\theta = - \frac{1}{4\pi\varepsilon_0} \frac{\omega^2}{c^2} \frac{\sin\theta}{r} p_0 e^{-i\omega t'} \quad ,$$

$$H_\phi = - \frac{1}{4\pi} \frac{\omega^2}{c} \frac{\sin\theta}{r} p_0 e^{-i\omega t'} \quad ,$$

$$E_r = E_\phi = H_r = H_\theta = 0 \quad . \tag{4.65}$$

It is convenient to introduce a *propagation vector* $\mathbf{k}$ defined as a vector in the direction of $\mathbf{r}$ of magnitude ω/c. The radiation field vectors can then be written as

$$\mathbf{E}_{rad} = \frac{(\mathbf{k} \times \mathbf{p}) \times \mathbf{k}}{4\pi\varepsilon_0 r} \quad ,$$

$$\mathbf{H}_{rad} = \frac{\omega \mathbf{k} \times \mathbf{p}}{4\pi r} \quad . \tag{4.66}$$

It can be seen that the electric and the magnetic vector of the radiation field are mutually perpendicular and are perpendicular to $\mathbf{r}$, whose direction, as will be seen, is also the direction of the Poynting vector. Thus the radiation is purely transverse. Furthermore, the field vectors are related in magnitude by

$$\sqrt{\varepsilon_0}\, |\mathbf{E}_{rad}| = \sqrt{\mu_0}\, |\mathbf{H}_{rad}| \quad ,$$

showing that the structure of the radiation field is that of a plane electromagnetic wave. Thus the radiation from a Hertzian dipole propagates through space as plane electromagnetic waves. In the vicinity of the source there are, in addition to the radiation field, static and induction fields which do not radiate energy.

Both $\mathbf{E}_{rad}$ and $\mathbf{H}_{rad}$ vanish on the axis $\theta = 0$ but become maximum in the equatorial plane $\theta = \pi/2$. Hence, no energy is radiated along the axis of the dipole and the radiated energy has maximum intensity in the equatorial plane. Finally, the radiation is completely polarized with the electric vector in the plane of the dipole and the direction of propagation since $\mathbf{E}_{rad}$ is perpendicular to $\mathbf{k} \times \mathbf{p}$.

4.10 Radiation from a Hertzian Dipole

While $\mathbf{E}$ and $\mathbf{H}$ are written as complex vectors it is the real part that has physical significance. Thus the Poynting vector is given by

$$\mathbf{N} = \mathrm{Re}\mathbf{E} \times \mathrm{Re}\mathbf{H} = \frac{1}{4}(\mathbf{E} + \mathbf{E}^*) \times (\mathbf{H} + \mathbf{H}^*)$$

$$= \frac{1}{4}(\mathbf{E} \times \mathbf{H} + \mathbf{E}^* \times \mathbf{H} + \mathbf{E} \times \mathbf{H}^* + \mathbf{E}^* \times \mathbf{H}^*) \quad , \qquad (4.67)$$

where the asterisks denote complex conjugates.

We are interested in the mean value of $\mathbf{N}$ over one cycle of oscillation since the time of measurement is usually much longer than a period, i.e., in

$$\langle \mathbf{N} \rangle = \frac{1}{T}\int_0^T \mathbf{N}\, dt' \quad ,$$

where T is the period of oscillation, being equal to $2\pi/\omega$. The terms $\mathbf{E} \times \mathbf{H}$ and $\mathbf{E}^* \times \mathbf{H}^*$ in the expression for $\mathbf{N}$ contain exponential time factors $\exp(\mp 2i\omega t')$. As

$$\frac{1}{T}\int_0^T e^{\mp 2i\omega t'} dt' = 0 \quad ,$$

these terms do not contribute to $\langle \mathbf{N} \rangle$. On the other hand, the remaining terms $\mathbf{E}^* \times \mathbf{H}$ and $\mathbf{E} \times \mathbf{H}^*$ do not contain any time factor and are not affected by averaging. Hence

$$\langle \mathbf{N} \rangle = \frac{1}{4}(\mathbf{E}^* \times \mathbf{H} + \mathbf{E} \times \mathbf{H}^*)$$

$$= \frac{1}{4}\{(\mathbf{E} \times \mathbf{H}^*)^* + \mathbf{E} \times \mathbf{H}^*\} = \frac{1}{2}\,\mathrm{Re}(\mathbf{E} \times \mathbf{H}^*) \quad . \qquad (4.68)$$

Applying this result to the electromagnetic field of a Hertzian dipole we have

$$\langle \mathbf{N} \rangle = \frac{1}{2}\,\mathrm{Re}(E_\theta H_\phi^*)\mathbf{i}_r - \frac{1}{2}\,\mathrm{Re}(E_r H_\phi^*)\mathbf{i}_\theta$$

$$= \frac{\omega^4 p_0^2}{32\pi^2 \varepsilon_0 c^3} \frac{\sin^2\theta}{r^2}\,\mathbf{i}_r \quad . \qquad (4.69)$$

Note that the θ-component does not contribute and the average Poynting vector is the same as that for the radiation field alone. This confirms the general conclusion reached in Sec. 4.6 that only the r^{-1}-component contributes to energy radiation. Energy is not radiated from the other component fields although they contribute to the storage of energy in the electromagnetic field.

The instantaneous Poynting vector is obtained from the real parts of the radiation field vectors $\mathbf{E}_{rad}$ and $\mathbf{H}_{rad}$ as

$$\mathbf{N} = \frac{\omega^4 p_0^2 \sin^2\theta \cos^2\omega t'}{16\pi^2 \varepsilon_0 c^3 r^2}\,\mathbf{i}_r = \frac{\ddot{p}^2 \sin^2\theta}{16\pi^2 \varepsilon_0 c^3 r^2}\,\mathbf{i}_r \quad . \qquad (4.70)$$

If we take the view that the dipole is formed by a stationary charge and a mobile charge separated by an oscillating displacement $\boldsymbol{\ell}$, then $\ddot{\mathbf{p}} = q\ddot{\boldsymbol{\ell}} = q\dot{\mathbf{u}}$ and

$$\mathbf{N} = \frac{q^2 \dot{u}^2}{16\pi^2 \varepsilon_0 c^3} \frac{\sin^2\theta}{r^2}\,\mathbf{i}_r \quad ,$$

where $\dot{u}$ is the acceleration of the mobile charge. This expression is identical with Eq. (4.54) for the radiation field of an accelerated charge, showing that the entire radiation from a Hertzian dipole may be attributed to the mobile charge which alone undergoes acceleration. The stationary charge does not radiate energy, but provides a centre for the restoring force.

It might be noted that the assumption that the period of the dipole is large compared with the time taken for electromagnetic effects to travel across the length of the dipole, i.e. $\frac{\ell}{c} \ll T$, or $\frac{\ell}{T} \ll c$, means that the average velocity of the mobile charge is much smaller than the velocity of light. Thus the approximation involved here is the same as the assumption of small velocities for an accelerated charge.

The total radiation emitted per unit time from a Hertzian dipole averaged over one cycle is

$$-\frac{dW}{dt'} = \frac{1}{32\pi^2\varepsilon_0}\frac{\omega^4}{c^3}p_0^2\int_0^\pi \frac{\sin^2\theta}{r^2}2\pi r^2\sin\theta\, d\theta$$

$$= \frac{1}{12\pi\varepsilon_0}\frac{\omega^4}{c^3}p_0^2 = \frac{\mu_0}{12\pi}\frac{\omega^4}{c}p_0^2 \quad . \tag{4.71}$$

In free space $\mu_0 = 4\pi\times 10^{-7}$ Wb $A^{-1}m^{-1}$, $c = 3\times 10^8 m\ s^{-1}$, the above gives

$$-\frac{dW}{dt'} = \frac{1}{9}\times 10^{-15}\omega^4 p_0^2 \quad \text{watts} \quad .$$

If the Hertzian dipole is formed by an oscillating current I in a short conductor ℓ, then

$$\dot{p} = I\ell = -\omega p_0 \sin\omega t'$$

so that

$$\langle I^2\rangle \ell^2 = \frac{1}{2}\omega^2 p_0^2 \quad .$$

Equation (4.71) can be written as

$$-\frac{dW}{dt'} = \frac{\mu_o}{6\pi}\frac{\omega^2}{c}\ell^2 < I^2 > = R < I^2 > \quad ,$$

where

$$R = \frac{\mu_o}{6\pi}\frac{\omega^2\ell^2}{c} = 80\pi^2 \left(\frac{\ell}{\lambda}\right)^2 \Omega \quad ,$$

as

$$\sqrt{\frac{\mu_o}{\varepsilon_o}} = 120\pi\ \Omega \quad .$$

R is called the *radiation resistance* of the short conductor, defined as the ohmic resistance which would cause the same rate of dissipation of energy as the radiation of the dipole for the same root-mean-square current.

4.11 Field of an Oscillating Magnetic Dipole

A closed loop carrying an oscillating current acts as an oscillating magnetic dipole. Such a loop radiates electromagnetic radiation and is also known as a *loop antenna*. For simplicity we shall consider a circular loop C of radius a carrying a sinusoidal current

$$[I] = I_o e^{-i\omega t'}$$

in vacuum. If r is the distance of the field point P from the centre O of the loop, $\lambda = 2\pi c/\omega$ is the wavelength of an electromagnetic wave of the same frequency, then, for mathematical simplicity, we shall assume the conditions to be such that $a \ll \lambda \ll r$ as in the case of the Hertzian dipole.

Choose a coordinate system with the origin at O and the z-axis along the axis of the loop as shown in Fig. 4.6, and let $[I]d\mathbf{r}'$ be a current element at $\mathbf{r}'$. The vector potential at P is

$$\mathbf{A}(\mathbf{r}, t) = \frac{\mu_o}{4\pi}\oint_C \frac{[I]d\mathbf{r}'}{\bar{r}} = \frac{\mu_o}{4\pi r}\oint_C [I]d\mathbf{r}' + \frac{\mu_o}{4\pi r^3}\oint_C [I]\mathbf{r}\cdot\mathbf{r}'d\mathbf{r}' + \ldots \quad , \tag{4.72}$$

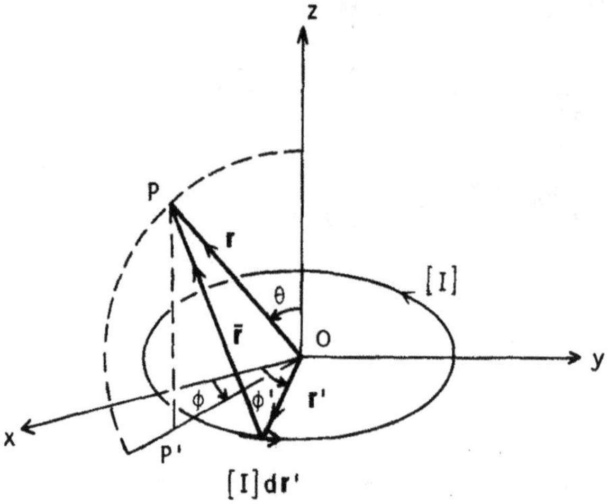

Fig. 4.6 A circular loop carrying a current [I] lies on the xy-plane with its centre at the origin. The meridian plane containing the field point P intercepts the plane of the loop on the line OP'.

where $\bar{r}^{-1} = |\mathbf{r} - \mathbf{r}'|^{-1}$ has been expanded as in Eq. (3.27). As the integrand is not a full differential, the first term in the series does not vanish but remains the dominant term.

The exponent in [I], namely $\exp[-i\omega(t - \bar{r}/c)]$, can be expanded in a similar way. As

$$\bar{r} = (r^2 - 2\mathbf{r}\cdot\mathbf{r}' + r'^2)^{\frac{1}{2}} = r(1 - \frac{\mathbf{r}\cdot\mathbf{r}'}{r^2} + \ldots) \quad ,$$

the exponent can be written as

$$-\, i\omega(t - \frac{r}{c} + \frac{\mathbf{r}\cdot\mathbf{r}'}{rc} + \ldots) = i(kr - \omega t - \mathbf{k}\cdot\mathbf{r}' + \ldots) \quad , \tag{4.73}$$

where $k = \frac{\omega}{c} = \frac{2\pi}{\lambda}$ is the magnitude of the propagation vector **k**, which is in the direction of **r**.

As $r' = a \ll r$ we shall retain only the dominant terms in the above expansions. This in effect restricts our treatment to the radiation field which dominates at large distances. Note that for the first integral in Eq. (4.72) to be finite it would be necessary to retain the $\mathbf{k} \cdot \mathbf{r}'$ term in Eq. (4.73). The vector potential is then, to the first approximation,

$$\mathbf{A} = \frac{\mu_o I_o}{4\pi r} e^{i(kr - \omega t)} \oint_C e^{-i\mathbf{k} \cdot \mathbf{r}'} d\mathbf{r}' \quad . \tag{4.74}$$

Let the spherical coordinates of $\mathbf{r}$ and $\mathbf{r}'$ be (r, θ, ϕ) and $(a, \pi/2, \phi')$ respectively, then

$$\mathbf{k} \cdot \mathbf{r}' = ka \sin\theta \cos(\phi' - \phi) = \rho \cos\psi \quad ,$$

where $\psi = \phi' - \phi$ is the azimuth angle of $\mathbf{r}'$ with respect to the intercept OP' of the meridian plane through the field point P on the plane of the loop, and $\rho = ka \sin\theta$.

The contribution of the current element $[I]d\mathbf{r}'$ to $\mathbf{A}$ is parallel to $d\mathbf{r}'$. As can be seen from Fig. 4.7, for every current element $[I]|d\mathbf{r}'| = [I]ad\phi' = [I]ad\psi$ at angle ψ there exists an element at angle $-\psi$. If their contributions are resolved into two components, one parallel to OP' and the other perpendicular to it, then on summation over all current elements the former components will cancel out. The resultant vector potential will therefore be perpendicular to OP', i.e. in the direction of $\mathbf{i}_\phi$. Hence the vector potential has the following components in the spherical coordinates

$$A_r = A_\theta = 0 \quad ,$$

$$A_\phi = \frac{\mu_o I_o}{4\pi} \frac{e^{i(kr - \omega t)}}{r} \int_0^{2\pi} e^{-i\rho\cos\psi} \cos\psi \, ad\psi$$

$$= \frac{\mu_o a}{4\pi} \frac{I_o e^{i(kr - \omega t)}}{r} \int_0^{2\pi} (1 - i\rho\cos\psi - \frac{\rho^2}{2!}\cos^2\psi$$

$$+ \frac{i\rho^3}{3!}\cos^3\psi + \ldots)\cos\psi \, d\psi \quad .$$

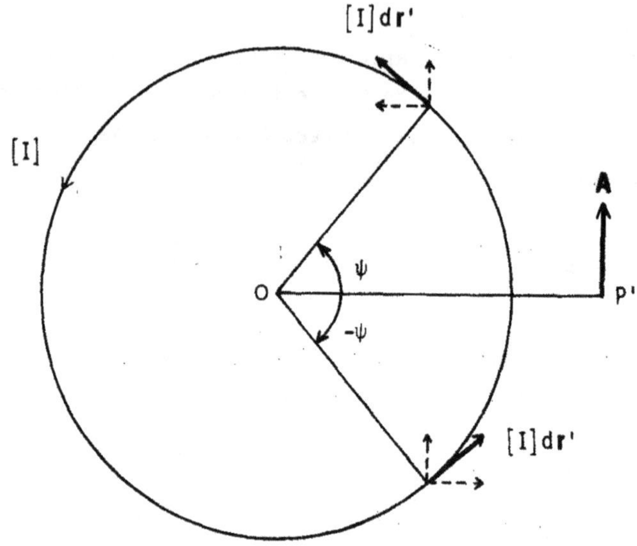

Fig. 4.7 To every current element at angle ψ there exists one at angle $-\psi$. As d**A** is parallel to $[I]d\mathbf{r}'$ for each current element, the component of the vector potential parallel to OP' vanishes when summation is carried out over all current elements.

Since for a non-negative integer n we have

$$\int_0^{2\pi} \cos^{2n+1}\psi d\psi = 0 \quad ,$$

$$\int_0^{2\pi} \cos^{2n}\psi d\psi = \frac{(2n-1)(2n-3)\ldots 3.1}{2n(2n-2)\ldots 4.2}\, 2\pi \quad ,$$

the above becomes

$$A_\phi \simeq -\frac{i\mu_0}{4\pi}\frac{\rho a}{r} I_0 e^{i(kr-\omega t)} \int_0^{2\pi} \cos^2\psi d\psi$$

$$= -i\frac{\mu_0 k}{4}\frac{a^2}{r}\sin\theta\, I_0 e^{i(kr-\omega t)} \quad . \qquad (4.75)$$

neglecting terms involving $\rho^3 \sim (\frac{a}{\lambda})^3$ and higher order magnitudes which give rise to the higher multipole potentials.

The scalar potential Φ can be evaluated by means of the Lorentz condition. As

$$-\mu_o\varepsilon_o \frac{\partial\Phi}{\partial t} = \nabla \cdot \mathbf{A} = \frac{1}{r^2}\frac{\partial}{\partial r}(r^2 A_r) + \frac{1}{r\sin\theta}\frac{\partial}{\partial\theta}(A_\theta \sin\theta) + \frac{1}{r\sin\theta}\frac{\partial A_\phi}{\partial\phi} = 0 \quad ,$$

Φ can be taken as zero, neglecting any possible stationary potential.

The field vectors are then obtained by straightforward differentiation of the vector potential. The relations

$$\mathbf{E} = -\frac{\partial \mathbf{A}}{\partial t} \quad , \qquad \mathbf{H} = \frac{1}{\mu_o}\nabla \times \mathbf{A}$$

give

$$E_r = E_\theta = 0 \quad ,$$

$$E_\phi = \frac{\mu_o}{4}\frac{a^2}{r}\omega k \sin\theta \, I_o e^{i(kr-\omega t)} \quad ,$$

$$H_r = -\frac{ia^2}{2r^2} k\cos\theta \, I_o e^{i(kr-\omega t)} \quad ,$$

$$H_\theta = -\frac{a^2}{4r}k^2 \sin\theta \, I_o e^{i(kr-\omega t)} \quad ,$$

$$H_\phi = 0 \quad . \tag{4.76}$$

It should be noted that in the above derivation only the r^{-1}-dependent terms have been retained from the beginning; consequently only such terms are significant. These terms represent the *radiation field* of an oscillating magnetic dipole, which alone is responsible for energy dissipation by radiation.

The radiation field vectors have non-zero components E_ϕ and H_θ only. Thus, compared with the field of the Hertzian dipole, the roles of **E** and **H** are reversed. Here **H** lies in the meridian plane of the dipole while the lines of **E** form coaxial circles about the dipole axis (note that $\nabla \cdot \mathbf{E} = 0$). The condition

$$\sqrt{\varepsilon_o}\ |\mathbf{E}_{rad}| = \sqrt{\mu_o}\ |\mathbf{H}_{rad}|$$

is again satisfied, indicating that the radiation is propagated in the form of plane electromagnetic waves. The radiation is polarized with **E** perpendicular to the meridian plane containing the direction of propagation.

The Poynting vector is in the radial direction and has an average magnitude

$$\langle N \rangle = -\frac{1}{2} E_\phi H_\theta^* = \frac{\mu_o}{32} \frac{a^4}{r^2} \omega k^3 I_o^2 \sin^2\theta$$

$$= \frac{\mu_o}{32} \frac{a^4}{r^2} c k^4 I_o^2 \sin^2\theta \qquad . \qquad (4.77)$$

Thus the intensity of radiation is maximum in the plane of the loop and zero along the axis.

The average radiation loss per unit time is

$$-\frac{dW}{dt'} = \int_0^\pi 2\pi r^2 \langle N \rangle \sin\theta \, d\theta = \frac{\pi}{12} \mu_o c k^4 a^4 I_o^2 = \frac{1}{2} R I_o^2 \quad ,$$

where $R = (\pi/6)(\mu_o/\varepsilon_o)^{1/2} k^4 a^4$ is the *radiation resistance* of the loop antenna and is defined as the ohmic resistance which would cause the same rate of energy dissipation for the same root-mean-square current. As $(\mu_o/\varepsilon_o)^{1/2} = 120\pi\ \Omega$, we find that

$$R = 320\pi^4 \left(\frac{\pi a^2}{\lambda^2}\right)^2 = 320\pi^4 \left(\frac{S}{\lambda^2}\right)^2 \Omega \qquad . \qquad (4.78)$$

where $S = \pi a^2$ is the area of the loop. The last expression for R also holds approximately for a planar loop of any shape provided its linear dimensions are much smaller than the wavelength of the radiation.

4.12 Multipole Fields

In the above we have considered the fields of oscillating electric and magnetic dipoles. In each case we started from the retarded vector potential with some degree of approximation. To put the treatments in the proper perspective we shall consider a general expansion of the retarded vector potential, assuming a sinusoidal time variation for the current density:

$$\mathbf{J}(\mathbf{r}', t') = \mathbf{J}(\mathbf{r}')e^{-i\omega t'} \quad , \tag{4.79}$$

where t' is the retarded time. It is further assumed that the current distribution is confined to a small region of linear dimensions $\ell \ll \lambda$, where $\lambda = 2\pi c/\omega$ is the wavelength of an electromagnetic wave of the same frequency.

Choose for the origin of coordinates a fixed point in the region of current distribution. The vector potential at a distant field point $P(\mathbf{r})$ is

$$\mathbf{A}(\mathbf{r}, t) = \frac{\mu_0}{4\pi}\int \frac{\mathbf{J}(\mathbf{r}')e^{-i\omega t'}}{\bar{r}} dV' \quad , \tag{4.80}$$

where $\bar{r} = |\mathbf{r} - \mathbf{r}'|$, $t' = t - \bar{r}/c$, and the integral is over the entire distribution. As $r' \leq \ell \ll r$, the potential may be expanded as a series in r'/r. The expansion will be greatly simplified if we first make the following approximation:

$$\bar{r} \simeq r - \hat{\mathbf{r}} \cdot \mathbf{r}' \tag{4.81}$$

where $\hat{\mathbf{r}}$ is a unit vector parallel to $\mathbf{r}$. Note that this approximation is equivalent to replacing the term r'^2 in $\bar{r}^2$ with $(\hat{\mathbf{r}} \cdot \mathbf{r}')^2$, and is therefore a second order approximation. Defining the propagation vector $\mathbf{k}$ as a vector of magnitude ω/c in the direction of $\mathbf{r}$, we can write (4.80) as

$$\mathbf{A}(\mathbf{r},\ t) \simeq \frac{\mu_0}{4\pi}\frac{e^{i(kr-\omega t)}}{r}\int \left(1-\frac{\mathbf{k}\cdot\mathbf{r}'}{kr}\right)^{-1} \mathbf{J}(\mathbf{r}')e^{-i\mathbf{k}\cdot\mathbf{r}'}dV' \quad . \tag{4.82}$$

The integrand may be expanded as a series in $\mathbf{k}\cdot\mathbf{r}'$:

$$\mathbf{J}(\mathbf{r}')\left\{1+\frac{(-i)}{1!}(\mathbf{k}\cdot\mathbf{r}')+\frac{(-i)^2}{2!}(\mathbf{k}\cdot\mathbf{r}')^2+\frac{(-i)^3}{3!}(\mathbf{k}\cdot\mathbf{r}')^3+\ldots\right\}$$

$$\times\left\{1+\frac{\mathbf{k}\cdot\mathbf{r}'}{kr}+\left(\frac{\mathbf{k}\cdot\mathbf{r}'}{kr}\right)^2+\ldots\right\}$$

$$=\mathbf{J}(\mathbf{r}')\left\{1+\left[\frac{(-i)}{1!}+\frac{1}{kr}\right](\mathbf{k}\cdot\mathbf{r}')+\left[\frac{(-i)^2}{2!}+\frac{(-i)}{1!kr}+\frac{1}{(kr)^2}\right](\mathbf{k}\cdot\mathbf{r}')^2\right.$$

$$\left.+\left[\frac{(-i)^3}{3!}+\frac{(-i)^2}{2!kr}+\ldots+\frac{1}{(kr)^3}\right](\mathbf{k}\cdot\mathbf{r}')^3+\ldots\right\} \quad .$$

Thus

$$\mathbf{A} = \sum_{s=0}^{\infty}\mathbf{A}_s \tag{4.83}$$

with

$$\mathbf{A}_s = \frac{\mu_0}{4\pi}\frac{e^{-i\omega t'}}{r}\sum_{m=0}^{s}\frac{(-i)^{s-m}}{(s-m)!(kr)^m}\int(\mathbf{k}\cdot\mathbf{r}')^s\mathbf{J}(\mathbf{r}')dV' \quad , \tag{4.84}$$

where we have re-defined the retarded time as $t' = t - r/c$.

As we shall be interested in the radiation field alone, we need only retain the asymptotic form of $\mathbf{A}_s$ which holds for $\frac{r}{r'}\to\infty$:

$$\mathbf{A}_s \sim \frac{\mu_0}{4\pi}\frac{(-i)^s}{s!}\frac{e^{-i\omega t'}}{r}\int(\mathbf{k}\cdot\mathbf{r}')^s\mathbf{J}(\mathbf{r}')dV' \quad . \tag{4.85}$$

As this contains an exponent linear in r, the dominant terms for the field vectors will also vary as r^{-1} for large r, corresponding to a radiation field. Thus irrespective of s, each $\mathbf{A}_s$ will give rise to a radiation field. On the other hand, for successive values of the integer s, $\mathbf{A}_s$ is reduced in magnitude by a factor $kr' \sim \ell/\lambda$. Under our assumption $\ell \ll \lambda$, the radiation emitted by the current distribution will be due mainly to the first non-vanishing term in the series (4.83).

For $s = 0$, we have

$$\mathbf{A}_0 \sim \frac{\mu_0}{4\pi}\frac{e^{-i\omega t'}}{r}\int \mathbf{J}(\mathbf{r}')dV' \quad . \tag{4.86}$$

This is the vector potential of a *Hertzian dipole* as we have seen before. The integral can be written in a more familiar form if we make use of the continuity equation

$$\nabla' \cdot \mathbf{J}(\mathbf{r}', t') = -\frac{\partial \rho}{\partial t'} = i\omega\rho(\mathbf{r}', t') \tag{4.87}$$

and the identity

$$\nabla' \cdot (x_i'\mathbf{J}) = \frac{\partial}{\partial x_j'}(x_i'J_j) = J_i + x_i'\nabla' \cdot \mathbf{J} \quad .$$

Consider the integral over the current distribution

$$\int_V J_i dV' = \oint_S x_i'\mathbf{J} \cdot d\mathbf{S}' - \int_V x_i'\nabla' \cdot \mathbf{J}\, dV' \quad .$$

The surface integral vanishes if we take the volume of integration slightly larger than the volume of the distribution. Hence

$$\int \mathbf{J}(\mathbf{r}', t')dV' = -\int \mathbf{r}'\nabla' \cdot \mathbf{J}dV' = -i\omega\int \mathbf{r}'\rho dV' = -i\omega\mathbf{p} \quad ,$$

where $\mathbf{p} \simeq \mathbf{p}_0\exp\{i(kr - \omega t)\}$ is the electric dipole moment of the current distribution. $\mathbf{A}_0$ now has the asymptotic form given in Sec. 4.9:

$$\mathbf{A}_0 \sim -i\frac{\mu_0}{4\pi}\frac{\omega}{r}\mathbf{p}_0 e^{i(kr - \omega t)} = \frac{\mu_0}{4\pi}\frac{[\dot{\mathbf{p}}]}{r} \quad .$$

If $\mathbf{A}_0$ vanishes, we look to the approximation of the next order, i.e., $s=1$, for which

$$\mathbf{A}_1 \sim -\frac{i\mu_0}{4\pi}\frac{e^{-i\omega t'}}{r}\int (\mathbf{k}\cdot\mathbf{r}')\mathbf{J}(\mathbf{r}')dV' \quad .$$

This is essentially the same as the vector potential (Eq. (4.74)) given for a closed current loop under the approximation $r' \ll r$.

For a general distribution we can write the integrand as the sum of two parts, one antisymmetric and one symmetric with respect to $\mathbf{r}'$ and $\mathbf{J}$:

$$(\mathbf{k}\cdot\mathbf{r}')\mathbf{J} = \frac{1}{2}\{(\mathbf{k}\cdot\mathbf{r}')\mathbf{J} - (\mathbf{k}\cdot\mathbf{J})\mathbf{r}'\} + \frac{1}{2}\{(\mathbf{k}\cdot\mathbf{r}')\mathbf{J} + (\mathbf{k}\cdot\mathbf{J})\mathbf{r}'\}$$

$$= \frac{1}{2}(\mathbf{r}'\times\mathbf{J})\times\mathbf{k} + \frac{1}{2}\{(\mathbf{k}\cdot\mathbf{r}')\mathbf{J} + (\mathbf{k}\cdot\mathbf{J})\mathbf{r}'\} \quad . \quad (4.88)$$

By Eq. (3.74) we recognize $\frac{1}{2}\mathbf{r}'\times\mathbf{J}e^{-i\omega t'}$ as the magnetization $\mathbf{M}$. Thus the antisymmetric part gives rise to a potential

$$\mathbf{A}_{1m} \sim -\frac{i\mu_0}{4\pi}\frac{\mathbf{m}\times\mathbf{k}}{r} \quad , \quad (4.89)$$

where

$$\mathbf{m} = \int \mathbf{M}dV'$$

is the magnetic dipole moment of the current distribution. It may be noted that $\mathbf{A}_{1m}$ is the same as the vector potential of an oscillating circular current loop as given by Eq. (4.75).

Consider next the symmetric part of Eq. (4.88). The i-th component is

$$\frac{1}{2}\{(\mathbf{k}\cdot\mathbf{r}')J_i + (\mathbf{k}\cdot\mathbf{J})x_i'\} = \frac{1}{2}\nabla'\cdot\{(\mathbf{k}\cdot\mathbf{r}')x_i'\mathbf{J}\} - \frac{1}{2}(\mathbf{k}\cdot\mathbf{r}')x_i'(\nabla'\cdot\mathbf{J}) \quad .$$

On integration, the divergence term gives rise to a surface integral which vanishes for a finite current distribution and we have, using relation (4.87),

$$\mathbf{A}_{1e} \sim -\frac{\mu_0}{4\pi}\frac{\omega}{2r} e^{-i\omega t'} \int (\mathbf{k}\cdot\mathbf{r}')\mathbf{r}'\rho dV' \quad . \tag{4.90}$$

This represents the vector potential of an *electric quadrupole* as we shall see when we evaluate its field vectors.

If we restrict ourselves to the radiation field alone, i.e. to the $\frac{1}{r}$ -dependent term only, then

$$\mathbf{B}_{rad} = \nabla\times\mathbf{A}_{1e} = -\frac{\mu_0}{4\pi}\frac{\omega}{2}\nabla\left\{\frac{e^{i(kr-\omega t)}}{r}\right\}$$

$$\times \int (\mathbf{k}\cdot\mathbf{r}')\mathbf{r}'\rho dV' \sim i\mathbf{k}\times\mathbf{A}_{1e} \quad , \tag{4.91}$$

where, as the propagation vector $\mathbf{k}$ is parallel to $\mathbf{r}$, we have put $kr = \mathbf{k}\mathbf{r}$. Thus

$$\mathbf{B}_{rad} = -i\frac{\mu_0}{4\pi}\frac{\omega k^2}{2r^3} e^{-i\omega t'} \int \mathbf{r}\times\mathbf{r}'(\mathbf{r}\cdot\mathbf{r}')\rho dV' \quad .$$

The i-th component of the integral is

$$\varepsilon_{ijk}x_j x_\ell \int x'_k x'_\ell \rho dV' = \frac{1}{3}\varepsilon_{ijk}x_j x_\ell \int (3x'_k x'_\ell - r'^2\delta_{k\ell})\rho dV'$$

$$= \frac{1}{3}\varepsilon_{ijk}x_j x_\ell Q_{k\ell} \quad ,$$

since $\varepsilon_{ijk}x_j x_\ell \delta_{k\ell} = \varepsilon_{ijk}x_j x_k = (\mathbf{r}\times\mathbf{r})_i = 0$. If we define a vector $\mathbf{Q}$ by

$$Q_i = \frac{1}{r}Q_{ij}x_j \quad , \tag{4.92}$$

we can write the magnetic intensity as

$$\mathbf{H}_{rad} = -\frac{i}{4\pi}\frac{\omega k}{6r} e^{-i\omega t'}\mathbf{k}\times\mathbf{Q} \quad . \tag{4.93}$$

The fact that Q_{ij} are the components of the electric quadrupole moment tensor defined by Eq. (3.28) shows that the field is that of an electric quadrupole.

The electric vector is given by Eq. (4.63) and a calculation similar to that set out in Eq. (4.91) gives

$$\mathbf{E}_{rad} = \frac{i}{\omega\varepsilon_0} \nabla \times \mathbf{H}_{rad} = -\frac{\mathbf{k}\times\mathbf{H}}{\omega\varepsilon_0} \quad .$$

The average Poynting vector of the radiation field is therefore

$$\langle \mathbf{N} \rangle = \frac{1}{2}\,\mathrm{Re}(\mathbf{E}_{rad}\times\mathbf{H}^*_{rad}) = \frac{1}{2\omega\varepsilon_0}\, H_{rad}H^*_{rad}\,\mathbf{k} = \frac{1}{4\pi\varepsilon_0}\,\frac{ck^3}{288\pi r^2}\,|\mathbf{k}\times\mathbf{Q}|^2\,\mathbf{k} \quad , \tag{4.94}$$

where we have made use of the fact that $\mathbf{H}_{rad}$ is transverse to the direction of propagation.

The angular distribution of the quadrupole field is rather complicated and will not be discussed here. The interested reader may refer to the more advanced texts. It suffices to remark that while the dipole radiation intensity varies as ω^4, the quadrupole radiation intensity varies as ω^6.

The higher order terms in the expansion of $\mathbf{A}$ correspond to the fields of the higher multipoles. In general, each $\mathbf{A}_s$ can be similarly divided into electric and magnetic multipole potentials. Mathematical manipulation becomes increasing complex, however, and other methods must be used for the calculation of the higher multipole radiations.

In this and the previous chapter we have taken a survey of the electromagnetic fields under the various approximations. Under static conditions, the electric and the magnetic fields may be treated separately, often along parallel lines. *This is no longer possible when time variation is involved as the interaction between the two fields sets in.* In the study of quasistationary fields we are concerned mostly with currents and applied electromotive forces. The rapidly varying fields are important in that they are the sources of electromagnetic radiation. At large distances from the source and for limited regions of space the radiation may be considered as plane electromagnetic waves. In the following chapters we shall consider the propagation of such waves in matter.

PROBLEMS

1. A circular wire loop of radius a and a long straight wire lie in the same plane such that the centre of the loop is at a distance h from the wire, where $h > a$. Show that the coefficient of mutual induction of the wires is $\mu_0(h - \sqrt{h^2 - a^2})$.
Hint: Assume that the straight wire has a large but finite length $2\ell \gg h$.

2. Two parallel, coaxial circular wire loops of radii a_1 and a_2 are separated by a distance h. Show that their mutual inductance may be expressed as an integral

$$L_{12} = \frac{\mu_0}{2} a_1 a_2 \int_0^{2\pi} \frac{\cos\theta \, d\theta}{(h^2 + a_1^2 + a_2^2 - 2a_1a_2\cos\theta)^{\frac{1}{2}}} ,$$

where θ denotes the angle between two length elements $d\ell_1$ and $d\ell_2$.

Hence show that, in terms of the elliptical integrals

$$K = \int_0^{\pi/2} \frac{d\psi}{(1 - k^2\sin^2\psi)^{\frac{1}{2}}} , \qquad E = \int_0^{\pi/2} (1 - k^2\sin^2\psi)^{\frac{1}{2}} d\psi ,$$

L_{12} may be expressed as

$$L_{12} = \mu_0(a_1a_2)^{\frac{1}{2}}\left[\left(\frac{2}{k} - k\right)K - \frac{2}{k}E\right] ,$$

where $k^2 = \dfrac{4a_1a_2}{(a_1 + a_2)^2 + h^2}$, $\qquad \psi = \dfrac{1}{2}(\pi - \theta)$.

3. Show that the coefficient of self-induction of a circuit carrying a steady current I is given by

$$L = \frac{1}{I^2}\int \mathbf{H}\cdot\mathbf{B}\,dV \quad ,$$

integrating over all space.

The *internal self-inductance* is the part of L which arises from the magnetic field energy contained inside the conductor. Show that the internal self-inductance per unit length of a straight, long homogeneous conductor of permeability μ is $\frac{\mu}{8\pi}$.

4. A charge q moves with a uniform velocity $\mathbf{u}$ in a dielectric medium in which the phase velocity of electromagnetic waves is v. Show that the retardation time $t - t'$ is given by the equation

$$(u^2 - v^2)(t - t')^2 + 2\mathbf{u}\cdot\mathbf{r}(t - t') + r^2 = 0 \quad ,$$

where $\mathbf{r}$ is the radius vector from the "present" position of the charge to the field point. Hence show that (a) there is one and only one positive real solution for $t - t'$ if $u < v$, and that (b) there are no positive real solutions for $t - t'$ if $u > v$, except in the backward cone with vertex at the present position of q of half angle ψ_0 given by

$$\psi_0 = \arccos\left(1 - \frac{v^2}{u^2}\right)^{\frac{1}{2}} \quad .$$

Then there are two positive real solutions for each value of the angle between $\mathbf{u}$ and $\mathbf{r}$, corresponding to two retarded positions.

5. Show that the z-component of $\mathbf{E}$ at a point on the z-axis due to a point charge describing a fixed circle on the xy-plane with the origin as centre is independent of the motion of the charge.

6. Two point charges q and $-q$ move in a circle of radius a centred at the origin in the xy-plane in such a way that they are always at the ends of a diameter, which turns with a constant angular velocity ω and coincides with the x-axis at $t' = 0$. The magnitudes are such that $2aq \to p_0$, a constant, as $a \to 0$ and $q \to \infty$. Write down the vector potential in cylindrical coordinates and show that at a distant point of spherical coordinates (r, θ, ϕ)

$$H_\phi = \frac{\mu_0}{4\pi} \frac{p_0 \omega^2}{cr} \cos\theta \cos\{\omega(t - \frac{r}{c}) - \phi\} \quad .$$

Hence find the electric field at the same point. Hint: Time dependence for the complex field vectors is $\exp\{-i\omega(t - \frac{r}{c})\}$.

7. A charge is accelerated in the direction of motion. Show that the directions of maximum radiation intensity make an angle θ_{max} with the direction of motion given by

$$\cos\theta_{max} = \frac{(1 + 15\beta^2)^{\frac{1}{2}} - 1}{3\beta} \quad ,$$

where $\beta = u/c$, u being its velocity. Show also that as $\beta \to 1$

$$\theta_{max} \to \frac{1}{2}(1 - \beta^2)^{\frac{1}{2}} \quad .$$

Plot (a) θ_{max} vs β, (b) the relative maximum intensity vs β.

8. A particle of mass m and charge q moves with a speed $u \ll c$ in a uniform magnetic field of induction $\mathbf{B}$. Find the rate of radiation if the orbit is circular. Show that for a helical orbit the rate of radiation is reduced by a factor $\cos^2\delta$, where δ is the pitch angle defined as the arctangent of the ratio of the axial component to the azimuthal component of the particle velocity.

9. An electron makes a head-on collision with a nucleus of charge Ze and is scattered back along the line of incidence. If the interaction is purely Coulomb and the electron has an initial

velocity $u \ll c$, show that the electron will suffer an energy loss of approximately

$$\Delta E = \frac{8E_0}{45Z}\left(\frac{u}{c}\right)^5$$

in the collision, where E_0 is the rest energy mc^2 of an electron.

10. A charge q moves in a circular orbit of radius a with a uniform angular velocity ω about the centre. Using a coordinate system with the origin at the instantaneous location of the charge and the x-axis parallel to the axis of the orbit, show that the rate of radiation loss per unit solid angle of the charge is given by

$$-\frac{dw(\theta, \phi)}{dt'} = \frac{q^2a^2\omega^4}{16\pi^2\varepsilon_0 c^3}\left\{\frac{(1-\beta\cos\theta)^2 - (1-\beta^2)\sin^2\theta\sin^2\phi}{(1-\beta\cos\theta)^5}\right\},$$

where θ and ϕ are the polar and azimuth angles respectively and $\beta = a\omega/c$.

Hence show that the total rate of radiation is

$$-\frac{dW}{dt'} = \frac{q^2a^2\omega^4}{6\pi\varepsilon_0 c^3}\frac{1}{(1-\beta^2)^2}.$$

11. Show that the Hertz vector as defined in Prob. (2.8) for a Hertzian dipole $\mathbf{p}_0 e^{-i\omega t'}$ is

$$\boldsymbol{\pi} = \frac{1}{4\pi\varepsilon_0}\frac{\mathbf{p}_0 e^{-i\omega t'}}{r}.$$

Hence obtain its electromagnetic field.

12. Two identical Hertzian dipoles oscillating in phase, each of moment $\mathbf{p}_0 e^{-i\omega t'}$, are placed along the z-axis with their centres separated by $a = n\lambda$, where λ is the wavelength of the radiation and n an integer. Show that the average intensity of the

radiation emitted by the system at an angle θ with the z-axis is increased by a factor $2\{1+\cos(2n\pi\cos\theta)\}$ from that of the radiation emitted by either dipole oscillating alone. Hence show that the total radiation emitted is smaller by a factor $1-3/(2n\pi)^2$ when compared with that of the two dipoles if their radiations do not interfere.

.3. Two identical Hertzian dipoles, each of moment $\mathbf{p}_0 e^{-i\omega t}$, lie along the z-axis and oscillate out of phase. Show that if the distance a between their centres is small compared with the wavelength of the radiation, the average rate of radiation per unit solid angle emitted at an angle θ from the z-axis is

$$-\frac{dw}{dt'} = \frac{1}{4\pi\varepsilon_0}\frac{p_0^2 a^2 \omega^6}{32\pi c^5}\sin^2 2\theta \quad .$$

In this case the dipole moment of the system vanishes and the radiation is due to the quadrupole moment as can be seen by the ω^6-dependence of the radiation loss. Sketch a graph showing the angular distribution of the radiation.

14. A linear antenna of length 2ℓ lies along the z-axis with its centre at the origin and carries a current

$$I(z, t') = I_0\cos(kz)e^{-i\omega t'} \quad ,$$

where I_0 is a constant and $k = \frac{\omega}{c} = \frac{\pi}{2\ell}$. By considering the antenna as a system of Hertzian dipoles, each of moment $\mathbf{p}$ given by $\dot{\mathbf{p}} = I(z, t')dz\mathbf{i}_z$ aligned along the antenna, show that the electric vector of the radiation field at a distant point of position vector $\mathbf{r}$ (where $r \gg \ell$) has only one component in spherical coordinates:

$$E_\theta(\mathbf{r}, t) = -\frac{i}{2\pi\varepsilon_0}\frac{I_0}{rc}e^{i(kr-\omega t)}\frac{\cos(\frac{\pi}{2}\cos\theta)}{\sin\theta} \quad .$$

Hence find the rate of radiation per unit solid angle emitted at an angle θ from the z-axis.

Chapter V

REFLECTION AND REFRACTION OF PLANE ELECTROMAGNETIC WAVES

When an electromagnetic wave passes through the boundary between two different media, reflection and refraction occur. The consequent change of direction, phase and intensity may all be derived from the boundary conditions governing the change of the associated field vectors.

5.1 Laws of Reflection and Refraction

We have seen in Sec. 2.8 that the electric field of a plane electromagnetic wave travelling parallel to the x-axis in a linear, isotropic and homogeneous medium of permittivity ε and permeability μ can be represented by either $E_y(x-vt)$ or $E_z(x-vt)$, or their vector sum, where E_y and E_z are arbitrary functions of $x-vt$, and $v=(\mu\varepsilon)^{-\frac{1}{2}}$ is the phase velocity of the wave in the medium. It is often desirable to have a more general representation. Let O be the origin and $\hat{\mathbf{k}}$ a unit vector in the direction of propagation, then for a point P in the path of the wave, $\vec{OP}=x\hat{\mathbf{k}}$. If O' is the new origin and we denote $\vec{OO'}=\mathbf{r}_0$, $\vec{O'P}=\mathbf{r}$, then as shown in Fig. 5.1 we have

$$x\hat{\mathbf{k}}=\mathbf{r}_0+\mathbf{r} \quad .$$

Hence

$$x-vt=\hat{\mathbf{k}}\cdot\mathbf{r}_0+\hat{\mathbf{k}}\cdot\mathbf{r}-vt=\hat{\mathbf{k}}\cdot\mathbf{r}_0+\frac{1}{k}(\mathbf{k}\cdot\mathbf{r}-\omega t) \quad ,$$

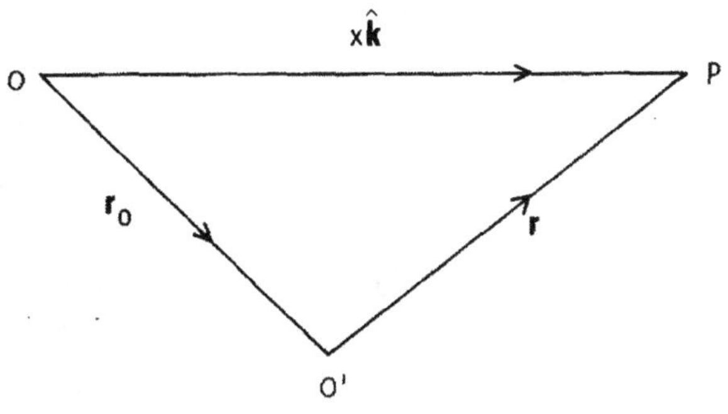

Fig. 5.1 P is a point in the path of the wave, and O' the new origin.

where $k = \omega/v$ and the vector $\mathbf{k} = k\hat{\mathbf{k}}$ is the *propagation vector*. Since $\hat{\mathbf{k}} \cdot \mathbf{r}_0$ and k are both constants, the wave can be represented by an arbitrary vector function $\mathbf{E}(\mathbf{k} \cdot \mathbf{r} - \omega t)$ whose direction is transverse to $\mathbf{k}$. The magnetic vector associated with the wave is then obtained by the relation

$$\mathbf{H} = \sqrt{\frac{\varepsilon}{\mu}}\, \hat{\mathbf{k}} \times \mathbf{E} = \frac{1}{\mu\omega} \mathbf{k} \times \mathbf{E} \qquad . \tag{5.1}$$

Almost all common sources of radiation can be described in terms of harmonic oscillators at least in an approximate way. In any case a wave function can be decomposed into harmonic functions. We shall therefore confine ourselves to plane electromagnetic waves with the electric vector varying sinusoidally:

$$\mathbf{E} = \mathbf{E}_0 e^{i(\mathbf{k} \cdot \mathbf{r} - \omega t)} \qquad , \tag{5.2}$$

where $\mathbf{E}_0$ is a constant real vector. It is understood that only the real part of $\mathbf{E}$ is physically significant.

Consider such a wave travelling from medium 1 across the boundary to medium 2. If the wavelength is small compared with the extension of the boundary, we may consider the boundary surface to be plane provided we restrict ourselves to a beam of small cross section. The deflection at the boundary is governed by the boundary conditions for the electromagnetic field vectors. If no surface current is present, we require the tangential components of both the electric and magnetic vectors to be continuous across the interface. These conditions cannot be satisfied in general by a single progressive wave and it will be necessary to have in the first medium both the incident wave and a reflected wave.

We shall first consider the case where the media are dielectric, so that the propagation wave equation applies to both. Denoting quantities pertaining to the transmitted wave by symbols with a prime and those pertaining to the reflected wave by symbols with double primes, and assuming these to remain plane and sinusoidal the waves may be represented by

$$\mathbf{E}' = \mathbf{E}_0' e^{i(\mathbf{k}' \cdot \mathbf{r} - \omega' t)} \quad , \tag{5.3}$$

$$\mathbf{E}'' = \mathbf{E}_0'' e^{i(\mathbf{k}'' \cdot \mathbf{r} - \omega'' t)} \quad . \tag{5.4}$$

The conditions to be satisfied at the boundary are

$$\mathbf{E}_t + \mathbf{E}_t'' = \mathbf{E}_t' \tag{5.5}$$

and

$$\mathbf{H}_t + \mathbf{H}_t'' = \mathbf{H}_t' \quad , \tag{5.6}$$

where the suffix t denotes components tangential to the boundary surface.

Equations (5.5) and (5.6) can be satisfied at the boundary for all time only if all the exponents involved in the terms are the same. We therefore require that

$$\omega = \omega' = \omega'' \tag{5.7}$$

and

$$\mathbf{k} \cdot \mathbf{r} = \mathbf{k}' \cdot \mathbf{r} = \mathbf{k}'' \cdot \mathbf{r} \tag{5.8}$$

for any point $\mathbf{r}$ at the boundary.

Equation (5.7) shows that reflection and transmission do not change the frequency of a wave. For the interpretation of Eq. (5.8) we choose a coordinate system with the origin in the boundary plane. In particular the origin O is so chosen that the position vector $\mathbf{r}$ of the point O' where the wave strikes the boundary is perpendicular to $\mathbf{k}$. Equation (5.8) then becomes

$$\mathbf{k}' \cdot \mathbf{r} = \mathbf{k}'' \cdot \mathbf{r} = \mathbf{k} \cdot \mathbf{r} = 0 \quad . \tag{5.9}$$

Furthermore, if $\mathbf{n}$ is the normal to the boundary at O',

$$\mathbf{n} \cdot \mathbf{r} = 0 \quad . \tag{5.10}$$

Since the vectors $\mathbf{k}$, $\mathbf{k}'$, $\mathbf{k}''$ and $\mathbf{n}$ meet at a point O' as shown in Fig. 5.2, Eq. (5.9) means that $\mathbf{k}$, $\mathbf{k}'$ and $\mathbf{k}''$ are coplanar. Equation (5.10) furthermore states that their common plane is perpendicular to the boundary plane. Thus reflection and refraction take place in the normal plane containing the incident direction. For oblique incidence this plane is unique and is called the *plane of incidence*. For normal incidence the vectors $\mathbf{k}$, $\mathbf{k}'$, $\mathbf{k}''$ and $\mathbf{n}$ are collinear and any normal plane containing the incident direction is a plane of incidence.

To be more specific, choose a coordinate system with the origin at an arbitrary point O on the boundary plane, the x-axis parallel to the plane of incidence, the y-axis parallel to the normal $\mathbf{n}$, and let θ, θ' and θ'' respectively be the angles of incidence, refraction and reflection, measured from the normal. Then a point on the boundary will have position vector $\mathbf{r} = (x, 0, z)$ where x and z are arbitrary.

Now as

$$\mathbf{k} = k(\sin\theta, -\cos\theta, 0) \quad ,$$

$$\mathbf{k}' = k'(\sin\theta', -\cos\theta', 0) \quad ,$$

$$\mathbf{k}'' = k(\sin\theta'', \cos\theta'', 0) \quad ,$$

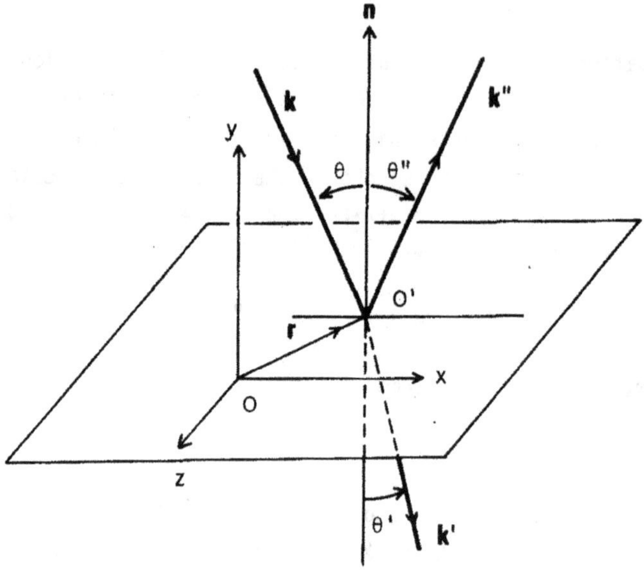

Fig. 5.2 The origin O is chosen in the boundary plane such that **r** is perpendicular to **k**.

Equation (5.8) gives for an arbitrary value of x

$$k \sin\theta = k' \sin\theta' = k \sin\theta'' \quad .$$

Note that $k'' = k$, since they refer to the same medium and frequency remains unaltered on reflection. As all the angles involved are by definition not greater than $\pi/2$, the above means that

$$\theta = \theta'' \tag{5.11}$$

and

$$\frac{\sin\theta}{\sin\theta'} = \frac{k'}{k} = \frac{v_1}{v_2} = \frac{n_2}{n_1} \quad , \tag{5.12}$$

where v is the phase velocity of the wave in a medium and n the *refraction index* of the medium defined as

$$n = \frac{c}{v} = \sqrt{\frac{\mu\varepsilon}{\mu_0\varepsilon_0}} \quad , \tag{5.13}$$

c being the velocity of light in free space. Equation (5.11) shows that the angles of incidence and reflection are equal. Equation (5.12) gives the relation between the angles of incidence and refraction and is known as *Snell's law of refraction.*

5.2 Fresnel's Formulas

To obtain the relative amplitudes of the reflected and the refracted wave it is convenient to consider linearly polarized waves. As we have seen in Sec. 2.8, an unpolarized wave may be considered as consisting of two linearly polarized components with equal amplitudes and mutually perpendicular planes of polarization. There is no loss of generality, therefore, to consider the incident wave as linearly polarized, in the first instance with the plane of polarization, i.e. the plane containing the $\mathbf{E}$ vector and $\mathbf{k}$, normal to the plane of incidence, and later with the plane of polarization parallel to it. If the incoming wave is already linearly polarized, it can still be considered as being made up of two components with the electric vectors parallel and perpendicular to the plane of incidence, their vector sum being equal to the electric vector of the resultant wave. Furthermore, if we exclude ferromagnetic materials we may assume to a good approximation that $\mu_1 \simeq \mu_2 \simeq \mu_0$. The refractive index of a medium of permittivity ε is then given by

$$n = \frac{c}{v} = \sqrt{\frac{\varepsilon}{\varepsilon_0}} \quad . \tag{5.14}$$

Case (a): Electric vector $\mathbf{E}$ *normal to the plane of incidence*

As $\mathbf{E}$ is tangential, the boundary condition may be written simply as

$$E + E'' = E' \quad . \tag{5.15}$$

For convenience, we shall relocate the origin of the coordinate system at O' where the incident wave strikes the boundary as shown in Fig. 5.3. As $\mathbf{E}, \mathbf{H}$ and $\mathbf{k}$ form a right-handed set, if we assume $\mathbf{E}, \mathbf{E}'$

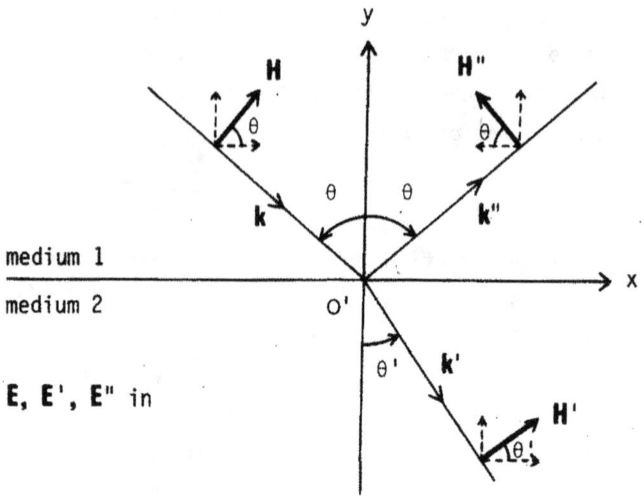

Fig. 5.3 Reflection and refraction at the boundary between two dielectric media. The incident wave is linearly polarized with **E** normal to the plane of incidence. If we assume the electric vectors **E**, **E**' and **E**" to be all pointing into the diagram, then as **E**, **H** and **k** form a right-handed set, the magnetic vectors **H**, **H**' and **H**" will have the directions shown.

and **E**" to have directions pointing into the diagram, the magnetic vectors **H**, **H**' and **H**" will have the directions shown. The boundary condition for **H** is then

$$H\cos\theta - H''\cos\theta = H'\cos\theta' \quad . \tag{5.16}$$

Using Eq. (5.1) and the approximation $\mu_1 \simeq \mu_2 \simeq \mu_0$, this may be written as

$$\sqrt{\varepsilon_1}\,(E - E'')\cos\theta = \sqrt{\varepsilon_2}\,E'\cos\theta' \quad ,$$

or applying Snell's law, as

$$E - E'' = \frac{\sin\theta\cos\theta'}{\sin\theta'\cos\theta}\,E' \quad . \tag{5.17}$$

Solving the simultaneous equations (5.15) and (5.17) we find

$$\frac{E'}{E} = \frac{2\sin\theta'\cos\theta}{\sin\theta'\cos\theta + \sin\theta\cos\theta'} = \frac{2\sin\theta'\cos\theta}{\sin(\theta' + \theta)} \quad , \tag{5.18}$$

$$\frac{E''}{E} = \frac{\sin\theta'\cos\theta - \sin\theta\cos\theta'}{\sin\theta'\cos\theta + \sin\theta\cos\theta'} = \frac{\sin(\theta' - \theta)}{\sin(\theta' + \theta)} \quad . \tag{5.19}$$

Case (b): Electric Vector **E** *in the plane of incidence*

In this case the magnetic vector **H** is tangential to the boundary plane. If we assume the **E** vectors to have the directions shown in Fig. 5.4, then the boundary conditions can be written as

$$H - H'' = H' \tag{5.20}$$

and

$$E + E'' = \frac{\cos\theta'}{\cos\theta} E' \quad . \tag{5.21}$$

Equation (5.20) may be written as

$$E - E'' = \sqrt{\frac{\varepsilon_2}{\varepsilon_1}}\, E' = \frac{\sin\theta}{\sin\theta'} E' \tag{5.22}$$

using Eq. (5.1) and Snell's law. Solving Eqs. (5.21) and (5.22), we find

$$\frac{E'}{E} = \frac{2\sin\theta'\cos\theta}{\sin(\theta' + \theta)\cos(\theta' - \theta)} \quad , \tag{5.23}$$

$$\frac{E''}{E} = \frac{\tan(\theta' - \theta)}{\tan(\theta' + \theta)} \quad . \tag{5.24}$$

The expressions giving the relative amplitude changes on reflection and refraction are known as *Fresnel's formulas*. They were first derived by A. Fresnel on the hypothesis that light consisted of undulations in elastic ether.

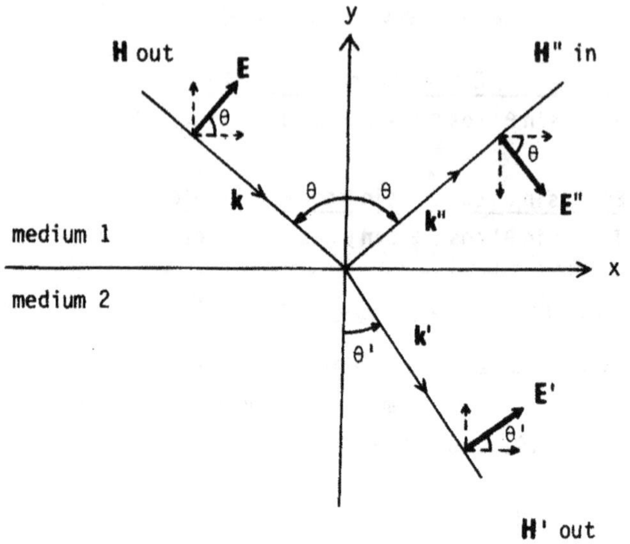

Fig. 5.4 Reflection and refraction at the boundary between two dielectric media. The incident wave is linearly polarized with the plane of polarization coinciding with the plane of incidence.

The intensity of a plane electromagnetic wave is given by the magnitude of the average Poynting vector, which by Eq. (4.68) is

$$\langle \mathbf{N} \rangle = \frac{1}{2}\,\mathrm{Re}(\mathbf{E}\times\mathbf{H}^*) = \frac{1}{2}\,\mathbf{E}\times\mathbf{H}^* = \frac{1}{2}\,EH^*\hat{\mathbf{k}} \quad , \tag{5.25}$$

where $\hat{\mathbf{k}}$ is a unit vector in the direction of propagation. The *transmission* and *reflection coefficients* are defined as the fraction of the incident energy that is transmitted or reflected at the boundary surface. Consider a small area $\mathbf{S}$ at the interface, then the energy incident on or leaving the area per unit time is $|\langle \mathbf{N} \rangle \cdot \mathbf{S}|$. Hence the transmission and reflection coefficients are respectively given by

$$T = \frac{E'H'^*|\hat{\mathbf{k}}'\cdot\mathbf{S}|}{EH^*|\hat{\mathbf{k}}\cdot\mathbf{S}|} = \sqrt{\frac{\varepsilon_2}{\varepsilon_1}}\,\frac{E'E'^*}{EE^*}\,\frac{\cos\theta'}{\cos\theta} = \frac{E'E'^*}{EE^*}\,\frac{\sin\theta\cos\theta'}{\sin\theta'\cos\theta} \quad , \tag{5.26}$$

$$R = \frac{E''H''^*|\hat{\mathbf{k}}'' \cdot \mathbf{S}|}{EH^*|\hat{\mathbf{k}} \cdot \mathbf{S}|} = \frac{E''E''^*}{EE^*} \quad , \tag{5.27}$$

where we have made use of the laws of reflection and refraction.

Let the symbols $\perp$ and $\parallel$ refer to the two cases (a) and (b) respectively. Fresnel's formulas give

$$T_\perp = \frac{\sin 2\theta' \sin 2\theta}{\sin^2(\theta' + \theta)} \quad , \tag{5.28}$$

$$R_\perp = \frac{\sin^2(\theta' - \theta)}{\sin^2(\theta' + \theta)} \quad , \tag{5.29}$$

$$T_\parallel = \frac{\sin 2\theta' \sin 2\theta}{\sin^2(\theta' + \theta)\cos^2(\theta' - \theta)} \quad , \tag{5.30}$$

$$R_\parallel = \frac{\tan^2(\theta' - \theta)}{\tan^2(\theta' + \theta)} \quad . \tag{5.31}$$

It may be readily shown that $T + R = 1$ in each case[a] as required by conservation of energy.

The above equations show that transmission and reflection coefficients are different for different polarizations. This means that if the incident wave is unpolarized the reflected and the transmitted waves will in general become partially plane polarized. Now for reflection and refraction to occur at a boundary, the two media must have different optical properties with the result that θ' and θ will not be equal. Equation (5.31) then shows that when $\theta' + \theta = \pi/2$, $R_\parallel = 0$, which means that the component whose electric vector is in the plane of incidence will be totally refracted. The reflected wave will then

---

[a] Some authors define T and R as ratios of intensities, in which case this relation obviously does not hold in general.

consist of the component with **E** perpendicular to the plane of incidence alone. It follows that the reflected wave is totally plane polarized. Snell's law allows us to express the condition for $R_{\parallel}$ to vanish in terms of the incident angle θ_B. As

$$n_1 \sin\theta_B = n_2 \sin\theta' = n_2 \cos\theta_B \quad ,$$

we have

$$\tan\theta_B = \frac{n_2}{n_1} \quad . \tag{5.32}$$

The angle θ_B is called the *polarizing angle* or *Brewster's angle* and is determined by the refractive indices of the two media.

Fresnel's formulas also make definite predictions on the phase relationship at the boundary.

In case (a) where the incident wave is linearly polarized with **E** normal to the plane of incidence, if the second medium is optically denser, i.e. $n_1 < n_2$, then $\theta > \theta'$, and **E**" and **E** will have opposite signs. Thus there will be a phase change of π on reflection. On the other hand, if the first medium is optically denser no phase change will occur on reflection.

In case (b) where the incident wave is polarized with **E** in the plane of incidence, the situation is more complicated. For $n_1 < n_2$ or equivalently $\theta > \theta'$: **E**" and **E** will have opposite phases if $\theta + \theta' < \pi/2$, i.e., if the incident angle is $\theta < \theta_B$; **E**" and **E** will have the same phase if $\theta > \theta_B$. For $n_1 > n_2$ or $\theta < \theta'$, **E**" and **E** will have the same phase if $\theta < \theta_B$, and a phase difference of π if $\theta > \theta_B$.

To find the phase relationship for the magnetic field vector, consider for example the case where **E** is normal to the plane of incidence. If we draw **H**" in the same way as we draw **E**" in Fig. 5.4, then **E**" must be drawn opposite to **E**. This means that if **H**" and **H** have the same phase, **E**" and **E** will have opposite phases, and vice versa[b]. *The same conclusion holds for the case where* **E** *is in the plane of incidence.*

In a similar way we see that **E**' and **E**, **H**' and **H** will always have the same phase at the boundary. In other words, refraction does not incur a change of phase.

In optics, the polarization of a plane light wave is defined in terms of an optical vector. It is found experimentally that light whose optical vector lies in the plane of incidence is not reflected if it is incident at Brewster's angle, showing that the optical vector is to be identified with the electric vector **E**. Hence the phase relationship between the reflected and incident light waves at the boundary is the same as that between their electric vectors.

The special case of normal incidence can be obtained by letting $\theta, \theta' \to 0$. Snell's law now becomes

$$n_1\theta = n_2\theta' \quad , \tag{5.33}$$

and Fresnel's formulas can be given in terms of the refractive indices:

$$\frac{E'}{E} = \frac{2n_1}{n_1 + n_2} \quad , \quad \frac{H'}{H} = \frac{2n_2}{n_1 + n_2} \quad ,$$

$$\frac{E''}{E} = \frac{n_1 - n_2}{n_1 + n_2} \quad , \quad \frac{H''}{H} = \frac{n_2 - n_1}{n_1 + n_2} \quad .$$

These are the same for both the cases considered, as would be expected since the plane of incidence is arbitrary for normal incidence. The transmission and reflection coefficients follow:

---

b That it is the tangential components, not the normal components, that determine the relative signs of H and H" can be seen by taking the limit $\theta \to 0$.

$$T = \frac{4n_1n_2}{(n_1+n_2)^2} \quad ,$$

$$R = \left(\frac{n_1-n_2}{n_1+n_2}\right)^2 \quad .$$

5.3 Total Reflection

Snell's law imposes a condition for refraction

$$\sin\theta = \frac{n_2}{n_1}\sin\theta' \leq \frac{n_2}{n_1} \quad .$$

If $n_2 > n_1$, this is trivial. If $n_1 > n_2$, this condition is not satisfied for angles of incidence greater than a *critical angle*

$$\theta_0 = \arcsin\left(\frac{n_2}{n_1}\right) \quad . \tag{5.34}$$

Refraction will not occur then and the wave is *totally reflected.*

In total reflection the reflected and the incident wave will have the same amplitude. A treatment of the phenomenon is interesting mainly in the aspect of phase change at the reflecting surface. A theory can be most conveniently formulated using complex angles.

If we admit complex values for the angle of refraction θ' we are not restricted by the condition $|\sin\theta'| \leq 1$. Since the waves are represented by complex functions anyway, the use of complex angles does not introduce anything new. The condition imposed by Snell's law can now be always satisfied and Fresnel's formulas will still apply formally. The method of complex angles is justified if the coefficient of reflection turns out to be unity, or equivalently, if the reflected wave has the same amplitude as the incident wave.

For convenience we introduce the relative refractive index $n = \frac{n_2}{n_1}$ with the understanding that $n < 1$ and consider an incident angle $\theta > \theta_0$. We can eliminate the complex angle θ' by expressing it as a complex function of the real angle of incidence θ, which under the conditions of total reflection is greater than arcsin n

$$\sin\theta' = \frac{1}{n}\sin\theta \quad ,$$

$$\cos\theta' = \pm\sqrt{1-\sin^2\theta'} = \pm\frac{i}{n}\sqrt{\sin^2\theta - n^2} \quad . \tag{5.35}$$

The choice of the sign in the expression for $\cos\theta'$ is determined by the requirement that $\mathbf{E}'$ must be finite in the second medium, which is assumed to extend infinitely. Using the coordinate system shown in Fig. 5.3, we have

$$\mathbf{E}' = \mathbf{E}'_0 \exp\{i(k'\sin\theta' x - k'\cos\theta' y - \omega t)\}$$

$$= \mathbf{E}'_0 \exp\left(\pm\frac{k'}{n}\sqrt{\sin^2\theta - n^2}\, y\right)\exp\left\{i\left(\frac{k'}{n}\sin\theta x - \omega t\right)\right\} \quad . \tag{5.36}$$

As the field penetrates into the second medium, $y \to -\infty$. Then, for $\mathbf{E}'$ to remain finite, we must choose the positive sign in Eq. (5.35)[c]. Thus the wave in the second medium is propagating in the x-direction with a modified phase velocity $\frac{n\omega}{k'\sin\theta}$ and has an amplitude which attenuates exponentially with increasing penetration. Consider now the reflected wave for the two cases considered above.

Case (a): $\mathbf{E}$ *normal to the plane of incidence*

Substitution of Eq. (5.35) in Eq. (5.19) gives

$$\frac{E''}{E} = \frac{\cos\theta - i\sqrt{\sin^2\theta - n^2}}{\cos\theta + i\sqrt{\sin^2\theta - n^2}} = \rho_\perp e^{-i\phi_\perp} \quad ,$$

---

[c] Consider a second medium of finite thickness t. The wave in the medium will consist of components corresponding to the two signs. It can be shown that with $t \to \infty$ only the component with finite amplitude remains.

where $\rho_\perp$, the modulus of the complex function E''/E, has the value unity and

$$\phi_\perp = -\arg\left(\frac{E''}{E}\right) = -2\arg(\cos\theta - i\sqrt{\sin^2\theta - n^2})$$

$$= 2\arctan\left(\frac{\sqrt{\sin^2\theta - n^2}}{\cos\theta}\right) \quad . \tag{5.37}$$

Case (b): **E** *in the plane of incidence*

Substitution of Eq. (5.35) in Eq. (5.24) gives

$$-\frac{E''}{E} = \frac{n^2\cos\theta - i\sqrt{\sin^2\theta - n^2}}{n^2\cos\theta + i\sqrt{\sin^2\theta - n^2}} = \rho_{\parallel} e^{-i\phi_{\parallel}} \quad ,$$

where

$$\rho_{\parallel} = 1$$

and

$$\phi_{\parallel} = 2\arctan\left(\frac{\sqrt{\sin^2\theta - n^2}}{n^2\cos\theta}\right) \quad . \tag{5.38}$$

We see that in each case $R = \rho^2 = 1$ satisfying the physical requirement for total reflection. The conservation of energy then insures that no transmission occurs. This is also seen from the fact that T as given by Eq. (5.26) is imaginary. It is clear from Eqs. (5.26) and (5.27) that only the component of the energy flow normal to the boundary is involved in the definitions of T and R. The result that $\mathrm{Re}\,T = 0$ means that the time-average of the normal component of the energy flow into the second medium vanishes. The instantaneous normal energy flow, which has a real part, need not vanish however. The time-average of the component parallel to the boundary, on the other hand, does not vanish as it is real but attentuates exponentially with the depth of penetration [Prob. (5.8)].

In total reflection, the phase change is no longer either 0 or π but can have any intermediate value. Furthermore, the phase changes are different for different polarizations giving rise to some interesting consequences.

If the incident wave is linearly polarized with the electric vector **E** neither in the plane of incidence nor perpendicular to it, it can be decomposed into two such components. On total reflection, each will undergo a different phase change and the two components will in general recombine into an elliptically polarized wave. Let ψ be the angle between the electric vector **E** of the incident wave and the plane of incidence. The components of **E** corresponding to cases (a) and (b) after total reflection are

$$E''_{\perp} = E_0 \sin\psi \cos(\mathbf{k}'' \cdot \mathbf{r} - \omega t - \phi_{\perp}) \quad ,$$

and

$$E''_{||} = - E_0 \cos\psi \cos(\mathbf{k}'' \cdot \mathbf{r} - \omega t - \phi_{||}) \quad . \tag{5.39}$$

In general, $\phi_{\perp}$ and $\phi_{||}$ will not be equal and the terminus of the resultant electric vector $\mathbf{E}''$ at a fixed point on the path of the reflected wave will trace out an ellipse in the plane normal to the direction of propagation [Prob. (5.7)]. Thus the totally reflected wave is elliptically polarized. In the special case of $\psi = \pi/4$ and $\phi_{\perp} - \phi_{||} = \pm\pi/2$, the totally reflected wave is circularly polarized. An example of the application of this effect is *Fresnel's rhomb*, which is a parallelepiped of crown glass. A beam of light incident normally on one face of the rhomb undergoes internal total reflection twice and emerges normally from the opposite face. If the incident light is linearly polarized with the plane of polarization making an angle $\pi/4$ with the plane of internal reflection, and if the material of the rhomb and the angle of reflection are such that the relative phase change $\phi_{\perp} - \phi_{||}$ is $\pm\pi/4$ on each reflection, the light emerging from the rhomb will be circularly polarized.

Another interesting aspect of the theory is the prediction that some penetration of the electromagnetic field into the second medium will occur even under the conditions of total reflection as can be seen from Eq. (5.36). The tacit assumption made in choosing the sign of $\cos\theta'$ which prevents the amplitude in the second medium from becoming infinitely large is that the thickness of the second medium is large compared with the wavelength. If this condition is not satisfied, both signs are to be used; the wave can penetrate through the second medium and total reflection will not occur. For example, total reflection does not occur when light is incident on a thin layer of air sandwiched between two pieces of optically flat glass even though the incident angle exceeds the critical angle.

5.4 Refraction in a Conducting Medium

Consider now the case where medium 2 is a linear, isotropic and homogeneous conductor of permittivity ε_2, permeability μ_2 and conductivity σ_2. For a conducting medium the general wave equation is to be used. For waves with a sinusoidal time-dependence,

$$\mathbf{E} = \mathbf{E}_o(\mathbf{r})e^{-i\omega t} \quad ,$$

where $\mathbf{E}_o$ is complex, the general wave Eq. (2.39) can be written as

$$\nabla^2\mathbf{E} - \mu_2\varepsilon_2\left(1 + \frac{i\sigma_2}{\omega\varepsilon_2}\right)\ddot{\mathbf{E}} = 0 \quad . \tag{5.40}$$

This has the form of the propagation wave equation

$$\nabla^2\mathbf{E} - \mu\varepsilon\ddot{\mathbf{E}} = 0$$

in a dielectric of permittivity ε and permeability μ so that the results obtained for a dielectric medium can be formally taken over, provided we make the substitution

$$\mu\varepsilon = \mu_2\varepsilon_2\left(1 + \frac{i\sigma_2}{\omega\varepsilon_2}\right) \quad . \tag{5.41}$$

For convenience we shall assume the approximation $\mu_2 \simeq \mu_o$. In the case where this approximation is not satisfactory the correct results may be obtained by the re-substitution $\varepsilon_o \to \varepsilon_o(\mu_o/\mu_2)$. The refractive index of the conducting medium is then

$$n_2 = \sqrt{\frac{\varepsilon_2}{\varepsilon_o}\left(1 + \frac{i\sigma_2}{\omega\varepsilon_2}\right)} = \pm(\eta + i\kappa) \quad , \tag{5.42}$$

where η and κ are positive real numbers given by

$$\eta^2 - \kappa^2 = \frac{\varepsilon_2}{\varepsilon_0} \quad , \qquad 2\eta\kappa = \frac{\sigma_2}{\omega\varepsilon_0} \quad ,$$

or more explicitly by

$$\eta^2 = \frac{1}{2}\frac{\varepsilon_2}{\varepsilon_0}\left(\sqrt{1 + \left(\frac{\sigma_2}{\omega\varepsilon_2}\right)^2} + 1\right)$$

and

$$\kappa^2 = \frac{1}{2}\frac{\varepsilon_2}{\varepsilon_0}\left(\sqrt{1 + \left(\frac{\sigma_2}{\omega\varepsilon_2}\right)^2} - 1\right) \quad . \tag{5.43}$$

The sign to be used for the right-hand side of Eq. (5.42) is to be determined by the requirement that the amplitude of the wave should remain finite as the conducting medium is traversed. Suppose the plus sign is to be chosen, then the propagation vector has a magnitude

$$k' = \frac{\omega n_2}{c} = (\eta + i\kappa)k_0 \quad , \tag{5.44}$$

where $k_0 = \omega/c$. The wave in the conducting medium can then be represented, using the coordinate system shown in Fig. 5.3, by

$$\mathbf{E}' = \mathbf{E}_0' \, e^{i(k'x\sin\theta' - k'y\cos\theta' - \omega t)} \tag{5.45}$$

and

$$|\mathbf{H}'| = \frac{1}{\mu_2\omega}|\mathbf{k}' \times \mathbf{E}'| = \frac{(\eta + i\kappa)}{\mu_2\omega} k_0 |\mathbf{E}'| \quad . \tag{5.46}$$

Equation (5.45) means, as will be shown below, that the amplitude of the wave is not constant but attenuates exponentially. Equation (5.46) shows that the electric and the magnetic field are no longer in phase but have a phase difference of $\arctan(\kappa/\eta)$.

Snell's law and Fresnel's formulas can still be applied formally. The former now gives

$$\sin\theta' = \frac{n_1}{n_2}\sin\theta = \frac{n_1 \sin\theta}{\eta + i\kappa} \tag{5.47}$$

so that the angle of refraction θ' is complex. As

$$k'\sin\theta' = (\eta + i\kappa)k_o \frac{n_1}{\eta + i\kappa}\sin\theta = n_1 k_o \sin\theta \quad , \tag{5.48}$$

$$\begin{aligned} k'\cos\theta' &= \sqrt{k'^2 - k'^2\sin^2\theta'} = k_o\sqrt{\eta^2 - \kappa^2 - n_1^2\sin^2\theta + 2\eta\kappa i} \\ &= p + iq \quad , \end{aligned} \tag{5.49}$$

where p and q are positive real numbers given by

$$p^2 - q^2 = (\eta^2 - \kappa^2 - n_1^2\sin^2\theta)k_o^2 \tag{5.50}$$

and

$$pq = \eta\kappa k_o^2 \quad ,$$

we find

$$\mathbf{E}' = \mathbf{E}_o' \, e^{qy} \, e^{i(n_1 k_o \sin\theta \, x - py - \omega t)} \tag{5.51}$$

The equation for $\mathbf{H}'$ can be similarly written. Apart from a constant phase difference the exponent of $\mathbf{H}'$ is identical with that of $\mathbf{E}'$.

For the interpretation of the expression (5.51) we refer to Fig. 5.5.

The amplitude of $\mathbf{E}'$ has an attenuation factor $\exp(qy)$ so that the surfaces of constant amplitude are the planes $qy = \text{constant}$, which are parallel to the boundary surface. On the other hand the surfaces of constant phase are given by

$$n_1 k_o \sin\theta \, x - py - \omega t = \text{constant} \quad , \tag{5.52}$$

which at any given instant of time are a family of slanting parallel planes. In general the surfaces of constant amplitude and

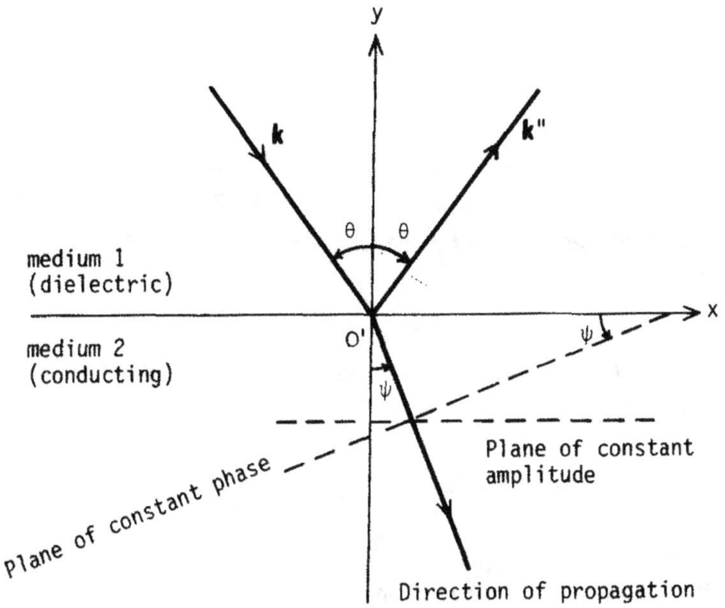

Fig. 5.5 For a wave refracted into a conducting medium the surfaces of constant amplitude and constant phase do not coincide as in a dielectric medium.

constant phase do not coincide as is the case of waves in a dielectric. This is an example of non-uniform plane waves, since on a wavefront (i.e. plane of constant phase) the amplitude varies from point to point.

The direction of propagation is normal to the planes of constant phase and the angle ψ it makes with the y-axis is the effective angle of refraction. As can be seen from Fig 5.5, ψ is also the angle between a plane of constant phase and the x-axis. Hence

$$\tan\psi = \left(\frac{\partial y}{\partial x}\right)_t = \frac{n_1 k_o \sin\theta}{p} \quad ,$$

which may be written as

$$n_1 \sin\theta = n_2' \sin\psi \quad , \tag{5.53}$$

with

$$n_2' = \frac{1}{k_o}(p^2 + n_1^2 k_o^2 \sin^2\theta)^{\frac{1}{2}} \quad . \tag{5.54}$$

As Eq. (5.53) has the form of Snell's law, we may consider n_2' as the *effective* refractive index. It should be noted that in addition to the constants of medium 2, n_2' depends on the angle of incidence as well as the frequency.

The phase velocity of the refracted wave is the rate of advance in the direction of propagation of a plane of constant phase. This may be obtained most simply from Eq. (5.54)

$$v_2' = \frac{c}{n_2'} = \frac{\omega}{k_o n_2'} \quad , \tag{5.55}$$

which is seen also to depend on the direction of incidence. Note that as the field vectors have components along the direction of propagation the refracted wave is not strictly transverse. This can be most easily seen by noting that $\mathbf{k}'$, which formally gives the direction of the plane of $\mathbf{E}'$ and $\mathbf{H}'$, has no definable direction as θ' is complex.

If the medium is a good conductor, which is the case of all metallic conductors, then for frequencies up to the optical range, $\sigma_2 \gg \omega\varepsilon_2$, so that Eqs. (5.43) have the approximate solutions

$$\eta \simeq \kappa \simeq \sqrt{\frac{\sigma_2}{2\omega\varepsilon_o}} \quad . \tag{5.56}$$

Equations (5.50) then yield the solution

$$q^2 \simeq \frac{1}{2} k_o^2 \left(\sqrt{4\eta^4 + n_1^4 \sin^4\theta} + n_1^2 \sin^2\theta\right) \quad ,$$

and

$$p^2 \simeq \frac{1}{2} k_o^2 \left(\sqrt{4\eta^4 + n_1^4 \sin^4\theta} - n_1^2 \sin^2\theta\right) \quad .$$

As η, $\kappa \gg 1$ and n_1 is of the order unity, the above can again be approximated by

$$q \simeq p \simeq \eta k_o \simeq \sqrt{\frac{\sigma_2\mu_2}{2\omega\varepsilon_o\mu_o}}\sqrt{\omega^2\mu_o\varepsilon_o} \simeq \sqrt{\frac{\omega\sigma_2\mu_2}{2}} \quad . \tag{5.57}$$

Equation (5.51) shows that the amplitudes of the fields inside the conducting medium are attenuated with a factor $\exp(\sqrt{\omega\sigma_2\mu_2/2}\, y)$, so that the fields will fall to e^{-1} of their surface values at a depth

$$-y = \sqrt{\frac{2}{\omega\sigma_2\mu_2}} \equiv \delta \quad . \tag{5.58}$$

The quantity δ measures the depth of penetration and is called the *skin depth*. Thus the greater the conductivity of the medium, or the greater the frequency of the wave, the smaller is the penetration. In the limit of a perfect conductor, $\sigma \to \infty$, $\delta \to 0$, showing that oscillating electromagnetic fields cannot penetrate into a perfect conductor.

According to Eqs. (5.46) and (5.56), the electric and the magnetic field of an electromagnetic wave in a good conductor will have a phase difference of $\pi/4$. Furthermore, the electromagnetic energy is no longer equally shared by the two fields since

$$\frac{\mu_2 H'H'^*}{\varepsilon_2 E'E'^*} = \frac{\mu_2}{\varepsilon_2}\left(\frac{\sqrt{2}\eta}{\mu_2\omega}\frac{\omega}{c}\right)^2 = \frac{\sigma_2}{\omega\varepsilon_2} \quad ,$$

which is a large number. In a good conductor, therefore, the electric field energy is negligible compared with the magnetic field energy. In other words, a good conductor is practically impenetrable to the oscillating electric fields for which the condition $\sigma \gg \omega\varepsilon$ is satisfied. Penetration of electromagnetic waves through a conducting medium is possible only if the thickness is of the same order of magnitude as the skin depth. The disparity in the field energies and the attenuation in the conductor are both due to the *Joule effect* whereby the electric energy of the wave is progressively dissipated as heat.

For a good conductor, the effective refractive index as given by Eq. (5.54) is approximately

$$n_2' \simeq \frac{p}{k_o} \simeq \eta \simeq \sqrt{\frac{\sigma_2}{2\omega\varepsilon_o}} \tag{5.59}$$

so that Snell's law takes the form

$$\sin\psi \simeq \sqrt{\frac{2\omega\varepsilon_1}{\sigma_2}} \sin\theta \quad . \tag{5.60}$$

This means that ψ is usually small for a good conductor, and tends to zero as the conductivity increases. Thus the plane of constant phase tends to coincide with the plane of constant amplitude and the refracted wave tends to propagate normal to the surface for a good conductor whatever the angle of incidence.

5.5 Reflection from a Conducting Surface

The reflection of a plane electromagnetic wave from the surface of a conducting medium may be similarly treated. We already have from Eqs. (5.48) and (5.49)

$$\sin\theta' = \frac{1}{k'} n_1 k_o \sin\theta \quad ,$$

and

$$\cos\theta' = \frac{1}{k'}(p + iq) \quad .$$

For the case where **E** is normal to the plane of incidence, the amplitude of the reflected wave is given by Eq. (5.19)

$$\frac{E''}{E} = \frac{\sin\theta'\cos\theta - \sin\theta\cos\theta'}{\sin\theta'\cos\theta + \sin\theta\cos\theta'} = \frac{(n_1 k_o \cos\theta - p) - iq}{(n_1 k_o \cos\theta + p) + iq} = \rho_\perp' e^{-i\phi_\perp} \quad , \tag{5.61}$$

where the modulus is given by

$$\rho_\perp'^2 = \frac{(n_1 k_o \cos\theta - p)^2 + q^2}{(n_1 k_o \cos\theta + p)^2 + q^2} \quad , \tag{5.62}$$

and the phase change is given by

$$\phi'_{\perp} = \arctan \frac{q}{n_1 k_o \cos\theta - p} + \arctan \frac{q}{n_1 k_o \cos\theta + p}$$

or

$$\tan \phi'_{\perp} = \frac{2 n_1 k_o q \cos\theta}{(n_1 k_o \cos\theta)^2 - (p^2 + q^2)} \quad . \qquad (5.63)$$

For a good conductor, for which $\sigma_2 >> \omega\varepsilon_2$, we may use the approximations

$$q \simeq p \simeq \eta k_o \quad ,$$

and

$$\eta \simeq \sqrt{\frac{\sigma_2}{2\omega\varepsilon_o}} >> 1$$

so that

$$\rho'^2_{\perp} \simeq \frac{(n_1 \cos\theta - \eta)^2 + \eta^2}{(n_1 \cos\theta + \eta)^2 + \eta^2} = 1 - \frac{4\alpha \cos\theta}{(1 + \alpha\cos\theta)^2 + 1} \simeq 1 - 2\alpha\cos\theta$$

$$(5.64)$$

and

$$\tan \phi'_{\perp} \simeq \frac{2 n_1 \eta \cos\theta}{n_1^2 \cos^2\theta - 2\eta^2} = \frac{2\alpha\cos\theta}{\alpha^2\cos^2\theta - 2} \simeq -\alpha\cos\theta \quad ,$$

where

$$\alpha = \frac{n_1}{\eta} \simeq \sqrt{\frac{2\omega\varepsilon_1}{\sigma_2}} << 1 \quad . \qquad (5.65)$$

For the case where **E** is in the plane of incidence, the calculation is more tedious and we shall assume the approximation of a good conductor from the start, for which

$$\sin\theta' = \frac{k_o n_1 \sin\theta}{k'} \simeq \frac{n_1 \sin\theta}{\eta(1+i)} = \frac{\alpha \sin\theta}{1+i} \quad ,$$

$$\cos\theta' = \frac{k_o \eta(1+i)}{k'} = 1 \quad .$$

Equation (5.24) then becomes

$$-\frac{E''}{E} = \frac{\sin\theta\cos\theta - \sin\theta'\cos\theta'}{\sin\theta\cos\theta + \sin\theta'\cos\theta'} \simeq \frac{(\cos\theta - \alpha) + i\cos\theta}{(\cos\theta + \alpha) + i\cos\theta} = \rho'_{||}e^{-i\phi'_{||}} \quad ,$$

where (5.66)

$$\rho'^2_{||} = \frac{(\cos\theta - \alpha)^2 + \cos^2\theta}{(\cos\theta + \alpha)^2 + \cos^2\theta} \tag{5.67}$$

and

$$\phi'_{||} = \arctan\frac{\cos\theta}{\alpha - \cos\theta} + \arctan\frac{\cos\theta}{\alpha + \cos\theta} = \arctan\frac{2\alpha\cos\theta}{\alpha^2 - 2\cos^2\theta} \quad ,$$

or

$$\tan\phi'_{||} = \frac{2\alpha\cos\theta}{\alpha^2 - 2\cos^2\theta} \quad . \tag{5.68}$$

The quantities $\rho_\perp^2$ and $\rho_{||}^2$ give the coefficients of reflection for the respective cases. It can be seen that both are close to unity and essentially independent of the angle of incidence. Furthermore, the greater the values of p and q, the smaller will be the value of α, and the nearer will be the values of $R_\perp$ and $R_{||}$ to unity. Hence wavelengths that are strongly absorbed (for which q is large) are also strongly reflected. Thus the colours of a thin metallic film by transmitted light and by reflected light are complementary. For example, a thin gold foil appears blue by transmitted light.

As in the case of total reflection, the change of phase on reflection from a metallic surface depends on the polarization of the incident wave. Consequently, an incident wave which is linearly polarized with the plane of polarization neither normal to the plane of incidence nor parallel to it will be reflected elliptically polarized.

PROBLEMS

1. Obtain Fresnel's formulas for the magnetic vector **H**. Hence obtain the coefficients of transmission and reflection.

2. Obtain the coefficients of transmission and reflection for normal incidence directly from the boundary conditions for the field vectors.

3. Show that linearly polarized electromagnetic waves remain linearly polarized after reflection and refraction at the boundary between two dielectric media except in the case of total reflection. If the plane of polarization of the incident wave makes an angle ψ with the plane of incidence and the angle of incidence is θ, show that the angles δ and β which the electric vectors of the refracted and reflected waves make with the plane of incidence are respectively given by

$$\tan\beta = \frac{\cos(\theta' - \theta)}{\sin(\theta' + \theta)}\tan\psi \quad ,$$

$$\tan\delta = \cos(\theta' - \theta)\tan\psi \quad ,$$

where θ' is the angle of refraction.

4. A plane electromagnetic wave linearly polarized with the electric vector at $\pi/4$ to the plane of incidence is incident from a dielectric medium of refractive index n_1 on the plane surface of a dielectric medium of refractive index n_2. Show that if the angle of incidence is such that the reflected wave is polarized normal to the plane of incidence, the refracted wave will be polarized at an angle $\operatorname{arccot}\left\{\frac{1}{2}\left(\frac{n_1}{n_2} + \frac{n_2}{n_1}\right)\right\}$ to the plane of incidence.

5. If the approximation $\mu_2 \simeq \mu_1$ is not valid, show that at the boundary between two dielectric media the reflection coefficients are

$$R_\perp = \left[\frac{\sqrt{\frac{\varepsilon_2}{\mu_2}}\cos\theta' - \sqrt{\frac{\varepsilon_1}{\mu_1}}\cos\theta}{\sqrt{\frac{\varepsilon_2}{\mu_2}}\cos\theta' + \sqrt{\frac{\varepsilon_1}{\mu_1}}\cos\theta}\right]^2$$

and

$$R_{||} = \left[\frac{\sqrt{\frac{\mu_2}{\varepsilon_2}}\cos\theta' - \sqrt{\frac{\mu_1}{\varepsilon_1}}\cos\theta}{\sqrt{\frac{\mu_2}{\varepsilon_2}}\cos\theta' + \sqrt{\frac{\mu_1}{\varepsilon_1}}\cos\theta}\right]^2 .$$

Find the transmission coefficients and discuss whether a polarizing angle (Brewster's angle) exists for each polarization.

6. The degree of polarization P is defined as

$$P = \left|\frac{I_{||} - I_{\perp}}{I_{||} + I_{\perp}}\right| ,$$

where $I_{||}$ and $I_{\perp}$ are respectively the intensities of the polarized components of an electromagnetic wave with **E** parallel and normal to the plane of incidence. Show that when an unpolarized wave travelling in air is incident on the plane surface of a dielectric of refractive index n at Brewster's angle, the refracted wave is polarized to a degree

$$P = \frac{(1+n^2)^2 - 4n^2}{(1+n^2)^2 + 4n^2} .$$

7. Show that at a fixed point on the path of the totally reflected wave Eq. (5.39) may be written as

$$\frac{E_{\perp}''^2}{a^2} + \frac{E_{||}''^2}{b^2} - \frac{2\cos\delta}{ab} E_{\perp}'' E_{||}'' = \sin^2\delta ,$$

where $\delta = \phi_{||} - \phi_{\perp}$, $a = E_0 \sin\psi$ and $b = -E_0\cos\psi$.
Hence show that in a coordinate system with the z-axis along the direction of propagation and the x-axis at an angle ψ to plane of incidence, the above may be written as

$$\frac{E_x''^2}{A^2} + \frac{E_y''^2}{B^2} = 1 \quad ,$$

where A and B are constants. Thus it is clear that the terminus of the vector **E**" describes an ellipse in a plane normal to the direction of propagation.

8. A plane electromagnetic wave is totally reflected at the boundary surface between two dielectric media. Show that the normal component of the average energy flow across the surface is purely imaginary so that the transmission coefficient T vanishes. Show that, on the other hand, the component of the average energy flow along the interface and just inside the second medium is real. Hint: Calculate $(\mathbf{E}' \times \mathbf{H}'^*) \cdot \mathbf{S}$ and $(\mathbf{E}' \times \mathbf{H}'^*) \times \mathbf{S}$.

9. A linearly polarized electromagnetic wave travelling in a dielectric medium of refractive index n_1 is totally reflected from the plane surface of a dielectric medium of refractive index n_2. If the plane of polarization of the incident wave makes an angle $\pi/4$ with the plane of incidence, show that for the totally reflected wave to become circularly polarized the refractive indices must be such that

$$\frac{n_2}{n_1} \leq \sqrt{2} - 1 \quad .$$

10. A plane electromagnetic wave is totally reflected in a prism which it enters and leaves normally. The material of the prism has a refractive index $n > 1$. (a) Show that the intensity of the emergent beam is $16n^2(1+n)^{-4}I_0$, where I_0 is the intensity of the incident wave. (b) If the incident wave is polarized at 45° to the plane of incidence, show that the emergent beam is elliptically polarized with a phase difference ϕ between $\mathbf{E}_{\parallel}$ and $\mathbf{E}_{\perp}$ given by

$$n \tan \frac{1}{2}\phi = \cot\theta \csc\theta\,(n^2\sin^2\theta - 1)^{\frac{1}{2}} \quad ,$$

where θ is the angle of incidence on the back of the prism, multiple reflections being neglected.

11. A plane electromagnetic wave travelling in a medium of constants ε_1 and μ_1 is reflected from a medium of constants $\varepsilon_2 = \varepsilon_1(1+2\eta)$, where $\eta \ll 1$, and $\mu_2 = \mu_1$. Show that, provided the angle of incidence θ is not near $\pi/2$, the ratio E":E is approximately $-\frac{1}{2}\eta\sec^2\theta$ for **E** perpendicular to the plane of incidence and $-\eta(1-\tan^2\theta)$ for **E** in the plane of incidence. What are the conditions for total reflection to take place?

12. Defining a positive number $\delta = \theta_0 - \theta$, θ_0 and θ being the critical angle and the incident angle respectively, show that as $\delta \to 0$,

$$R_\perp \simeq 1 - a_\perp\sqrt{\delta}$$

and $$R_{||} \simeq 1 - a_{||}\sqrt{\delta} \quad ,$$

where $a_\perp$ and $a_{||}$ are constants depending on the refractive indices of the media.

13. A plane electromagnetic wave travelling in a dielectric medium of permittivity ε_1 falls normally on the plane surface of a conductor of permittivity ε_2 and conductivity σ. Show that the coefficient of reflection R in the limiting case of low conductivity is

$$R = \left(\frac{\sqrt{\varepsilon_1}-\sqrt{\varepsilon_2}}{\sqrt{\varepsilon_1}+\sqrt{\varepsilon_2}}\right)^2 + \frac{1}{(\sqrt{\varepsilon_1}+\sqrt{\varepsilon_2})^4}\sqrt{\frac{\varepsilon_1}{\varepsilon_2}}\left(\frac{\sigma}{\omega}\right)^2 \quad .$$

Find also the phase change of the wave on reflection.

14. A plane electromagnetic wave travelling in a dielectric medium of refractive index n_1 falls normally on its boundary with a conducting medium of complex refractive index $\eta + i\kappa$. Show that the reflected wave suffers a phase change ϕ given by

$$\tan\phi = \frac{2n_1\kappa}{n_1^2 - (\eta^2 + \kappa^2)} .$$

Hence show that if the second medium is a good conductor the phase change is always nearly π.

15. A plane electromagnetic wave travelling in a dielectric medium of permittivity ε_1 is reflected from the plane surface of a good conductor of permittivity ε_2 and conductivity σ_2. For oblique incidence, show that the energy is reduced by a factor approximately equal to $1 - 2\sec\theta(2\omega\varepsilon_1/\sigma_2)^{\frac{1}{2}}$ if **E** is in the plane of incidence and if the incident angle is not near $\pi/2$. For normal incidence, show that the energy is reduced by a factor approximately equal to $1 - 2(2\omega\varepsilon_1/\sigma_2)^{\frac{1}{2}}$.

Chapter VI

PROPAGATION OF PLANE ELECTROMAGNETIC WAVES IN MATTER

In traversing matter electromagnetic waves suffer scattering and absorption. These, as well as the phenomenon of dispersion, are related to the molecular structure of matter, so that a proper treatment of these phenomena must be carried out employing the principles of quantum mechanics. A reasonably satisfactory qualitative account can however be given on the basis of classical electrodynamics employing a simple mechanical model first proposed by Maxwell and Sellmeier independently, and later extended by Lorentz.

6.1 Basic Ideas Underlying the Classical Theory

Microscopically, dielectric matter consists of neutral molecules or atoms which are made up of oppositely charged particles: ions and ions or electrons and ions bound together with certain definite binding energies. According to the classical theory, the opposite charges in a system are displaced parallel and antiparallel to the external electric field, when one is applied, forming dipoles. In the oscillating field of a traversing electromagnetic wave these molecular or electronic dipoles are forced to oscillate with the frequency of the wave. As a charge in a molecular or atomic system has a definite binding energy, the dipoles formed are endowed with definite natural or characteristic frequencies if they are approximated by harmonic oscillators. Thus, a dielectric medium may be represented in the first approximation by an ensemble of harmonic oscillators which undergo forced oscillations as an electromagnetic wave passes by. In a conducting medium, it is the free electrons that participate in the forced oscillations. This case will be considered later.

The oscillating dipoles emit radiation, while they absorb energy required for sustaining the vibration from the traversing wave. In the classical theory, the secondary radiation has the same frequency as the primary wave but is emitted in various directions. The traversing wave is therefore said to be *scattered*. The corresponding attenuation of the radiation in the primary direction is called *absorption*. The coherent interference of the forward-scattered radiation and the primary wave modifies the velocity of propagation, giving rise to a refractive index which depends on frequency. This dependence is known as *dispersion*. For the treatment of dispersion, it is far more convenient, however, to consider the polarization of the medium in the field of the traversing wave, which results in a dielectric constant depending on frequency.

The force suffered by a charge q in the fields of the traversing wave is the sum of the electric and magnetic forces given by Eqs. (1.1) and (1.30), i.e.

$$\mathbf{F} = q(\mathbf{E} + \mathbf{u} \times \mathbf{B}) \quad ,$$

and is called the *Lorentz force*[a]. The magnitudes of the electric and magnetic fields of a plane electromagnetic wave are related by

$$|\mathbf{B}| = \sqrt{\mu_0 \varepsilon_0}\, |\mathbf{E}| = \frac{1}{c}\, |\mathbf{E}| \quad ,$$

---

[a] The Lorentz force equation as assumed here should be regarded as an additional postulate since the fields under consideration are not stationary. It is shown to be a direct consequence of the Lorentz transformation in Sec. 8.11.

showing that, unless the speed u of the charge reaches a magnitude comparable to the velocity of light c, the magnetic force may be neglected.

A *molecular oscillator* consists of oppositely charged ions of comparable masses, so that its motion is equivalent to that of a particle with a mass equal to the reduced mass of the system and a displacement from the centre of mass equal to the separation of the ions. In an *atomic* or *electronic oscillator*, the negative charge is an electron while the positive charge is an atomic nucleus or an ion having a much greater mass, so that only the motion of the electron needs to be considered. In either case we are dealing with a mobile charge, say a negative charge, oscillating with respect to a fixed centre of the opposite charge. In general, both types of oscillation may be present, but owing to the much greater effective mass of the molecular oscillator only the effects of the electronic oscillators are important.

More than one electron of an atom may participate in the oscillation, in which case each is to be represented by an electronic oscillator. While all such oscillators are forced to vibrate with the same frequency, the amplitudes depend on their characteristic frequencies. These amplitudes in turn determine their importance in affecting the propagation of the traversing wave. Usually, however, only the motion of one or two electrons in an atom plays a significant role in the process.

6.2 Forced Oscillation of an Electronic Oscillator

Since an electronic oscillator loses energy mainly by radiation, it will be necessary to introduce in the equation of motion a radiation reaction in addition to the electrical force exerted by the applied field and the Coulomb restoring force. We shall assume the speed u of the electron to remain much smaller than the velocity of light c throughout the oscillation. Then as the radiation loss can be accounted for entirely by the acceleration $\dot{\mathbf{u}}$ of the electron, the instantaneous rate of energy loss of the oscillator is given by Eq. (4.56):

$$-\frac{dW}{dt} = \frac{e^2\dot{u}^2}{6\pi\varepsilon_0 c^3} \quad .$$

The *radiation reaction* $\mathbf{f}_r$ is the force which, if applied against the motion of the electron, would cause the same average rate of energy dissipation as the loss through radiation. It is therefore given by

$$\int_0^T \mathbf{f}_r \cdot \mathbf{u}\, dt = -\int_0^T \frac{e^2\dot{u}^2}{6\pi\varepsilon_0 c^3}\, dt$$

$$= -\left[\frac{e^2\dot{\mathbf{u}}\cdot\mathbf{u}}{6\pi\varepsilon_0 c^3}\right]_0^T + \int_0^T \frac{e^2\ddot{\mathbf{u}}\cdot\mathbf{u}}{6\pi\varepsilon_0 c^3}\, dt \quad ,$$

where T is the period of oscillation. If the loss of energy by radiation is small, the oscillation may be assumed to be harmonic and $\dot{\mathbf{u}}\cdot\mathbf{u}$ will repeat itself after one cycle. The first term on the lower right-hand side will thus vanish and we may represent the radiation reaction by

$$\mathbf{f}_r = \frac{e^2\ddot{\mathbf{u}}}{6\pi\varepsilon_0 c^3} \quad . \tag{6.1}$$

The radiation reaction acts as a damping force. Under the assumption of harmonic oscillations, $\ddot{\mathbf{u}} = -\omega^2\mathbf{u}$, where ω is the angular frequency of oscillation. $\mathbf{f}_r$ can then be written in the usual velocity-dependent form as $-m\gamma\mathbf{u}$, where m is the electron mass and γ is known as the *damping coefficient*. An estimate of γ is obtained from Eq. (6.1) by replacing ω with ω_0, the characteristic angular frequency of the oscillator:

$$\gamma \approx \frac{1}{6\pi\varepsilon_0}\frac{e^2}{mc^2}\frac{\omega_0^2}{c} \quad . \tag{6.2}$$

The radiation damping $|m\gamma\mathbf{u}|$ is usually much smaller than the restoring force $|m\omega_0\mathbf{u}|$ and the above estimate should be considered an upper limit.

In the oscillation of free electrons in a conducting medium, the damping is however due primarily to collisions with the ions or atoms present. To a first approximation, however, the damping force will still have the same form, $-m\gamma\mathbf{u}$, as will be shown in Sec. 6.7. In the following we shall simply regard γ as the damping coefficient of whatever origin so that the effect of collisions is also included. Furthermore we shall assume that damping is small so that $\omega_0 >> \gamma$.

Consider a plane electromagnetic wave $\mathbf{E} = \mathbf{E}_0 e^{i(\mathbf{k}\cdot\mathbf{r} - \omega t)}$ incident on an electronic oscillator located at $\mathbf{r}$. As we are interested in the field at a fixed point, we may for convenience put $\mathbf{r} = 0$ and write for the electric field $\mathbf{E} = \mathbf{E}_0 e^{-i\omega t}$. The equation of the motion of the electron of the oscillator is then

$$\ddot{\mathbf{r}}' + \omega_0^2 \mathbf{r}' = -\frac{e}{m}\mathbf{E}_0 e^{-i\omega t} - \gamma\dot{\mathbf{r}}' \quad , \tag{6.3}$$

where $\mathbf{r}'$ is the displacement from the fixed positive centre. Note that we have assumed the neighbouring oscillators to be so far away that their contribution to the local field is negligible. In the steady state the electron oscillates with the same frequency ω as the field. Its displacement from the neutral position can therefore be assumed to be

$$\mathbf{r}' = \mathbf{r}_0' e^{-i\omega t} \quad .$$

Substituting this in Eq. (6.3) we find

$$\mathbf{r}' = -\frac{e}{m}\frac{\mathbf{E}_0 e^{-i\omega t}}{(\omega_0^2 - \omega^2 - i\omega\gamma)} \quad , \tag{6.4}$$

which means that the oscillator has an instantaneous dipole moment

$$\mathbf{p} = -e\mathbf{r}' = \frac{e^2}{m}\frac{\mathbf{E}_0 e^{-i\omega t}}{(\omega_0^2 - \omega^2 - i\omega\gamma)} = \mathbf{p}_0 e^{-i(\omega t + \delta)} \quad ,$$

where $\mathbf{p}_0$ is real and δ is a constant phase angle. The oscillator thus behaves like a Hertzian dipole of amplitude

$$p_0 = \frac{e^2}{m} \frac{E_0}{\{(\omega_0^2 - \omega^2)^2 + \omega^2\gamma^2\}^{\frac{1}{2}}} \quad . \tag{6.5}$$

Taking the location of the oscillator as the origin, the average radiation emitted by the oscillator crossing a point of position vector **r** per unit time per unit area normal to $\hat{\mathbf{r}}$ is given by Eq. (4.69) to be

$$<\mathbf{N}> = \frac{e^4 E_0^2 \sin^2\theta}{32\pi^2\varepsilon_0 c^3 m^2 r^2} \frac{\omega^4}{(\omega_0^2 - \omega^2)^2 + \omega^2\gamma^2} \hat{\mathbf{r}} \quad , \tag{6.6}$$

where θ is the angle between the directions of radiation and oscillation.

6.3 Scattering by a Bound Electron

By definition $<N>$ is the intensity of the scattered wave at **r**. As the intensity of the primary wave averaged over one period is

$$I_0 = \frac{1}{2} \text{Re} \, |\mathbf{E} \times \mathbf{H}^*| = \frac{1}{2} \varepsilon_0 c E_0^2 \quad ,$$

the intensity of the scattered radiation may be written as

$$I_s = <N> = \left(\frac{1}{4\pi\varepsilon_0} \frac{e^2}{mc^2}\right)^2 \frac{\omega^4}{(\omega_0^2 - \omega^2)^2 + \omega^2\gamma^2} \frac{I_0}{r^2} \sin^2\theta \quad . \tag{6.7}$$

This expression obviously holds for any harmonic oscillator provided m is interpreted as the reduced mass of the system. Then as I_s varies as m^{-2}, scattering by the much heavier molecular oscillators is negligible compared with that by the electronic oscillators. For an electron, the quantity

$$r_0 = \frac{1}{4\pi\varepsilon_0} \frac{e^2}{mc^2} \tag{6.8}$$

is known as its *classical radius* and has the value of 2.8178×10^{-15}m.

The total energy scattered per unit time by an electron is

$$-\frac{dW}{dt} = \int_0^{2\pi} I_s r^2 2\pi \sin\theta \, d\theta = \frac{8}{3}\pi r_o^2 I_o \frac{\omega^4}{(\omega_o^2 - \omega^2)^2 + \omega^2\gamma^2} \quad .$$

The quantity

$$-\frac{1}{I_o}\frac{dW}{dt} = \frac{8}{3}\pi r_o^2 \frac{\omega^4}{(\omega_o^2 - \omega^2)^2 + \omega^2\gamma^2} \equiv \sigma \qquad (6.9)$$

has the dimensions of an area and is called the *total scattering cross section* per electron.

If the frequency of the primary radiation is much greater than the characteristic frequency, i.e. $\omega >> \omega_o$, the electron is said to be weakly bound and may be considered as essentially free. As we then have $\omega >> \gamma$ also, the total cross section is

$$\sigma_o = \frac{8}{3}\pi r_o^2 \quad , \qquad (6.10)$$

which is 8/3 times the "geometrical" cross section πr_o^2 of the electron and is known as the *Thomson scattering cross section*.

On the other hand, for low frequencies or strong binding, we have $\omega_o >> \omega$. As $\omega_o >> \gamma$ also, the total cross section is

$$\sigma_R = \left(\frac{\omega}{\omega_o}\right)^4 \sigma_o \quad . \qquad (6.11)$$

The cross section for strong binding is thus inversely proportional to the fourth power of the wavelength. This inverse fourth-power law was first derived by Rayleigh in his investigation of the blueness of the sky and is known as *Rayleigh's law*.

To give an indication of the validity of the various approximations, it may be noted that for all elements the lowest characteristic frequency occurs at $\omega_o \simeq 10^{16} - 10^{17}\ s^{-1}$. The coefficient of radiation damping will then be $\gamma \leq 10^{-8}\omega_0$ as can be seen from Eq. (6.2).

In the above cases where ω is far away from ω_0, the damping term $\gamma^2\omega^2$ in the denominator of Eq. (6.9) has been neglected. This term becomes important however if $\omega \approx \omega_0$. Under such conditions

$$\sigma \approx \frac{\omega^2}{4(\omega_0 - \omega)^2 + \gamma^2}\sigma_0 \quad .$$

In particular, the total cross section is a maximum at $\omega = \omega_0$ and has the value

$$\sigma_m = \left(\frac{\omega_0}{\gamma}\right)^2 \sigma_0 \quad . \tag{6.12}$$

The large increase of the scattering cross section in the neighbourhood of the characteristic frequency of the electron is known as resonance, and ω_0 is often referred to as the *resonance frequency*.

The cross section per unit solid angle $d\sigma/d\Omega$ is called the *differential scattering cross section*. If the incident radiation is linearly polarized, the oscillators will vibrate along the fixed direction of the electric field. The angle θ in Eq. (6.7) is therefore the angle between the direction of polarization of the primary wave at the dipole and the direction of propagation of the scattered wave. As the solid angle subtended at the origin by unit area at a distance r is r^{-2}, the differential cross section for scattering is given by Eq. (6.7) to be

$$\frac{d\sigma}{d\Omega} = \frac{I_s}{I_0} r^2 = \frac{r_0^2\omega^4\sin^2\theta}{(\omega_0^2 - \omega^2)^2 + \omega^2\gamma^2} \quad . \tag{6.13}$$

If the primary wave is unpolarized, the direction of oscillation of the electron is random and the differential cross section has to be averaged over all possible directions of polarization. Choose a Cartesian coordinate system with the origin at the location of the oscillator, the z-axis along the direction of the primary wave, the x-axis in the plane formed by the z-axis and the direction $\hat{\mathbf{r}}$ of the scattered wave as shown in Fig. 6.1. Let ϕ be the angle between the electric field $\mathbf{E}$ of the primary wave and the x-axis. As the oscillator vibrates parallel to $\mathbf{E}$, the angle between $\mathbf{E}$ and $\hat{\mathbf{r}}$ is the angle θ which appears in Eq. (6.13). Then as $\mathbf{E}$ is perpendicular to the z-axis we have

$$\cos\theta = \cos\phi \sin\psi \quad ,$$

where ψ is the scattering angle, i.e. the angle between the directions of the primary and scattered waves. For random polarization we have to average $\sin^2\theta$ over the angle of polarization ϕ:

$$< \sin^2\theta > = \frac{1}{2\pi} \int_0^{2\pi} (1 - \cos^2\phi \sin^2\psi) d\phi = \frac{1}{2} (1 + \cos^2\psi) \quad .$$

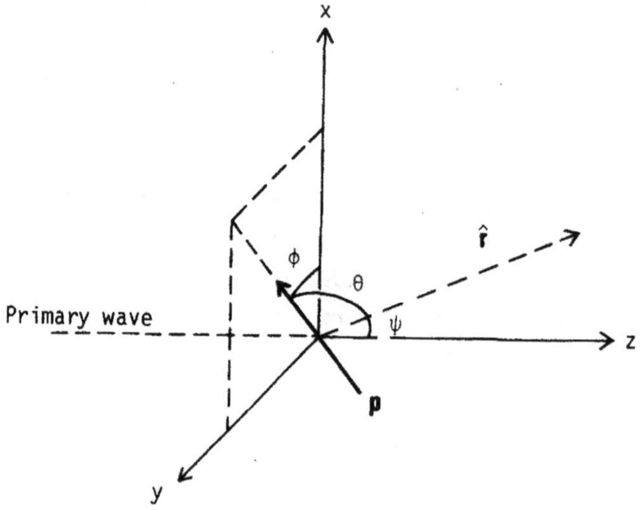

Fig. 6.1 An electromagnetic wave incident along the z-direction forces the oscillator $\mathbf{p}$ to vibrate in the direction of its electric field $\mathbf{E}$. The x-axis is chosen to lie in the plane containing the directions of the primary and scattered waves. It makes an angle ϕ with $\mathbf{p}$.

The differential cross section for the scattering of unpolarized electromagnetic waves by an electron is then

$$\frac{d\sigma}{d\Omega} = \frac{r_o^2}{2} \frac{\omega^4}{(\omega_o^2 - \omega^2)^2 + \omega^2\gamma^2} (1 + \cos^2\psi) \quad . \tag{6.14}$$

The classical theory therefore predicts maximum scattering in the forward and backward directions of the unpolarized primary wave.

If the primary wave is linearly polarized, the scattered wave will also be linearly polarized. Since the electron will oscillate in the plane of polarization of the primary wave, the wave scattered in any direction will be linearly polarized with **E** in the plane containing the direction of the electric vector of the primary wave at the oscillator and the direction of propagation. This has the consequence that, for a given scattering angle ψ, the intensity of the scattered radiation will be a maximum if the scattering occurs in a plane perpendicular to the plane of polarization of the primary wave as then $\theta = \pi/2$, a minimum if it occurs in this plane where $\theta = \pi/2 - \psi$.

If the primary radiation is unpolarized, the secondary wave will still be completely plane polarized if it is scattered into a direction normal to the primary radiation, for then the responsible oscillator must oscillate normal to both the primary and secondary directions. Any oscillator parallel to the scattered direction cannot contribute. Between the transverse and longitudinal directions of the primary radiation, waves will be scattered partially polarized [Prob. (6.2)].

In considering the effect of the presence of matter on the propagation of electromagnetic waves, reference must be made to the microscopic nature of matter. It is then questionable whether the classical theories of electrodynamics and mechanics serve as any approximation at all. It will therefore be interesting to compare the results of the classical theory of scattering with the predictions of the *quantum* theory. In the quantum theory the incident waves are considered as consisting of photons of energy $\hbar\omega$, where $\hbar$ is Planck's constant divided by 2π. If the photon energy is much larger than the binding energy $\hbar\omega_0$ of the electron, the latter may be considered as being free and initially at rest. The scattering of a photon by a free electron is known as the *Compton scattering*. The quantum theory predicts a frequency ω' for the scattered photon which is in general different from the frequency ω of the primary photon, the two being related by

$$\omega' = \omega \left\{ 1 + \frac{\hbar\omega}{mc^2} (1 - \cos\psi) \right\}^{-1} .$$

Note that in the non-relativistic approximation, $\hbar\omega \ll mc^2$, and the above gives $\omega' \approx \omega$, i.e., the frequency of the radiation remains essentially unaltered as in the classical theory.

The cross section for scattering in the quantum theory is given by the *Klein-Nishima formula*. In the non-relativistic limit, both the differential and total scattering cross sections again reduce to the classical Thomson cross sections. The angular distribution of the Compton-scattered photons for several values of $\hbar\omega/mc^2$ is shown in Fig. 6.2. It may be seen that the distribution tends to that for the Thomson scattering as $\hbar\omega/mc^2 \to 0$. Furthermore the quantum and classical cross sections are essentially the same in the forward direction $(\cos\psi \approx 1)$ although the former is much reduced for large scattering angles. Regarding the validity of the quantum theory it may be stated that the Klein-Nishima formula fits the experimental data extermely well and may be regarded as proven. What is rather pleasantly surprising is that the classical theory serves rather well when only non-relativistic energies are involved. This gives us some confidence in the classical treatment of dispersion, at least as a crude model for correlating the experimental data.

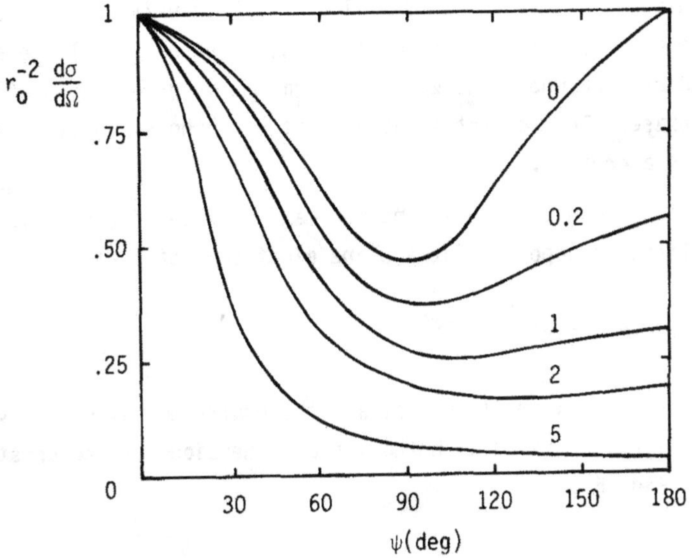

Fig. 6.2 The angular distribution for the electromagnetic radiation scattered by a free electron according to the Klein-Nishima formula. The primary radiation is unpolarized. The numbers shown above the curves are values of $\hbar\omega/mc^2$. For $\hbar\omega/mc^2 \to 0$, the distribution reduces to that obtained in the classical theory.

6.4 Attenuation of Traversing Radiation

The energy required for maintaining the vibration of the oscillators is abstracted from the primary radiation which consequently suffers attenuation as the medium is being traversed. If damping is due mainly to radiation reaction this energy is re-emitted as scattered radiation. If however the effect of collision is important, a substantial part of the vibrational energy is dissipated as heat and real absorption occurs. Scattered waves emitted in the primary direction will combine coherently with the primary waves, thus modifying the effective velocity of propagation in the medium. Wavelets scattered in any other direction will superpose on one another, so that the resultant disturbance will depend on the phase difference between these wavelets. The problem is then one of interference. If the oscillators are regularly arranged as in a crystal, they will act collectively as a three-dimensional diffraction grating and maximum intensity will be expected in any direction where the wavelets from the neighbouring scatterers are in phase. This effect is demonstrated by the Bragg diffraction of X-rays in a crystal.

To obtain the conditions for maximum intensity in crystal diffraction, we suppose that a plane electromagnetic wave

$$\mathbf{E} = \mathbf{E}_0 e^{i(\mathbf{k}\cdot\mathbf{r} - \omega t)}$$

is incident upon a small crystal and the scattered waves observed at a distance large compared with the linear dimensions of the crystal as shown in Fig. 6.3.

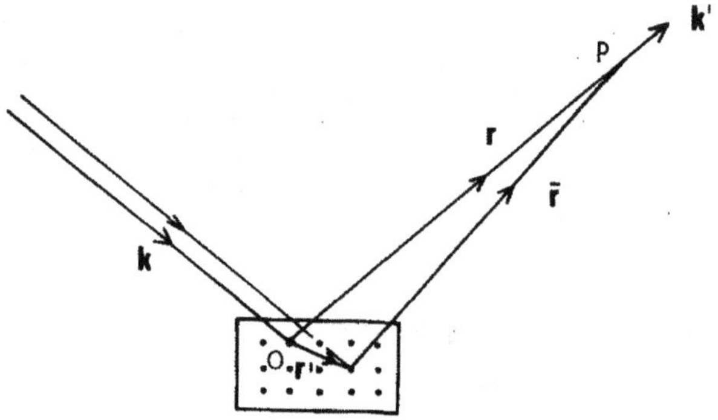

Fig. 6.3 Plane electromagnetic waves of propagation vector **k** fall upon a crystal and the scattered waves are observed at P, at a distance r large compared with the linear dimensions of the crystal.

Take the origin O at a lattice point of the crystal and consider the secondary waves arriving at a distant point P of radius vector $\mathbf{r}$ at a particular instant t from an oscillator at O and another at $\mathbf{r}'$. The electric vector $\mathbf{E}'$ of a scattered wave will have an amplitude which, according to Eq. (6.7), is proportional to the amplitude of the incident wave and inversely proportional to the distance $\bar{r} = |\mathbf{r} - \mathbf{r}'|$ from the scattering centre. Furthermore, the frequency ω, and hence the magnitude of the propagation vector, is not altered by scattering. Thus, the amplitude of the secondary wave from 0 is proportional to

$$\frac{E_0}{\bar{r}} \exp\{i(kr - \omega t)\}$$

and that of the secondary wave from $\mathbf{r}'$ is proportional to

$$\frac{E_0}{\bar{r}} \exp\{i(\mathbf{k} \cdot \mathbf{r}' + k\bar{r} - \omega t)\} \quad .$$

In addition, there is a factor depending on the angle of scattering.

As observations are made at a large distance away, $r \gg r'$, the angle of scattering may be taken to be the same for all scattering centres in the crystal. Furthermore, in the relation above, we may use the approximations (see Sec. 4.12)

$$\bar{r} \simeq r - \mathbf{r}' \cdot \hat{\mathbf{r}}$$

and $$\bar{r}^{-1} \simeq r^{-1} \quad .$$

The last exponent is then

$$\begin{aligned} i(\mathbf{k} \cdot \mathbf{r}' + kr - k\mathbf{r}' \cdot \hat{\mathbf{r}} - \omega t) &= i(kr - \omega t) - i\mathbf{r}' \cdot (\mathbf{k}' - \mathbf{k}) \\ &= i(kr - \omega t) - i\mathbf{r}' \cdot \Delta\mathbf{k} \quad , \end{aligned}$$

where $\mathbf{k}' = k\hat{\mathbf{r}}$ may be taken to be the propagation vector of the radiation scattered by the crystal. The phase of a secondary wave from a scattering centre is determined by $\mathbf{r}' \cdot \Delta\mathbf{k}$.

A crystal may be considered as atoms arranged in a lattice defined by three fundamental vectors $\mathbf{a}, \mathbf{b}, \mathbf{c}$, such that every lattice point is given by

$$\mathbf{r}' = \ell\mathbf{a} + m\mathbf{b} + n\mathbf{c} \quad ,$$

where ℓ, m and n are integers, provided the origin O is itself a lattice point. To each of such points is attached a basis of atoms, which for our purpose may be considered as a number of electronic oscillators arranged in some definite manner.

The amplitude of the resultant wave at $\mathbf{r}$ is proportional to

$$\sum_{\ell}\sum_{m}\sum_{n} \exp\{-i(\ell\mathbf{a}+m\mathbf{b}+n\mathbf{c})\cdot\Delta\mathbf{k}\}$$

$$= \sum_{\ell} \exp(-i\ell\mathbf{a}\cdot\Delta\mathbf{k}) \sum_{m} \exp(-im\mathbf{b}\cdot\Delta\mathbf{k}) \sum_{n} \exp(-in\mathbf{c}\cdot\Delta\mathbf{k}) \quad .$$

Each set of the integers ℓ, m, n defines a lattice point and the sums are to be taken over all such points in the crystal. For simplicity we shall assume that there are N^3 lattice points and let ℓ, m, n each run from 0 to N - 1. We then have

$$\sum_{\ell=0}^{N-1} \exp(-i\ell\mathbf{a}\cdot\Delta\mathbf{k}) = \exp\left\{-\frac{i(N-1)\mathbf{a}\cdot\Delta\mathbf{k}}{2}\right\}\frac{\sin(\frac{1}{2}N\mathbf{a}\cdot\Delta\mathbf{k})}{\sin(\frac{1}{2}\mathbf{a}\cdot\Delta\mathbf{k})} \quad .$$

The intensity of the scattered radiation at $\mathbf{r}$ is $\mathbf{E}'\cdot\mathbf{E}'^*$, which is proportional to the product of three factors like

$$\left|\sum_{\ell=0}^{N-1} \exp(-i\ell\mathbf{a}\cdot\Delta k)\right|^2 = \frac{\sin^2(\frac{1}{2}N\mathbf{a}\cdot\Delta\mathbf{k})}{\sin^2(\frac{1}{2}\mathbf{a}\cdot\Delta\mathbf{k})} \quad .$$

The above has the maximum value N^2 when $\mathbf{a}\cdot\Delta\mathbf{k}$ is an integral multiple of 2π, since each term of the left-hand side then takes the maximum value unity. We conclude therefore that constructive interference will occur for directions for which

$$\mathbf{a}\cdot\Delta\mathbf{k} = 2\pi p \quad ,$$

$$\mathbf{b}\cdot\Delta\mathbf{k} = 2\pi q \quad ,$$

and

$$\mathbf{c} \cdot \Delta\mathbf{k} = 2\pi s \quad ,$$

where p, q and s are integers. These equations are known as the *Laue equations*, and may be shown to be equivalent to *Bragg's law*.

The Laue equations define the necessary conditions for constructive interference. However, each scattering centre may consist of a number of atoms, and each atom a number of electronic oscillators. Summation must first be made over all the oscillators at each lattice point. If the atoms are identical and suitably situated, the secondary wavelet from the same scattering centre may interfere destructively with one another for certain Laue directions. Such directions are then forbidden and no bright spot will appear accordingly.

If, at the other extreme, the oscillators are distributed completely at random such as in a gas, the phase difference δ between any two wavelets arriving from a direction will be as likely positive as negative provided the direction of observation is sufficiently far removed from the primary direction. The resultant intensity is then proportional to

$$(\sum_s E_s \sin\delta_s)^2 = \sum_s E_s^2 \sin^2\delta_s + \sum_s \sum_{r \neq s} E_s E_r \sin\delta_s \sin\delta_r$$

$$= \sum_s E_s^2 \sin^2\delta_s \quad ,$$

where the cross terms cancel out as the signs and magnitudes of δ_s are random. Thus the total scattered intensity is simply the sum of the intensities of the scattered wavelets.

In a condensed medium such as a liquid or a solid the oscillators are arranged with some degree of regularity. The partial coherence of the secondary waves observed in a given direction results in a reduction of the scattered intensity. For this reason scattering in a dense medium need not be stronger than in a gas though the number of scatterers is very much greater.

The attenuation of the incoming radiation in a gas where all the oscillators are identical and randomly distributed can be readily calculated. Let S be the cross sectional area of the beam and N the number of oscillators per unit volume of the gas. In traversing a gas of thickness dx, the beam will have encountered $NSdx$ oscillators, each of which will in unit time scatter an amount of energy equal to $I\sigma$, where I is the intensity of the radiation and σ the scattering cross section per electron. The energy abstracted from the beam in traversing the layer of gas is therefore $I\sigma NSdx$. This will cause a decrease in intensity of dI given by

$$-\frac{1}{I}\frac{dI}{dx} = \sigma N \equiv \mu_s \quad . \tag{6.15}$$

Integrating we have

$$I(x) = I_0 e^{-\mu_s x} \quad ,$$

I_0 being the initial intensity and I the intensity after traversing a thickness x of the gas. The parameter μ_s is called the *attenuation constant for scattering.*

6.5 Dispersion in a Dilute Medium

To obtain an expression for the refractive index of a medium, it is simplest to consider the polarization of matter in the field of the traversing radiation. For a linear, isotropic and homogeneous dielectric medium with $\mu \approx \mu_0$, the refractive index n is given by

$$n^2 = \frac{\varepsilon}{\varepsilon_0} = \frac{D}{\varepsilon_0 E} = 1 + \frac{P}{\varepsilon_0 E} \quad .$$

Microscopically, the polarization $\mathbf{P}$ is related to the displacement $\mathbf{r}'$ of the oscillating electrons from their positive centres by

$$\mathbf{P} = -Ne\mathbf{r}' \quad ,$$

where N is the number of the electronic oscillators per unit volume.

In a dilute system such as a gas or vapour, neighbouring oscillators are far apart so that a good approximation is achieved by taking for the local field at an oscillator simply the field of the traversing wave. Equation (6.4) then gives

$$n^2 = 1 + \frac{Ne^2}{\varepsilon_0 m(\omega_0^2 - \omega^2 - i\omega\gamma)} = 1 + \frac{\alpha}{\omega_0^2 - \omega^2 - i\omega\gamma} \quad , \tag{6.16}$$

where

$$\alpha \equiv \frac{Ne^2}{\varepsilon_0 m} \quad , \tag{6.17}$$

assuming all the oscillators of the medium to be identical and have a characteristic angular frequency ω_0. This equation gives the dependence of the refractive index of the system on the frequency of the traversing wave and is called a *dispersion equation.* The phase velocity in the medium will likewise be dependent on the frequency.

The complex refractive index n may be written as $\eta + i\kappa$, where η and κ are positive real numbers given by

$$\eta^2 - \kappa^2 = 1 + \frac{\alpha(\omega_0^2 - \omega^2)}{(\omega_0^2 - \omega^2)^2 + \omega^2\gamma^2} \tag{6.18}$$

and

$$2\eta\kappa = \frac{\alpha\omega\gamma}{(\omega_0^2 - \omega^2)^2 + \omega^2\gamma^2} \quad . \tag{6.19}$$

The significance of a complex refractive index has been considered in Sec. 5.4. It is clear from Eq. (5.45) and the subsequent discussions that the imaginary part of the refractive index is responsible for the attenuation of the amplitude of the traversing wave. In a dilute system, the density of scatterers is low with the result that absorption is small and the effective refractive index is close to unity. Then, as $\eta >> \kappa$ and $\eta \approx 1$, Eqs. (6.18) and (6.19) have the approximate solution

$$\eta^2 \approx 1 + \frac{\alpha(\omega_0^2 - \omega^2)}{(\omega_0^2 - \omega^2)^2 + \omega^2\gamma^2} \quad , \tag{6.20}$$

$$\kappa \approx \frac{\alpha}{2} \frac{\omega\gamma}{(\omega_0^2 - \omega^2)^2 + \omega^2\gamma^2} \quad . \tag{6.21}$$

Within the same approximation the parameters p and q defined by Eq. (5.49) have the values

$$p^2 \approx (\eta^2 - n_1^2 \sin^2\theta)k_0^2 \quad , \tag{6.22}$$

$$q \approx \kappa k_0 \quad , \tag{6.23}$$

where $k_0 = \omega/c$, so that the effective refractive index as given by Eq. (5.54) is simply η, the real part of n.

If the frequency is sufficiently far-removed from the characteristic frequency so that $(\omega_0^2 - \omega^2)^2 >> \omega^2\gamma^2$, the above reduce to

$$\eta^2 \approx 1 + \frac{\alpha}{\omega_0^2 - \omega^2} \quad , \tag{6.24}$$

$$\kappa \approx \frac{\alpha}{2} \frac{\omega\gamma}{(\omega_0^2 - \omega^2)^2} \quad . \tag{6.25}$$

The first equation shows that η is greater than unity for $\omega < \omega_0$, and less than unity for $\omega > \omega_0$. In either case the refractive index increases with increasing frequency of the traversing wave (see Fig. 6.4). Such a dependence of the refractive index on the frequency is known as *normal dispersion*. Note that at very low frequencies, $\omega << \omega_0$, η is essentially independent of ω and no dispersion occurs. The second equation shows that absorption is generally small at frequencies far removed from the characteristic frequency.

For waves with a narrow frequency spread, we may speak of some average phase velocity v in the medium and write η^2 in terms of the wavelengths $\lambda = 2\pi v/\omega$, $\lambda_o = 2\pi v/\omega_o$:

$$\eta^2 - 1 \simeq \frac{\alpha}{\omega_o^2} \frac{\lambda^2}{\lambda^2 - \lambda_o^2} \quad . \tag{6.26}$$

This is known as *Sellmeier's formula* and works well for frequencies not too close to a characteristic frequency.

For frequencies lower than the characteristic frequency we may expand the above as a series in λ_o^2/λ^2:

$$\eta^2 - 1 \approx \frac{\alpha}{\omega_o^2}\left(1 + \frac{\lambda_o^2}{\lambda^2} + \frac{\lambda_o^4}{\lambda^4} + \ldots\right) \quad .$$

Over limited regions of frequencies, η does not vary considerably and this may be written as

$$\eta - 1 \approx A + \frac{B}{\lambda^2} + \frac{C}{\lambda^4} + \ldots \quad , \tag{6.27}$$

where $A, B, C, \ldots,$ are positive constants for a given medium over a given range of wavelengths. This expression is known as *Cauchy's formula*.

Near the characteristic frequency, $\omega\gamma$ may no longer be ignored and η has the approximation

$$\eta \approx 1 + \frac{\alpha}{\gamma^2} \frac{\omega_o - \omega}{\omega_o} \quad .$$

Thus as ω increases from a value smaller than ω_o to a value greater, η changes rapidly from a value greater than unity to a value smaller. The abruptness of the change depends strongly on the damping coefficient as can be seen from the differential

$$\left(\frac{d\eta}{d\omega}\right)_{\omega=\omega_o} \approx -\frac{\alpha}{\omega_o \gamma^2} \quad .$$

Thus the greater the damping coefficient the more gradual will be the change. Furthermore, as the differential is negative, the refractive index decreases with increasing frequency in the neighbourhood of ω_0, an effect known as *anomalous dispersion.*

In the region of anomalous dispersion, κ has a sharp maximum at ω_0 of

$$\kappa_{max} \simeq \frac{\alpha}{2\omega_0\gamma} \quad .$$

This means that a medium becomes essentially opaque to waves of frequencies near the characteristic frequency. In other words, resonance absorption occurs. The behaviour of the effective refractive index and that of the absorption constant $q \approx \kappa k_0$ near the characteristic frequency ω_0 are shown in Fig. 6.4.

Hitherto we have for simplicity confined ourselves to a dilute medium having only one type of electronic oscillation. If more than one electron in an atom of the medium participate in the scattering process, each will act as an oscillator and may have a different characteristic frequency. Also, more than one type of atom may be present in the medium. To allow for these possibilities let N_s be the number of oscillators per unit volume having a characteristic frequency ω_s. The dipole moment per unit volume is then $\mathbf{P} = -\sum_s N_s e\mathbf{r}'_s$ and the dispersion formula becomes

$$n^2 = 1 + \frac{e^2}{\varepsilon_0 m} \sum \frac{N_s}{\omega_s^2 - \omega^2 - i\omega\gamma_s} \quad . \tag{6.28}$$

The effective refractive index η is then given by

$$\eta^2 = 1 + \sum \frac{\alpha_s(\omega_s^2 - \omega^2)}{(\omega_s^2 - \omega^2)^2 + \omega^2\gamma_s^2} \quad , \tag{6.29}$$

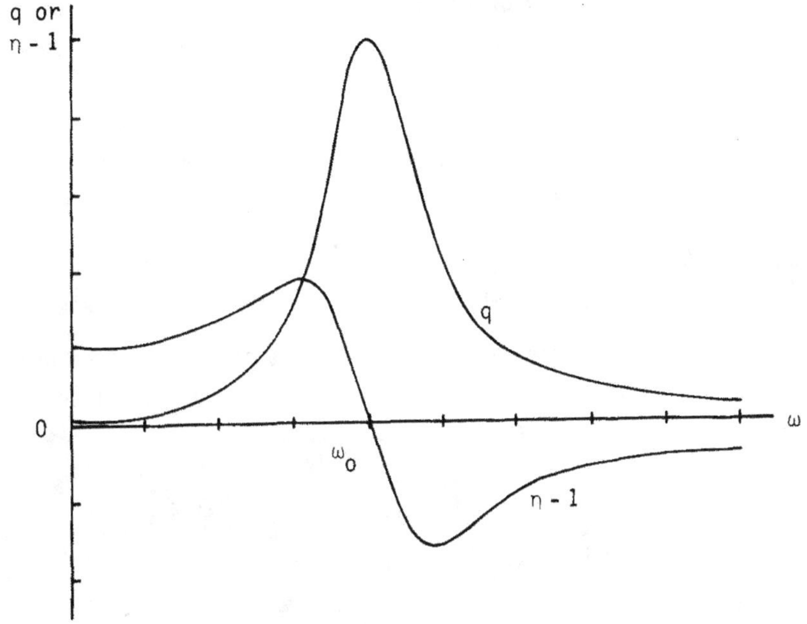

Fig. 6.4 The effective refractive index η and the absorption constant $q \approx \kappa k_o$ of a gas near its characteristic angular frequency ω_o. Note that η increases with ω when far removed from ω_o. Near ω_o, η decreases with increasing ω showing "anomalous" behaviour. The absorption constant has a sharp maximum at ω_o.

where for convenience we have put

$$\frac{N_s e^2}{\varepsilon_o m} \equiv \alpha_s \quad . \tag{6.30}$$

The terms in the summation of Eq. (6.29) are usually small except when ω is close to one of the characteristic frequencies, say ω_r. The term involving ω_r then becomes significant. Similarly, the corresponding term in κ will be the main contributor to the absorption coefficient. As a result, anomalous dispersion and resonance absorption will occur near a characteristic frequency in much the same way as for a gas with a single characteristic frequency.

If the frequency of the traversing wave is not close to any of the characteristic frequencies, damping may be neglected and Eq. (6.29) becomes

$$\eta^2 \approx 1 + \sum_s \frac{\alpha_s}{\omega_s^2 - \omega^2} \quad . \tag{6.31}$$

Then if $\omega < \omega_s$ for all s, each term in the summation may be expanded as a series in λ_s^2/λ^2 so that Cauchy's formula may again be applied.

If the frequency range of interest is intermediate between two characteristic frequencies, say ω_r and ω_{r+1} but not close to either, Eq. (6.31) may be written as a series:

$$\eta^2 \approx 1 + \sum_{s=r+1}^{S} \frac{\alpha_s}{\omega_s^2} + \sum_{s=r+1}^{S} \frac{\alpha_s}{\omega_s^4}\omega^2 + \sum_{s=r+1}^{S} \frac{\alpha_s}{\omega_s^6}\omega^4 + \ldots$$

$$- \sum_{s=1}^{r} \alpha_s \omega^{-2} - \sum_{s=1}^{r} \alpha_s \omega_s^2 \omega^{-4} - \ldots \quad , \tag{6.32}$$

where $\omega_1 < \omega_2 < \omega_3 \ldots$ and S is the total number of types of oscillator present in the medium. In optics the negative terms are usually negligible as evidenced by the fact that the refractive index is greater than unity for most transparent materials and, as η does not vary considerably for a limited range of frequencies, Cauchy's formula is still applicable. Improvement in fitting of experimental data can often be obtained by including in the leading low frequency term:

$$\eta - 1 \approx - A'\lambda^2 + A + \frac{B}{\lambda^2} + \frac{C}{\lambda^4} + \ldots \quad , \tag{6.33}$$

where A', A, B, C ... are positive constants.

For very high frequencies such as for X-rays, $\omega \gg \omega_s$ for all s and Eq. (6.31) becomes

$$\eta^2 = 1 - \frac{Ne^2}{\varepsilon_0 m \omega^2} \quad . \tag{6.34}$$

Thus the refractive index for X-rays is always less than unity and very close to it. Note that this is also the case of propagation of waves of any frequency in a volume distribution of free electrons with a concentration N, in which instance $\omega_0 = 0$.

6.6 Dispersion in a Dense Medium

The treatment of a dense medium, such as a solid or a liquid, follows exactly the same line. Here, however, polarization of the neighbouring atoms or molecules must be taken into account as this will give rise to an additional electric field at the location of an oscillator. To evaluate the effect imagine a spherical cavity of surface S, which is small macroscopically but large compared to atomic dimensions, with centre O at the location of the oscillator. The polarization of the neighbouring atoms will give rise to charges of surface density P_n on the inner surface of the sphere. In the neighbourhood of the oscillator we may assume the external field **E** and hence the polarization **P** to be uniform. Then by symmetry the field $\mathbf{E}_p$ of the induced surface charges is parallel to **P**. Referring to Fig. 6.5, we may express E_p by

$$E_p = \int_S \frac{\cos\theta\, \mathbf{P}\cdot d\mathbf{S}}{4\pi\varepsilon_0 r^2} = \frac{1}{4\pi\varepsilon_0}\int_0^{\pi} \frac{P\cos^2\theta\; 2\pi r^2 \sin\theta\, d\theta}{r^2} = \frac{P}{3\varepsilon_0} \quad .$$

The effective local field of an oscillator is therefore

$$\mathbf{E}_{eff} = \mathbf{E} + \frac{\mathbf{P}}{3\varepsilon_0} = \mathbf{E} - \frac{Ne\mathbf{r}'}{3\varepsilon_0} \quad . \qquad (6.35)$$

Its use in the equation of motion (6.3) gives

$$\mathbf{r}' = \frac{-e\mathbf{E}}{m(\omega_0^2 - \omega^2 - i\omega\gamma - \frac{\alpha}{3})} \quad ,$$

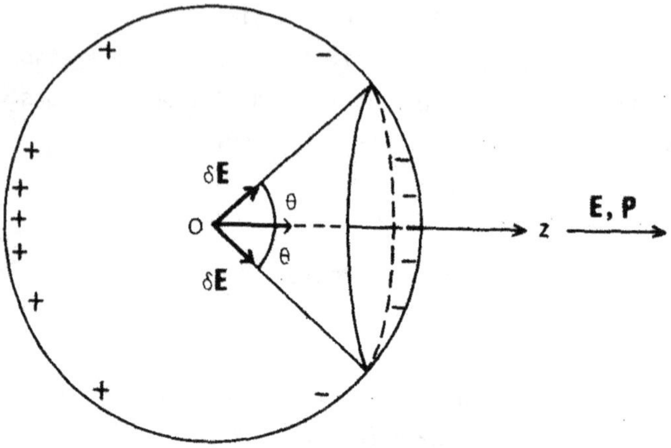

Fig. 6.5 A spherical cavity in a dielectric small macroscopically but large compared to atomic dimensions, centred at the location of an electronic oscillator. The polarization of the neighbouring atoms or molecules gives rise to charges on the inner surface of surface density P_n.

and hence

$$n^2 = 1 + \frac{\alpha}{\omega_0^2 - \omega^2 - i\omega\gamma - \frac{\alpha}{3}} , \qquad (6.36)$$

where

$$\alpha = \frac{Ne^2}{\varepsilon_0 m}$$

as before.

If more than one type of electronic oscillator is present it is necessary first to re-arrange the terms before summing up. Thus we obtain

$$\frac{n^2 - 1}{n^2 + 2} = \frac{e^2}{3\varepsilon_0 m} \sum_s \frac{N_s}{\omega_s^2 - \omega^2 - i\omega\gamma_s} = \frac{1}{3} \sum_s \frac{\alpha_s}{\omega_s^2 - \omega^2 - i\omega\gamma_s} . \qquad (6.37)$$

This expression may be considered an extension of the *Clausius-Mossotti equation* relating the relative permittivity $\varepsilon/\varepsilon_0$ to the density ρ of a non-polar and non-magnetic substance, for N_s is proportional to the total number of molecules per unit volume and hence to ρ. If damping may be neglected, the refractive index n becomes real and the relation is known as the *Lorentz-Lorenz equation*. It enables, for instance, the refractive index of a vapour to be calculated when that of its liquid is known.

Note that for a dilute system, $n^2 \approx 1$ and the equation becomes identical with Eq. (6.28). In general for a limited range of frequency, n, and hence η, does not change significantly with frequency so that Sellmeier's and Cauchy's formulas as well as Eq. (6.33) may also be employed for a liquid or a solid.

The classical theory of dispersion has been successful in fitting the optical refractive index-frequency data, in correlating the refractive indices of a substance in liquid and gaseous forms, and in explaining semi-quantitatively the phenomena of normal and anomalous dispersions. It is however incapable of predicting the characteristic frequencies of a substance. In general, an electron in an atomic system will have a number of quantum states, each of which corresponds to an ionization potential, which is the energy required to remove the electron to infinity. The ionization potential I_s is related to some "characteristic frequency" ω_s by $I_s = \hbar\omega_s$, where $\hbar$ is Planck's constant divided by 2π. Calculation of the ionization potential can only be made in quantum theory. It may also be remarked that the success of the classical theory suggests that atoms behave to a large extent like harmonic oscillators. It is for this reason that sinusoidal electromagnetic waves may traverse matter without deformation.

6.7 Dispersion in a Metallic Conductor

As a crude model, a metallic conductor may be considered as a fixed lattice of positive ions, each surrounded by a cloud of freely circulating electrons in such a way that the net charge within any macroscopic volume element is zero. Under the action of an applied electric field the free electrons will collectively drift in a direction opposite to the field. The drifting electrons collide with the ions resulting in a loss of momentum and hence in a damping force. The ions nevertheless remain essentially stationary apart from thermal vibrations. In addition to the free electrons, there may be bound electrons also. Their effect in modifying the velocity of the traversing electromagnetic wave may however be neglected. To the same approximation, the effective local field intensity may be taken simply as equal to the field of the traversing wave.

Let $\mathbf{r}'$ be the displacement of a free electron from some initial neutral position. The momentum transfer in a collision with a heavy ion varies from zero to $-2m\dot{\mathbf{r}}'$. If we assume the average momentum transfer to be $-m\dot{\mathbf{r}}'$ and the number of collisions suffered by a drifting electron per unit time to be ν, then the damping force acting on the electron will be $-\nu m\dot{\mathbf{r}}'$. Radiation damping is also present, but the total damping is due mainly to collisions. We shall write the damping force as $-m\gamma\dot{\mathbf{r}}'$ as in the case of bound electrons in a dielectric.

The equation of the motion of a free electron in the field of a sinusoidal electromagnetic wave is therefore

$$\ddot{\mathbf{r}}' = -\frac{e}{m}\mathbf{E}_0 e^{-i\omega t} - \gamma\dot{\mathbf{r}}' \quad . \tag{6.38}$$

This is the same as the equation of the motion of a bound electron Eq. (6.3) with $\omega_0 = 0$ so that the steady motion solution is

$$\mathbf{r}' = \frac{e}{m}\frac{1}{\omega(\omega + i\gamma)}\mathbf{E}_0 e^{-i\omega t} \quad .$$

If there are N free electrons per unit volume in the medium, the current density resulting from the drift motion of the free electrons over their random thermal motion is

$$\mathbf{J} = -Ne\dot{\mathbf{r}}' = \frac{Ne^2}{m}\frac{\mathbf{E}}{\gamma - i\omega} \quad .$$

The conductivity σ of the medium as defined by $\mathbf{J} = \sigma\mathbf{E}$ is therefore

$$\sigma = \frac{Ne^2}{m}\frac{1}{\gamma - i\omega} \quad . \tag{6.39}$$

For a conducting medium for which $\mu \approx \mu_0$, the refractive index is given by Eq. (5.42):

$$n = \sqrt{\frac{\varepsilon}{\varepsilon_0}\left(1 + \frac{i\sigma}{\omega\varepsilon}\right)} = \eta + i\kappa \quad . \tag{6.40}$$

Squaring both sides and substituting for σ the expression (6.39) we obtain

$$\eta^2 - \kappa^2 = \frac{\varepsilon}{\varepsilon_0} - \frac{Ne^2}{\varepsilon_0 m}\frac{1}{\omega^2 + \gamma^2} \tag{6.41}$$

and

$$2\eta\kappa = \frac{Ne^2}{\varepsilon_0 m}\frac{\gamma}{\omega(\omega^2 + \gamma^2)} \quad . \tag{6.42}$$

Equation (6.40) gives the dispersion of a metallic conductor[b].

At very low frequencies the conductivity is essentially independent of ω, for as $\omega \ll \gamma$ Eq. (6.39) becomes

$$\sigma_s \approx \frac{Ne^2}{m\gamma} \quad . \tag{6.43}$$

It has been found experimentally that the conductivity has approximately the value measured under static conditions up to the infrared frequencies

---

b The permittivity ε as defined by the equation $\mathbf{D} = \varepsilon\mathbf{E}$ is rather indefinite for a conductor since under static conditions an electric field cannot penetrate into a conductor. In fact the dispersion equation has been used as a basis for assigning ε to a conductor and it has been found that ε of a conductor usually has the same order of magnitude as ε_0.

of $\sim 10^{14}\ s^{-1}$. For higher frequencies the conductivity depends significantly on the frequency in a way roughly described by Eq. (6.39).

The plasma, which is a dilute ionized gas consisting of equal numbers of electrons and positive ions, is a special case of the conducting medium. In a plasma, the collision frequency is small and damping may be neglected. The conductivity is therefore practically purely imaginary

$$\sigma_p \approx \frac{iNe^2}{m\omega} \quad , \tag{6.44}$$

where N is the density of free electrons, the contribution of ions to the drift current being negligible in comparison. As $\varepsilon = \varepsilon_0$ essentially, the refractive index of a plasma is real:

$$n \approx \sqrt{1 - \frac{Ne^2}{\varepsilon_0 m}\frac{1}{\omega^2}} = \sqrt{1 - \frac{\omega_p^2}{\omega^2}} \quad , \tag{6.45}$$

where

$$\omega_p = \sqrt{\frac{Ne^2}{\varepsilon_0 m}} \tag{6.46}$$

is known as the *plasma frequency*.

6.8 Group Velocity

It was shown in Sec. 2.8 that the velocity v of the propagation of plane electromagnetic waves in a linear, isotropic and homogeneous medium of permittivity ε and permeability μ is given by $(\mu\varepsilon)^{-1/2}$. This is the velocity of a point of constant phase on a wave train and is known as the *phase velocity*. We have seen also that in a material medium ε and hence the phase velocity are frequency-dependent, as a consequence of the atomic nature of matter. A signal or a pulse which carries energy or information usually can be represented by a suitable combination of pure sinusoidal waves of differing frequencies. Thus in practice we almost always deal with a bundle of waves having a finite frequency spread. Then, since in most experimental situations different frequencies are involved, the phase velocity has little physical relevance if dispersion is present, although mathematically it is still convenient to speak of the phase velocity of a bundle of waves, meaning in reality some kind of a mean phase velocity for the range of frequencies involved.

To find a more realistic definition of the wave velocity in a dispersive medium, we consider two plane sinusoidal waves propagating in the same direction, say along the z-axis, having the same amplitude and polarization but slightly different frequencies: $\omega - \delta\omega$ and $\omega + \delta\omega$. Let the magnitudes of their propagation vectors be $k - \delta k$ and $k + \delta k$ respectively and represent these waves by

$$E_1 = E_o e^{i\{(k - \delta k)z - (\omega - \delta\omega)t\}} ,$$

and

$$E_2 = E_o e^{i\{(k + \delta k)z - (\omega + \delta\omega)t\}} .$$

The resultant wave is

$$E = E_1 + E_2 = 2E_o \cos(z\delta k - t\delta\omega) e^{i(kz - \omega t)} .$$

Thus, owing to the constructive and destructive interference of the two component waves, the resultant wave has a modulated amplitude $2E_o \cos(z\delta k - t\delta\omega)$. It therefore has an envelope which is cosine in shape with the peak given by

$$z\delta k - t\delta\omega = 0 .$$

This peak and indeed any point of constant phase on the envelope move with the velocity

$$v_g = \frac{dz}{dt} = \frac{\delta\omega}{\delta k} .$$

This velocity may be interpreted as the velocity with which the group of waves as a whole propagates. For this reason it is called the *group velocity*. The above analysis can be generalized to a continuous band of frequencies where the band width is small compared with the average frequency, such as is found in most practical cases. Hence we have in general

$$v_g = \frac{d\omega}{dk} \quad , \tag{6.47}$$

where the differentiation is to be taken at the average frequency. If the phase velocity v is defined in terms of the effective refractive index η by $v = c/\eta$, then $k = \eta\omega/c$ and

$$\frac{1}{v_g} = \frac{1}{c}\left(\eta + \frac{\omega d\eta}{d\omega}\right) \quad . \tag{6.48}$$

It is clear that for a non-dispersive medium, the group velocity coincides with the phase velocity.

Since a signal or a pulse of energy is always carried by a group of waves, the group velocity should be identical with the signal velocity. This is true for most media. Then, while the phase velocity may exceed the velocity of light c in free space, the group velocity is always smaller than c, consistent with Einstein's special theory of relativity (Sec. 8.9). For a medium that exhibits anomalous dispersion for the band of frequencies in question, however, the group velocity may be greater than the phase velocity, and may even exceed c. For such a case, it can nevertheless be shown that the group velocity is no longer identical with the physically significant signal velocity, and that the latter is always[c] less than c.

As an example of the phase velocity exceeding the velocity of light in free space consider plane sinusoidal waves in a plasma. Here

---

[c] J.A. Stratton, *Electromagnetic Theory* (McGraw-Hill, New York, 1941) Sec. 5.17.

the refractive index n is given by Eq. (6.45). Equation (6.48) then gives

$$v_g = cn \quad , \tag{6.49}$$

or

$$v_g v = c^2 \quad ,$$

showing that while v exceeds c, v_g is less than c.

6.9 Propagation in the Ionosphere

In the upper region of the earth's atmosphere, from about 50 km upwards, the constituent gases are partially ionized into free electrons and ions by radiations of extra-terrestrial sources such as solar ultra-violet rays, X-rays, solar particles and galactic cosmic rays, and by meteors. The ionosphere, as the ionized region is called, extends to the outer limit of the atmosphere. It can be divided into six layers, which from bottom to top are the C, D, E, F layers, the heliosphere and the protonosphere. The F layer can be further subdivided into F_1 and F_2 layers; and the C layer was until recently considered part of the D layer. In each of these layers the free electron and ion densities increase to a maximum and thereafter taper off, while the ionosphere as a whole remains very nearly neutral throughout.

The heights of the layers vary according to the time of the day, season and intensity of geomagnetic activity. Roughly speaking, the C layer starts at about 50 km above the earth's surface, the D layer at 70 km, the E layer at 90 km, and the F layer at 120 km. The spread of the F region and the heights of the heliosphere and the protonosphere are particularly highly variable. The outer limit of the protonosphere extends to several earth radii from the earth. In addition, regions of abnormally high ionization densities are frequently superimposed on these layers. They are known as *sporadic E* and *sporadic F* according to the heights where they occur.

Although ionization and re-combination processes are continually going on, typical values may be given for the electron densities. The values at noon are 10^8 electrons m^{-3} for the C layer, 10^9 to 10^{10} for D, 10^{11} for E and F_1, 10^{12} for F_2, 10^{10} for the heliosphere, and 10^9 to 10^{10} for the protonosphere. These densities vary with the time of day, season, solar and geomagnetic activities as well as with latitude. As a rule they are much lower at night. The F_2 density is particularly variable, ranging from 10^{10} to 10^{13} electrons m^{-3} under different conditions.

If the ions and the electrons were absent, the ionosphere would have constitutive constants essentially those of free space, i.e. permittivity ε_0, permeability μ_0 and zero conductivity. The presence of the charged particles gives rise to oscillating currents under the influence of the field of the traversing electromagnetic waves just as the free electrons in a metallic conductor. Again, owing to their much greater mass, the ions may be considered essentially stationary apart from the usual thermal motion. Then the dispersion equation (6.40) applies equally well to the ionosphere, provided we put $\varepsilon = \varepsilon_0$ and interpret N as the number of free electrons per unit volume.

The damping force arises mainly from collisions of the free electrons with the molecules and ions of the atmosphere, so that roughly we may take the damping coefficient γ to be equal to the collision frequency ν of an electron, ν depends on the thermal agitation, the numbers of air molecules and ions per unit volume, and thus on the temperature and pressure of the local atmosphere. It varies from about 10^{12} collisions per second at ground level to one collision per second at an altitude of 800 km.

The factor $e^2/\varepsilon_0 m$ for an electron is approximately 3×10^3 $m^3 s^{-2}$. At ground level, the free electron density N is roughly 5×10^8 electrons m^{-3}, so that $\eta^2 - \kappa^2 \approx 1$ and $2\eta\kappa \simeq 1.5\omega^{-1}$ for all radio frequencies ($\omega \approx 10^6 s^{-1} << \gamma$ at ground level). The ground level solution is therefore $\eta \simeq 1$, $\kappa \simeq 3/4\omega^{-1}$. This means that ground wave transmission is possible even at low radio frequencies. Furthermore, the refractive index is essentially real and equal to unity. On the other hand, N increases rapidly with increasing altitude, reaching 10^{11}-10^{12} electrons m^{-3} in regions of maximum ionization, while the collision frequency ν is reduced to a low level with the result that for low frequencies κ becomes large at some high altitudes. Such waves will therefore be quickly attenuated upon incidence on the ionized region.

At high altitudes, ν is small so that κ is small for sufficiently high frequencies. The effective refractive index is then approximately

$$\eta \approx \sqrt{1 - \frac{Ne^2}{\varepsilon_0 m\omega^2}} = \sqrt{1 - \frac{\omega_p^2}{\omega^2}} \quad , \qquad (6.50)$$

which is just the refractive index of a plasma, of plasma frequency $\omega_p = \frac{Ne^2}{\varepsilon_0 m}$. Electromagnetic waves in the megacycle range are capable of penetrating the intermediate atmosphere, where κ is appreciable, to reach the ionosphere. As a rule, the region of high absorption is confined to a relatively thin layer at the lower edge of the E or D layer. Comparatively little loss normally occurs higher up in the atmosphere. Transmission through the ionosphere is therefore possible for high frequency waves, to which we shall confine ourselves in the following discussion.

Consider the transmission of a high frequency signal through the atmosphere. It will travel with the group velocity given by Eq. (6.49)

$$v_g = c\sqrt{1 - \frac{Ne^2}{\varepsilon_0 m\omega^2}} \quad . \qquad (6.51)$$

First, suppose the signal is sent out vertically upwards from the ground. It will set out with the velocity of light in free space c. As it penetrates an ionized layer, its velocity v_g will gradually decrease as the concentration of free electrons increases, until a height is reached where the electron density N is such that

$$N = \frac{\varepsilon_o m\omega^2}{e^2} \quad , \tag{6.52}$$

when the group velocity vanishes and the signal is reflected back downwards. If the concentration of free electrons in the layer is not large enough for Eq. (6.52) to be satisfied, the signal will penetrate through the ionized layer, continuing on in its upward course. Thus for a given layer of ionized matter the maximum angular frequency which a vertically sounded wave may have and still be reflected back is

$$\omega_c = \sqrt{\frac{N_{max} e^2}{\varepsilon_o m}} \quad , \tag{6.53}$$

where N_{max} is the maximum free electron density of the layer. ω_c is called the *critical angular frequency* for the layer in question. Vertical soundings carried out over a wide range of frequency can be employed for determining the variation of the electron density with altitude.

For oblique incidence it is more convenient to apply Snell's law to relate the ground angle of incidence θ_o with the angle of refraction θ at any particular height where the electron density is N:

$$\sin\theta_o = \sqrt{1 - \frac{Ne^2}{\varepsilon_o m\omega^2}} \sin\theta \quad . \tag{6.54}$$

Note that the intermediate angles do not come in. As an ionized layer is being penetrated, N increases causing θ to increase according to Snell's law. When θ reaches $\pi/2$ the signal will be momentarily directed horizontally, and then reflected downwards. The maximum height or the point of reflection occurs therefore at an altitude where the electron density is

$$N = \frac{\varepsilon_o m}{e^2} \omega^2 \cos^2\theta_o \quad . \tag{6.55}$$

If the electron density of the layer is insufficient for this relation to be satisfied, the wave will penetrate through the layer.

For short wave communication, the angle of incidence θ_0 is determined by the height of the reflecting layer and the distance between the source and the receiving point. The maximum angular frequency that can be used for a given angle of incidence θ_0 is then

$$\omega_u = \omega_c \sec\theta_0 \quad , \tag{6.56}$$

which is known as the *maximum usable frequency*. Since attenuation decreases with increasing frequency, the optimum frequency is just under the maximum usable frequency.

6.10 Effect of Earth's Magnetic Field on Ionospheric Propagation

While the action of the force due to the magnetic field of the traversing radiation may be neglected in comparison with the electric force, the effect of the earth's magnetic field stands alone and must be taken into account. We shall consider the simple case of a linearly polarized plane electromagnetic wave propagating in the direction of the earth's magnetic field $\mathbf{B}_e$, which is taken as along the z-axis.

A linearly polarized plane wave may be decomposed into two circularly polarized components of the same amplitude rotating in opposite senses. This can be readily seen from the following. Consider a point z on the path of a sinusoidal plane wave of angular frequency ω which is linearly polarized in the x-direction. The electric field at the point is given by

$$E_x = \mathrm{Re}\{E_0 e^{i(kz-\omega t)}\} = E_0 \cos(\omega t - kz) \quad ,$$

and

$$E_y = E_z = 0 \quad ,$$

where E_0 is a real constant. This may be regarded as the sum of two rotating vectors $\mathbf{E}^+$ and $\mathbf{E}^-$ with components

$$E_x^+ = \frac{1}{2} E_o \cos(\omega t - kz) \quad , \quad E_y^+ = \frac{1}{2} E_o \sin(\omega t - kz)$$

and

$$E_x^- = \frac{1}{2} E_o \cos(-\omega t + kz) \quad , \quad E_y^- = \frac{1}{2} E_o \sin(-\omega t + kz) \quad ,$$

respectively. The wave with the electric vector $\mathbf{E}^+$ is said to have anticlockwise or right-handed circular polarization. (If a right-handed screw is turned in the same way as the rotation of $\mathbf{E}$, it will advance in the direction of propagation.) The wave with electric vector $\mathbf{E}^-$ is said to have clockwise or left-handed circular polarization.

Let $\mathbf{r}' = (x', y', z')$ be the displacement of a free electron in the ionosphere from some initial position z. Then as

$$\mathbf{E} = (E_x^\pm, E_y^\pm, 0) \quad , \quad \mathbf{B}_e = (0, 0, B_e) \quad ,$$

the equation of the motion of the electron

$$\ddot{\mathbf{r}}' + \gamma \dot{\mathbf{r}}' = -\frac{e}{m} (\mathbf{E} + \dot{\mathbf{r}}' \times \mathbf{B}_e)$$

has components

$$\ddot{x}' + \gamma \dot{x}' = -\frac{e}{m} (E_x^\pm + \dot{y}' B_e) \quad ,$$

$$\ddot{y}' + \gamma \dot{y}' = -\frac{e}{m} (E_y^\pm - \dot{x}' B_e) \quad ,$$

and

$$\ddot{z}' + \gamma \dot{z}' = 0 \quad .$$

The last of the above set of simultaneous equations has an integral $\dot{z}' \sim e^{-\gamma t}$ showing that there can be no motion in the z-direction after a transient period. By putting $x' + iy' = \xi$, $E_x^\pm + iE_y^\pm = \frac{1}{2} E_o e^{\pm i(\omega t - kz)} = E$, and $\frac{e}{m} B_e = \omega_g$, the first two equations may be combined into one complex equation:

$$\ddot{\xi} + \gamma\dot{\xi} = -\frac{e}{m}E + i\frac{e}{m}B_e\dot{\xi} \quad ,$$

or

$$\ddot{\xi} + (\gamma - i\omega_g)\dot{\xi} = -\frac{e}{m}E \quad . \tag{6.57}$$

It may be noted that ω_g is the angular frequency of precession of an electron moving in a plane perpendicular to a uniform magnetic field $\mathbf{B}_e$ and is usually called the *gyromagnetic* or *Larmor (angular) frequency.*

As E is sinusoidal in t, the steady state solution of Eq. (6.57) is

$$\xi = \xi_o e^{\pm i(\omega t - kz)} \quad .$$

Substitution of this in Eq. (6.57) gives

$$\xi = \frac{e}{m\omega}\frac{E}{\omega \mp \omega_g \mp i\gamma} \tag{6.58}$$

The motion of the electrons over their thermal motion gives rise to a current density $\mathbf{J} = -Ne(\dot{x}', \dot{y}', 0)$. We may then define a complex function J by

$$J = J_x + iJ_y = -Ne\dot{\xi} = -\frac{Ne^2}{m\omega}\frac{\dot{E}}{\omega \mp \omega_g \mp i\gamma} \quad . \tag{6.59}$$

Each component of the electric field of the propagation wave satisfies the general wave equation (2.39)

$$\nabla^2 E_j - \mu_o\varepsilon_o\ddot{E}_j - \mu_o\dot{J}_j = 0 \qquad (j = 1,2) \quad ,$$

so that the complex functions E and J satisfy

$$\nabla^2 E - \mu_o\varepsilon_o\ddot{E} - \mu_o\dot{J} = 0 \quad ,$$

or

$$\nabla^2 E - \mu_o \varepsilon_o \left\{ 1 - \frac{\omega_p^2}{\omega(\omega \mp \omega_g \mp i\gamma)} \right\} \ddot{E} = 0 \quad , \tag{6.60}$$

where $\omega_p = \sqrt{\frac{Ne^2}{\varepsilon_o m}}$ is the plasma frequency of the ionosphere. A comparison of this with the corresponding equation in a dielectric medium of permittivity ε and permeability μ_o,

$$\nabla^2 E - \mu_o \varepsilon \ddot{E} = 0 \quad ,$$

shows that the ionosphere must be assigned a complex refractive index

$$n_\pm = \sqrt{\frac{\varepsilon}{\varepsilon_o}} = \sqrt{1 - \frac{\omega_p^2}{\omega(\omega \mp \omega_g \mp i\gamma)}} = \eta \mp i\kappa \quad . \tag{6.61}$$

The damping coefficient γ is approximately equal to ν, the collision frequency. For high radio frequencies and the ionosphere we may thus assume $\omega >> \gamma$, so that the effective refractive index η and the factor κ in the absorption coefficient are approximately

$$\eta \approx \sqrt{1 - \frac{\omega_p^2}{\omega(\omega \mp \omega_g)}} \tag{6.62}$$

and

$$\kappa \approx \frac{\gamma}{2\omega} \frac{\omega_p^2}{(\omega \mp \omega_g)^2} \tag{6.63}$$

respectively.

It follows from Eq. (6.62) that the refractive index, and hence the velocity of propagation, is different for the two circular polarizations. The result is a split of the incident linearly polarized wave into two refracted waves, each circularly polarized but rotating in opposite senses.

Equation (6.63) shows that absorption is different for the two components. Thus travelling along the earth's magnetic field the component with right-handed circular polarization is more easily absorbed. As transmission progresses the wave will become more and more prominently circularly polarized with clockwise rotation.

In general the direction of propagation will not coincide with the direction of the geomagnetic field. In such cases, however, it is only the component of $\mathbf{B}_e$ which is longitudinal to the wave that leads to an effect of the first order. Thus to a first approximation if θ is the angle between the two directions, the above results are to be modified by replacing ω_g with $\omega_g\cos\theta$. This means that the refractive index will depend on the direction of propagation, so that the ionosphere actually behaves like an anisotropic medium. Furthermore, the component rays will have elliptical polarization with opposite rotations. The component whose behaviour is more closely similar to that of the wave in the absence of a magnetic field is often called the *ordinary wave* while the other the *extraordinary wave*.

PROBLEMS

1. The electric field of an elliptically polarized plane wave may be represented by

$$\mathbf{E} = \mathbf{E}_1 \cos\omega t + \mathbf{E}_2 \sin\omega t \quad ,$$

where $\mathbf{E}_1 \cdot \mathbf{E}_2 = 0$. Show that the scattering cross section of a free electron for such a wave is

$$\frac{d\sigma}{d\Omega} = \frac{r_0^2}{k^2}\,\frac{(\mathbf{E}_1 \times \mathbf{k})^2 + (\mathbf{E}_2 \times \mathbf{k})^2}{E_1^2 + E_2^2} \quad ,$$

where $\mathbf{k}$ is the propagation vector of the scattered wave and r_0 the classical radius of an electron.

Hint: Regard the incident wave as the superposition of two waves

$$\mathbf{E}_1 e^{-i\omega t} \quad \text{and} \quad \mathbf{E}_2 e^{-i(\omega t - \pi/2)} .$$

2. Show that the degree of polarization of an initially unpolarized plane electromagnetic wave scattered by a free electron defined as

$$P = \frac{|I_\perp - I_\parallel|}{I_\perp + I_\parallel} ,$$

where $I_\perp$ and $I_\parallel$ are respectively the intensities of the components of the scattered radiation polarized perpendicular to and in the plane containing the incident and scattering directions, is

$$P = \frac{\sin^2\psi}{1 + \cos^2\psi} ,$$

where ψ is the scattering angle. Discuss the variation of the degree of polarization with the direction of scattering.

3. A linearly polarized plane electromagnetic wave of wavelength λ and intensity I_0 is scattered by a gas in a small volume V containing N identical electronic oscillators per unit volume. Show that the energy scattered per unit time into a solid angle $d\Omega$ at an angle θ from the direction of the electric vector is

$$dw = \frac{V}{N} \pi^2 I_0 \frac{(n^2 - 1)^2}{n^4 \lambda^4} \sin^2\theta d\Omega ,$$

where n is the refractive index of the gas, assuming that the oscillators scatter the wave independently and that damping is negligible. This formula is called *Rayleigh's formula*.

4. In a conductor where there is no resistivity the equation of the motion of a conduction electron is $m\ddot{\mathbf{r}} = -e\mathbf{E}$. Show that

$$\dot{\mathbf{j}} = \frac{Ne^2}{m} \mathbf{E} ,$$

where N is the number of electrons per unit volume, is to be used in place of Ohm's law. This equation applies to a superconductor (*London's equation*).

Considering a superconductor as a linear, isotropic and homogeneous medium to which Maxwell's equations apply and incorporating the experimental fact that fields cannot be frozen in, deduce

$$\lambda_L^2 \nabla^2 \mathbf{H} = \mathbf{H} \quad ,$$

where $\lambda_L = \left(\frac{m}{Ne^2\mu}\right)^{\frac{1}{2}}$ is known as *London's penetration depth*.

Applying the above equation to a semi-infinite superconductor, show that it can be penetrated by an applied magnetic field to a depth of the order of λ_L only.

5. A long conductor carrying a steady current of density $\mathbf{J}$ is placed in a magnetic field of induction $\mathbf{B}$ which is perpendicular to the current. Show that an electric field of magnitude

$$E = -\frac{JB}{Ne} \quad ,$$

where N is the number of free electrons per unit volume of the conductor, is induced across the conductor, perpendicular to both the current and the magnetic field. This effect is known as the *Hall effect*.

(Hint: As no electron moves transverse to the conductor under steady-state conditions, the net transverse force must vanish.)

6. The refractive index of hydrogen at 0^oC and 76cm Hg (S.T.P.) for optical frequencies is given by the empirical formula

$$n = 1 + 1.358 \times 10^{-4} \left(1 + \frac{7.52 \times 10^{-11}}{\lambda^2}\right)$$

where λ is measured in cm. Assuming that each hydrogen molecule contains two identical electronic oscillators, find the characteristic frequency and estimate the value of the classical radius of the electron. It is known that one gram molecule of an ideal gas occupies 22.4 litres at S.T.P.

The ionization potential of hydrogen has been quoted as 13.59eV. Is this value consistent with the characteristic frequency deduced above?

7. Writing $n = \eta + i\kappa$ for the refractive index of a dense medium, find the equations relating the real numbers η and κ to the wave frequency and the characteristic parameters of the medium.

For a "transparent" dense medium we may assume that $\kappa^2 \ll \eta^2$. Show that for such a medium at frequencies far-removed from any of the characteristic frequencies we have the approximate relations

$$\frac{\eta^2 - 1}{\eta^2 + 2} \approx \frac{1}{3} \sum_s \frac{\alpha_s}{\omega_s^2 - \omega^2}$$

and

$$\frac{\eta\kappa}{(\eta^2 + 2)^2} \approx \frac{\omega}{18} \sum_s \frac{\alpha_s \gamma_s}{(\omega_s^2 - \omega^2)^2}$$

where the symbols are the same as those defined in Sec. 6.6.

8. The hydrogen gas at S.T.P. has a density 8.96×10^{-5} g cm^{-3} and a refractive index 1.000138 . The density of liquid hydrogen is 6.8×10^{-2} g cm^{-3}. Deduce its refractive index.

9. Carbon bisulfide is often used as a standard in determining refractive indices by a total reflection method. Its refractive indices at 0°C and at 20°C for nine wavelengths are given in the following table

λ (A)	2749	3612	3968	4861	5893	6708	7685
η at 0°C	2.0348	1.7572	1.7199	1.6713	1.6436	1.6328	1.6241
η at 20°C	2.0047	1.7381	1.7018	1.6547	1.6276	1.6168	1.6087

Its coefficient of cubical expansion may be represented by

$$\beta = a + bt \quad ,$$

where $a = 1.1398 \times 10^{-2}$ deg C^{-1} and $b = 2.74 \times 10^{-6}$ deg C^{-2}.

Fit the data to (a) a Cauchy type equation

$$\frac{\eta - 1}{\rho} = A + \frac{B}{\lambda^2} \quad ,$$

and (b) a Lorentz-Lorenz type equation

$$\frac{\eta^2 - 1}{\eta^2 + 2} \cdot \frac{1}{\rho} = A' + \frac{B'}{\lambda^2} \quad ,$$

with ρ expressed as a function of t.

Which expression fits the data better?

10. The experimental values of the refractive index of water at 20°C for various wavelengths are given below:

$\lambda(\times 10^{-5}$cm):	12.56	6.708	6.438	5.893	5.461	5.086
η :	1.3210	1.3308	1.3314	1.3330	1.3345	1.3360
$\lambda(\times 10^{-5}$cm):	4.800	4.047	3.034	2.144		
η :	1.3374	1.3428	1.3581	1.4032		

Assuming that

$$\frac{\eta^2 - 1}{\eta^2 + 2} = A + \frac{B}{\lambda^2} + \frac{C}{\lambda^4} \quad ,$$

estimate the values of the constants A, B and C from the data.

11. Anomalous dispersion occurs in a gas. Show that at frequencies at which the effective refractive index assumes maximum and minimum values the absorption constant is equal to one half of its maximum value.

12. Electromagnetic waves transmitted from an aircraft are received by another at a distance d away. Both aircrafts are at the same small height h above a lake. If the lake water has permittivity $\chi\varepsilon_0$ and permeability μ_0, show that the ratio of the intensity of the signal received by direct reflection to that of the emitted signal is

$$\left[\frac{\{(\chi-1)d^2+4\chi h^2\}^{\frac{1}{2}}-2\chi h}{\{(\chi-1)d^2+4\chi h^2\}^{\frac{1}{2}}+2\chi h}\right]^2 ,$$

assuming the waves to be polarized with the electric vector in the vertical plane containing the aircrafts and the air absorption to be negligible.

13. An electromagnetic wave of frequency 30 MHz propagates in the ionosphere. (a) If the smallest incident angle for reflection is $\theta_0 = 75^0$, find the maximum electron density. (b) In a region where the electron density is 3×10^{11} electrons m^{-3} and the collision frequency is 1 MHz, find the attenuation factor per kilometre.

14 (a) A plane electromagnetic wave travels in a uniform plasma. Show that the time-averaged Poynting vector vanishes for frequencies smaller than the plasma frequency.

(b) If the electron density in the interstellar space is one per cm^3, find the lower limit of the frequency of an electromagnetic wave that can be propagated there.

15. A linearly polarized plane electromagnetic wave travels through a plasma along the direction of the external magnetic field **B**. As the propagation vector **k** has different magnitudes for the two circularly polarized components, the plane of polarization of the wave will rotate as it progresses through the medium. Show that the rotation is clockwise when viewed along the direction of propagation, and that, if the plasma has a length ℓ and a plasma frequency ω_p, the angle of rotation is

$$\alpha = \frac{\omega \ell}{2c}\left[\sqrt{1 - \frac{\omega_p^2}{\omega(\omega+\omega_g)}} - \sqrt{1 - \frac{\omega_p^2}{\omega(\omega-\omega_g)}}\right] ,$$

where ω_g is the gyromagnetic frequency of an electron, when the wave emerges from the other end of the plasma. This effect is known as the *Faraday rotation*.

If the external magnetic field is weak, show that the Faraday rotation is proportional to $|\mathbf{B}|$.

16 (a) If the earth's magnetic field has induction 0.5×10^{-4} Wb m^{-2}, show that maximum selective absorption will occur in the ionosphere for radio waves of wavelength 212 m.

(b) Linearly polarized radiation of 10 cm wavelength from a galactic radio source has to pass through a region of ionized interstellar gas of diameter 10^{18}cm where the mean electron density is one per cm^3. If the magnetic field in the region has an induction 10^{-8} Wb m^{-2} and is parallel to the direction of propagation, find the magnitude of the Faraday rotation.

Chapter VII

SKIN EFFECT AND WAVEGUIDES

An oscillating electromagnetic field attenuates exponentially with penetration into a conductor. This gives rise to the skin effect, which confines high frequency currents to a thin layer beneath the surface of a conductor. For the same reason, high frequency electromagnetic waves cannot penetrate appreciably into a conductor and may thus be confined inside and guided along a hollow conducting tube. Standing waves may be set up if the conducting tube is closed at both ends. Waves may also be transmitted along two long parallel conductors.

7.1 High Frequency Currents in a Semi-Infinite Conductor

In the discussion of the refraction of electromagnetic waves into a conducting medium we have seen that high frequency fields can penetrate into a conductor only to a distance of the order of the skin depth defined by Eq. (5.58). If the conductor is ohmic we should expect that high frequency currents flowing in a conductor will also be confined mainly to a thin layer beneath the surface. This effect is known as the *skin effect*.

Charges introduced into the interior of a linear, isotropic and homogeneous medium of permittivity ε, permeability μ and conductivity σ will migrate to the surface in such a way that the charge density at an interior point decays exponentially and drops to e^{-1} of its initial value during the relaxation time ε/σ [Prob. (2.3)]. For all but the poorest conductors the relaxation time is exceedingly small. It is only $\approx 10^{-18}$ s for a pure metal and is still less than 10^{-6} s for distilled water. Thus in the interior of a good conductor the charge density may be assumed to be zero under steady state conditions. Furthermore, for a good conductor the displacement current may be neglected in comparison with the conduction current up to optical frequencies. Maxwell's equations for a good conducting medium are therefore

$$\nabla \times \mathbf{E} = -\dot{\mathbf{B}} \quad , \tag{7.1}$$

$$\nabla \times \mathbf{H} = \mathbf{J} \quad , \tag{7.2}$$

$$\nabla \cdot \mathbf{B} = 0 \quad , \tag{7.3}$$

$$\nabla \cdot \mathbf{D} = 0 \quad . \tag{7.4}$$

The general wave equation is now reduced to the diffusion equation

$$\nabla^2 \mathbf{E} - \mu\sigma\dot{\mathbf{E}} = 0 \quad , \tag{7.5}$$

or

$$\nabla^2 \mathbf{H} - \mu\sigma\dot{\mathbf{H}} = 0 \quad . \tag{7.6}$$

Ohm's law $\mathbf{J} = \sigma\mathbf{E}$

then gives

$$\nabla^2 \mathbf{J} - \mu\sigma\dot{\mathbf{J}} = 0 \quad . \tag{7.7}$$

For an alternating current $\mathbf{J}_o(\mathbf{r})e^{-i\omega t}$, it becomes

$$\nabla^2 \mathbf{J}_o + T^2 \mathbf{J}_o = 0 \quad . \tag{7.8}$$

with

$$T^2 \equiv i\mu\sigma\omega = \mu\sigma\omega \, e^{i\frac{\pi}{2}}$$

or

$$T = \sqrt{\mu\sigma\omega}\, e^{i\frac{\pi}{4}} = \frac{1+i}{\delta} \quad , \tag{7.9}$$

where

$$\delta \equiv \sqrt{\frac{2}{\mu\sigma\omega}} \tag{7.10}$$

is the *skin depth.*

Equation (7.8) can be readily solved for a semi-infinite plane conductor, that is, a plane conductor of infinite length, infinite width and infinite thickness. For symmetry the current may be assumed to flow parallel to the plane surface and vary only as a function of depth if we confine ourselves to a section of a length small compared with c/ω. In practice these conditions are satisfied to a good approximation for a long conductor of a length small compared with c/ω, a finite width, and a finite thickness, but without sharp surface curvatures, provided the skin depth is much smaller than all the dimensions involved. If we choose a Cartesian coordinate system with the origin at the surface, the x-axis along the inward normal and the z-axis along the direction of flow, then the current density will be given by $J_z = J_o(x)e^{-i\omega t}$ and Eq. (7.8) reduced to

$$\frac{d^2J_o}{dx^2} + T^2J_o = 0 \quad . \tag{7.11}$$

The general solution of the last equation is

$$J_o = Ae^{iTx} + Be^{-iTx} = Ae^{-\frac{x}{\delta}}e^{i\frac{x}{\delta}} + Be^{\frac{x}{\delta}}e^{-i\frac{x}{\delta}} \quad ,$$

where A and B are real constants. Since the current density in the conductor must be finite everywhere, particularly for $x \to \infty$, we require that $B = 0$. Then, if the current density at the surface has an amplitude J_s, we have

$$J_z = J_s e^{-\frac{x}{\delta}} e^{i(\frac{x}{\delta} - \omega t)} \quad . \tag{7.12}$$

Thus the amplitude of the current density decreases exponentially with increasing penetration into the conductor, reducing to e^{-1} of its surface value at the skin depth δ. On account of the rapid attenuation of the current density after the penetration of a few skin depths, the conditions of a semi-infinite plane conductor are satisfied by a finite conductor whose thickness and radius of surface curvature are each large compared with δ. With increasing depth of penetration there is also a gradual change of phase, reaching one radian at the skin depth.

For a metallic conductor, $\sigma \simeq 10^7$ ohm^{-1}m^{-1}. If $\mu \approx \mu_0$, we have $\delta \approx (2\pi\omega)^{-\frac{1}{2}}$ metres, so that skin effect is not significant at low frequencies. On the other hand, for ferromagnetic materials the permeability is relatively high, $\mu \approx 10^3 \mu_0$, and the effect can be quite appreciable even at relatively low frequencies.

7.2 High Frequency Currents in a Circular Wire

Consider now the more realistic case of an alternating current in a long uniform conductor of circular cross section of radius a. For reason of symmetry the current may be assumed to flow parallel to the axis and vary only radially. If we confine ourselves to a length of the conductor that satisfies the condition $\ell << c/\omega$, any axial variation may also be neglected. We may then assume a current density of the form $J_0(\rho)e^{-i\omega t}$, ρ being the radial distance from the axis. In cylindrical coordinates with the z-axis coinciding with the axis of the conductor, Eq. (7.8) becomes

$$\frac{d^2 J_0}{d\rho^2} + \frac{1}{\rho}\frac{dJ_0}{d\rho} + T^2 J_0 = 0 \quad , \tag{7.13}$$

or

$$\frac{d^2J_o}{d\zeta^2} + \frac{1}{\zeta}\frac{dJ_o}{d\zeta} + J_o = 0 \quad , \tag{7.14}$$

where $\zeta = T\rho$. The last equation is the zero-order Bessel equation with complex argument ζ and has the general solution

$$A\,\mathcal{J}_0(\zeta) + B\,\mathcal{Y}_0(\zeta) \quad ,$$

where A, B are constants, $\mathcal{J}_0$ and $\mathcal{Y}_0$ are zero-order Bessel functions of the first and the second kind respectively. For $\zeta \to 0$,

$$\mathcal{Y}_0(\zeta) \sim \frac{2}{\pi}\ln\zeta$$

so that the modulus of $\mathcal{Y}_0$ tends to ∞ as $\rho \to 0$. Since for a solid conductor J_z must be finite on the axis we have to put $B = 0$. Then, if the amplitude of the current density at the surface is J_s, we have

$$J_z = \frac{\mathcal{J}_0(T\rho)}{\mathcal{J}_0(Ta)} J_s e^{-i\omega t} \quad . \tag{7.15}$$

The argument $T\rho$ can be written as $\xi\sqrt{i}$, where $\xi = \sqrt{2}\,\rho/\delta$ is real. The real and imaginary parts of $\mathcal{J}_0(\xi\sqrt{i})$ can be found in standard mathematical tables[a]. Let $\mathcal{J}_0(T\rho) = u + iv$, $\mathcal{J}_0(Ta) = u_s + iv_s$, and we may write

$$J_z = \left(\frac{u^2 + v^2}{u_s^2 + v_s^2}\right)^{\frac{1}{2}} J_s e^{-i(\omega t - \phi)} \quad , \tag{7.16}$$

where

$$\phi = \arctan\left(\frac{vu_s - uv_s}{uu_s + vv_s}\right) \quad . \tag{7.17}$$

---

[a] For example E. Jahnke and F. Emde, *Table of Functions with Formulae and Curves*, Dover, New York.

The radial distribution of the current density for a given wire is thus a function of ρ/δ. The solid curves in Fig. 7.1 shows the relative amplitude $\frac{J_o}{J_s}$ against the radial distance ρ from the axis of a long round wire for three values of a/δ: 1.2, 3.8 and 6.6. As frequency increases, a/δ increases and the current becomes more and more concentrated near the surface of the conductor. For a copper wire of 0.5 mm diameter of which $\mu \approx \mu_0$ and $\sigma = 5.9 \times 10^7$ $ohm^{-1}m^{-1}$, the above values of a/δ correspond to the frequencies of 10^5, 10^6 and 3×10^6 Hz respectively. It can be seen that only at the lowest frequency is the current fairly uniformly distributed over the cross section of the wire. Equation (7.17) shows that for a given frequency the phase angle also varies with the radial distance.

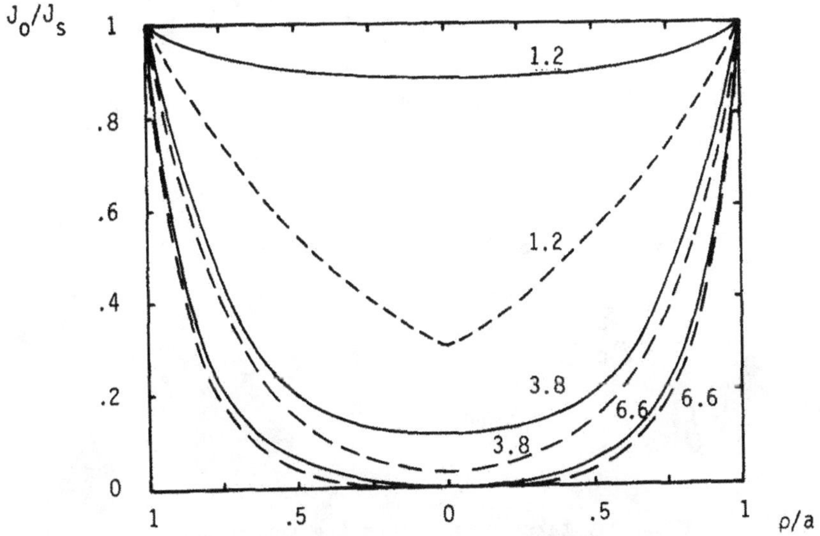

Fig. 7.1 Current distribution in a long wire of circular cross section of radius a for three values of a/δ, where δ is the skin depth. The solid curve is calculated from the exact solution, Eq. (7.16), and the dashed curve is obtained from the plane conductor approximation, Eq. (7.12). For $a/\delta \geq 10$ the two methods may be expected to give the same result for all practical purposes.

For large frequencies, the current flows mainly in a thin layer beneath the surface so that the surface curvature becomes unimportant. This means that for frequencies for which δ is much smaller than the radius of curvature of the surface of a conductor, the current distribution can be accurately predicted by taking out a layer of the conductor under the surface of a thickness large compared with δ and analysing it as a semi-infinite plane conductor. Consider for example the present case of a long cylindrical conductor. The depth x below the surface is given by $a-\rho$. For $|\zeta| \to \infty$, the zero-order Bessel function has the asymptotic form

$$\mathcal{J}_0(\zeta) \simeq \sqrt{\frac{2}{\pi\zeta}} \cos\left(\zeta - \frac{\pi}{4}\right)$$

$$= \frac{1}{\sqrt{2\pi T\rho}} \left\{ e^{-\frac{\rho}{\delta}} e^{i\left(\frac{\rho}{\delta} - \frac{\pi}{4}\right)} + e^{\frac{\rho}{\delta}} e^{-i\left(\frac{\rho}{\delta} - \frac{\pi}{4}\right)} \right\}$$

$$\approx \frac{1}{\sqrt{2\pi T\rho}} e^{\frac{\rho}{\delta}} e^{-i\left(\frac{\rho}{\delta} - \frac{\pi}{4}\right)} .$$

Since $\zeta = \frac{\rho}{\delta}(1+i)$ and $\rho >> \delta$, substitution in Eq. (7.15) results in

$$J_z \approx \sqrt{\frac{a}{\rho}}\, e^{-\frac{(a-\rho)}{\delta}} e^{i\frac{(a-\rho)}{\delta}} J_s e^{-i\omega t}$$

$$\approx J_s e^{-\frac{x}{\delta}} e^{i\left(\frac{x}{\delta} - \omega t\right)} ,$$

as for $x << a$ we may take $\rho \approx a$ under the square root sign. This expression is identical with Eq. (7.12) for a semi-infinite plane conductor as expected.

In Fig. 7.1 are also plotted the current distributions for the round wire for the values of a/δ indicated, calculated by treating it as a semi-infinite plane conductor. It is seen that agreement improves with increasing a/δ, and for values beyond about 10 the plane conductor approximation and the exact analysis may be expected to give the same result for all practical purposes.

7.3 Internal Impedance at High Frequencies

The concentration of a high frequency current in a thin surface layer must result in a large increase of the effective impedance of a conductor since the effective cross sectional area for conduction is greatly diminished. To find the total impedance of a circuit it will, however, be necessary to know the configuration of the conducting path since this affects the total inductance of the circuit. However, as a rough measure of the effect, we may use the internal impedance which takes account of the internal self-inductance only (defined in Prob. 4.3). For actual application to a circuit the total impedance must of course be employed.

For a semi-infinite plane conductor it is convenient to consider a cross section of unit width normal to the direction of current. The total flow of charge per unit time across this area is called the *linear current density* and is given by

$$I_z = \int_0^\infty J_z dx = \frac{J_s \delta e^{-i\omega t}}{1 - i} \ . \tag{7.18}$$

The internal impedance per unit width per unit length of flow is given by the ratio of the surface potential drop per unit path length to the linear current density. Since the former is simply the tangential electric field intensity at the surface, the impedance is

$$Z_s = \frac{J_s e^{-i\omega t}}{\sigma I_z} = \frac{1 - i}{\sigma \delta} \ . \tag{7.19}$$

As a surface area of unit width and unit length is involved, Z_s is known as the *internal impedance per unit square.*

It can be seen from Eq. (7.19) that the resistance and the

internal reactance are equal, each being

$$R_s = \frac{1}{\sigma\delta} = \sqrt{\frac{\mu\omega}{2\sigma}} \qquad (7.20)$$

per unit square. The internal impedance thus has a phase angle of $-\pi/4$ and a modulus that varies as $\omega^{1/2}$ if σ does not vary appreciably with frequency. R_s is called the *surface resistivity* of the conductor.

The results obtained here may be applied to long conductors of any geometry provided the radius of curvature at the surface is much greater than the skin depth δ. Consider for example the cylindrical conductor of the previous section. The total width of its cross section is the circumference $2\pi a$, so that the internal impedance per unit length is

$$Z = \frac{Z_s}{2\pi a} = \frac{(1-i)}{2\pi a}\sqrt{\frac{\mu\omega}{2\sigma}} \quad .$$

In the limiting case of a perfect conductor $\sigma \to \infty$, $\delta \to 0$, so that the electromagnetic field cannot penetrate into the interior; any excess charge or current must reside on the surface. As $Z_s \to 0$ also, the linear current density cannot remain finite unless the tangential component of the electric intensity at the surface vanishes. This is of course obvious from the boundary condition for $\mathbf{E}$, Eq. (2.9), if we consider one side of the transition layer as being just under the conductor surface. As the magnetic field is zero inside the conductor the boundary condition for $\mathbf{H}$, Eq. (2.11), requires the tangential component of $\mathbf{H}$ to be equal numerically to the linear current density I_ℓ. Vectorially, if we denote the outward normal to the conductor by $\mathbf{n}$, the boundary conditions for $\mathbf{E}$ and $\mathbf{H}$ at the surface of a perfect conductor may be written as

$$\mathbf{n} \times \mathbf{E} = 0 \quad , \qquad (7.21)$$

and

$$\mathbf{n} \times \mathbf{H} = \mathbf{I}_\ell \quad . \qquad (7.22)$$

Similarly, the boundary conditions for **D** and **B** are

$$\mathbf{n}\cdot\mathbf{D} = \sigma \quad , \tag{7.23}$$

and

$$\mathbf{n}\cdot\mathbf{B} = 0 \quad , \tag{7.24}$$

where σ is the surface charge density.

7.4 Waves Guided by Parallel Plane Conductors

On account of the skin effect an electromagnetic wave incident upon a good conductor will be reflected with very little loss. If the dielectric medium in which the wave propagates is bounded by two parallel plane conductors, we may expect the wave to be bounced back and forth and so guided down the dielectric without spreading into either conductor. To consider the transmission between parallel plane conductors quantitatively and obtain the conditions for free transmission, we assume both the dielectric and the conductors involved to be perfect.

Choose a Cartesian coordinate system with the x-axis perpendicular to the conductors and the xz-plane coinciding with the plane of reflection as shown in Fig. 7.2. The conductors are presented by the planes $x = 0$ and $x = a$. We shall assume that the electromagnetic wave may be represented by $\mathbf{E} = \mathbf{E}_0 e^{i(\mathbf{k}\cdot\mathbf{r} - \omega t)}$, where $\mathbf{E}_0$ is a real constant vector, and $\kappa = \omega\sqrt{\mu\varepsilon}$, ε and μ being respectively the permittivity and permeability of the dielectric.

Let the electric field intensity of the waves incident on and reflected from the conducting plane $x = 0$ be respectively

$$\mathbf{E}_0 e^{i(-kx\cos\theta + kz\sin\theta - \omega t)}$$

and

$$\mathbf{E}_0'' e^{i(kx\cos\theta + kz\sin\theta - \omega t)} \quad .$$

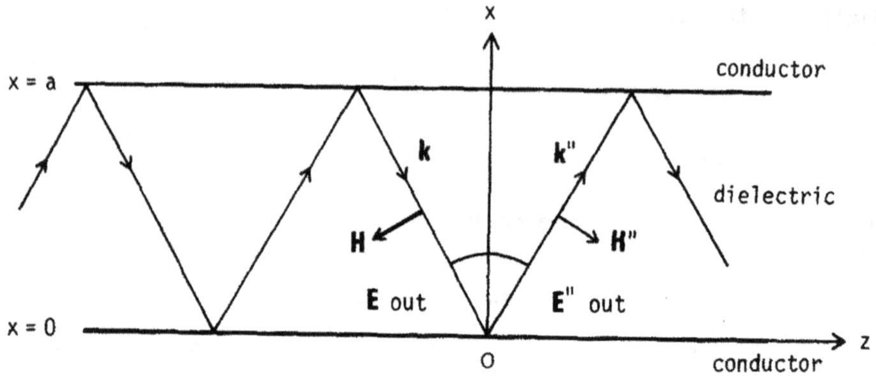

Fig. 7.2 Linearly polarized electromagnetic waves travelling between parallel plane conductors. The path of a ray is shown. The electric field is assumed to be perpendicular to the plane of reflection.

There is no loss of generality by assuming the incident electromagnetic wave to be linearly polarized. First we shall consider the wave as being polarized with the electric vector parallel to the y-axis. Then as polarization is not changed by reflection the resultant electric field between the conductors may be represented by

$$E_y = \left\{ E_0 e^{i(-kx\cos\theta + kz\sin\theta)} + E_0'' e^{i(kx\cos\theta + kz\sin\theta)} \right\} e^{-i\omega t} \quad .$$

The boundary condition Eq. (7.21) for **E** requires that $E_y = 0$ for $x = 0$, i.e. $E_0 + E_0'' = 0$, so that

$$E_y = -i2E_0 \sin(kx\cos\theta) e^{i(kz\sin\theta - \omega t)} \quad .$$

The boundary condition $E_y = 0$ must also be satisfied for $x = a$, which requires that

$$ka\cos\theta = m\pi \quad ,$$

or

$$\omega = \frac{1}{\sqrt{\mu\varepsilon}} \frac{m\pi}{a\cos\theta} \quad , \tag{7.25}$$

where m is a positive integer. If this condition is satisfied the resultant superposition of the incident and reflected waves is a wave propagating along the z-axis with a propagation vector **k**' of magnitude $k\sin\theta$. In this case the magnetic vector has in general a longitudinal component, i.e. component in the direction of propagation, as can be easily seen from Fig. 7.2. Only the electric field is transverse. The wave is therefore said to be a *transverse electric wave* (TE wave) or an M wave.

Consider next the incident wave as being polarized with the electric vector in the xz-plane. As can be seen from Fig. 7.3 the resultant electric field between the conductors has components

$$E_z = \left\{E_0 e^{i(-kx\cos\theta + kz\sin\theta)} - E_0'' e^{i(kx\cos\theta + kz\sin\theta)}\right\}\cos\theta e^{-i\omega t}$$

and

$$E_x = \left\{E_0 e^{i(-kx\cos\theta + kz\sin\theta)} + E_0'' e^{i(kx\cos\theta + kz\sin\theta)}\right\}\sin\theta e^{-i\omega t} \quad .$$

The boundary condition for **E** requires that $E_z = 0$ at $x = 0$, giving $E_0'' = E_0$ and

$$E_z = -i2E_0\cos\theta \sin(kx\cos\theta)e^{i(kz\sin\theta - \omega t)} \quad ,$$

$$E_x = 2E_0\sin\theta \cos(kx\cos\theta)e^{i(kz\sin\theta - \omega t)} \quad .$$

It also requires that $E_z = 0$ at $x = a$. Thus, if the same condition Eq. (7.25) is satisfied transmission along the z-axis is again possible. The transmitted wave now has a longitudinal electric field component and is said to be an E wave or a *transverse magnetic wave* (TM wave).

In the special case of $\theta = \pi/2$, the electromagnetic field has nonvanishing components E_x and H_y only. The waves propagated between the parallel conductors are transverse for both the electric and magnetic fields and are called *transverse electromagnetic waves* (TEM waves). Such waves travel down the guide without being affected by the presence of the conductors.

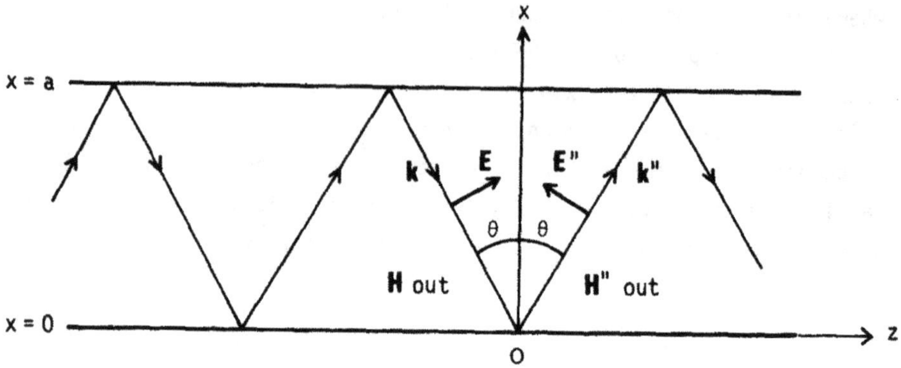

Fig. 7.3 Linearly polarized electromagnetic waves travelling between parallel plane conductors with the electric vector in the plane of reflection.

Thus it is possible to transmit electromagnetic waves between two parallel perfect plane conductors. The waves that are transmitted are prevented from spreading outside the region by the boundary condition $E_t = 0$. However, they are not prevented from spreading in the other transverse directions. We may make the protection complete by adding two more perfectly conducting planes at right angles to the conductors, thus enclosing the dielectric in which the waves propagate in a hollow tube. Such a system is called a *waveguide*.

The simplest and most commonly used guiding systems are hollow rectangular or circular cylindrical tubes enclosing a dielectric. Other cross sectional shapes may be used also, but do not usually offer much additional advantage and yet are more expensive to produce. We shall consider an ideal system of which the conductor and dielectric are both perfect. Condition for free transmission is obtained by applying the equation $E_t = 0$ at the conducting boundary to the solution of the wave equation for the dielectric. The configuration of the electromagnetic field may then be calculated using Maxwell's equations. Although the assumption of an ideal waveguide cannot be realized in practice it is a good approximation for nearly loss-free transmission.

The small resistivity of the conductor and the small conductivity of the dielectric are then treated as small perturbations for the estimation of the attenuation constant.

From the discussion on waves guided by parallel conductors we see that the wave transmitted by a guide will propagate in the axial direction, which is chosen as the direction of the z-axis. We shall then consider the propagation of a wave with a (z, t)-dependence of $e^{(\gamma z - i\omega t)}$ where ω is the angular frequency and γ a constant called the *propagation constant*. If γ is real, it must be chosen to have the appropriate sign for the amplitude to remain finite. The amplitude therefore attenuates exponentially and transmission will not be possible. If γ is purely imaginary, transmission will take place with no loss. In the dielectric of the waveguide, which is assumed linear, isotropic and homogeneous with permittivity ε and permeability μ, the equation for the transmitted wave may then be written in terms of $\mathbf{E}$ as

$$(\nabla^2 - \mu\varepsilon \frac{\partial^2}{\partial t^2})\mathbf{E} = (\frac{\partial^2}{\partial x^2} + \frac{\partial^2}{\partial y^2} + h^2)\mathbf{E} = 0 \quad , \tag{7.26}$$

or in terms of $\mathbf{H}$ as

$$(\frac{\partial^2}{\partial x^2} + \frac{\partial^2}{\partial y^2} + h^2)\mathbf{H} = 0 \quad , \tag{7.27}$$

where

$$h^2 = \gamma^2 + \mu\varepsilon\omega^2 \quad . \tag{7.28}$$

The field configuration is determined by Maxwell's equations

$$\nabla \times \mathbf{E} = i\mu\omega\mathbf{H} \quad , \tag{7.29}$$

$$\nabla \times \mathbf{H} = -i\varepsilon\omega\mathbf{E} \quad . \tag{7.30}$$

The above equations together with the boundary condition $E_t = 0$ at the conducting boundary completely determine the condition for propagation and characteristics of the guided wave.

7.5 Transmission by a Rectangular Waveguide

For a rectangular waveguide of internal width a and height b, choose Cartesian coordinates as shown in Fig. 7.4. The wave equations (7.26) and (7.27) are now to be solved subject to the boundary conditions

$$E_x = E_z = 0 \quad \text{at} \quad y = 0 \quad \text{and} \quad y = b \quad ,$$

$$E_y = E_z = 0 \quad \text{at} \quad x = 0 \quad \text{and} \quad x = a \quad .$$

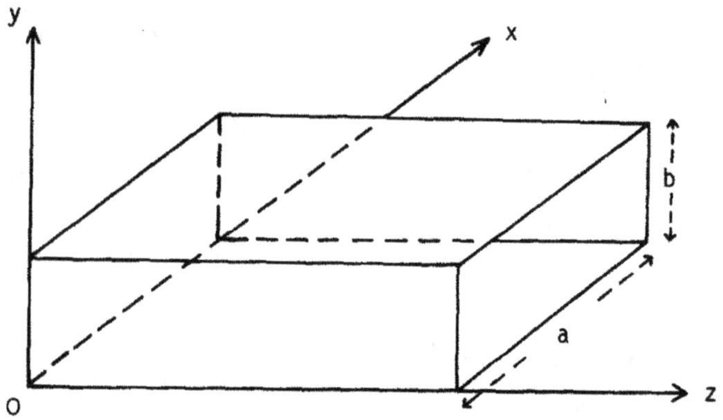

Fig. 7.4 A rectangular waveguide of internal width a and height b. By convention the x-axis is chosen along the larger transverse dimension.

Maxwell's Eqs. (7.29) and (7.30) for the fields in the dielectric of the guide may be written as

$$\frac{\partial E_z}{\partial y} - \gamma E_y = i\mu\omega H_x \quad , \tag{7.31}$$

$$\gamma E_x - \frac{\partial E_z}{\partial x} = i\mu\omega H_y \quad , \tag{7.32}$$

$$\frac{\partial E_y}{\partial x} - \frac{\partial E_x}{\partial y} = i\mu\omega H_z \quad , \tag{7.33}$$

$$\frac{\partial H_z}{\partial y} - \gamma H_y = -i\varepsilon\omega E_x \quad , \tag{7.34}$$

$$\gamma H_x - \frac{\partial H_z}{\partial x} = -i\varepsilon\omega E_y \quad , \tag{7.35}$$

$$\frac{\partial H_y}{\partial x} - \frac{\partial H_x}{\partial y} = -i\varepsilon\omega E_z \quad . \tag{7.36}$$

Either H_x or E_y may be eliminated from Eq. (7.31) and (7.35) to give

$$E_y = \frac{1}{h^2}\left(\gamma \frac{\partial E_z}{\partial y} - i\mu\omega \frac{\partial H_z}{\partial x}\right) \quad , \tag{7.37}$$

or

$$H_x = \frac{1}{h^2}\left(\gamma \frac{\partial H_z}{\partial x} - i\varepsilon\omega \frac{\partial E_z}{\partial y}\right) \quad . \tag{7.38}$$

Similarly from Eqs. (7.32) and (7.34) we obtain

$$E_x = \frac{1}{h^2}\left(\gamma \frac{\partial E_z}{\partial x} + i\mu\omega \frac{\partial H_z}{\partial y}\right) \quad , \tag{7.39}$$

$$H_y = \frac{1}{h^2}\left(\gamma \frac{\partial H_z}{\partial y} + i\varepsilon\omega \frac{\partial E_z}{\partial x}\right) \quad . \tag{7.40}$$

Equations (7.37)-(7.40) may be called *guide equations*.

It may be seen from these equations that if E_z and H_z are both zero all field components will vanish unless $h = 0$. If the latter is the case, then both **E** and **H** are transverse with components having amplitudes satisfying the two-dimensional Laplace's equation. For example, the equation for E_x or E_y has the form

$$\left(\frac{\partial^2}{\partial x^2} + \frac{\partial^2}{\partial y^2}\right) \begin{matrix} E_x \\ E_x \end{matrix} = 0 \quad .$$

If we define a potential function $V(x, y)e^{(\gamma z - i\omega t)}$ such that

$E_x = -\frac{\partial V}{\partial x} e^{(\gamma z - i\omega t)}$ and $E_y = -\frac{\partial V}{\partial y} e^{(\gamma z - i\omega t)}$, V can be easily seen to satisfy the two-dimensional Laplace's equation also. As the walls are perfectly conducting, **E** will be everywhere normal to the surface, i.e. the wall surface is an equipotential surface. Thus V is exactly the same as an electrostatic potential in a source-free volume which tends to a constant potential on the surfaces. In other words, the situation is the same as that in the interior of a conductor under static conditions for which the solution of Laplace's equation is $V = \text{constant}$, giving $\mathbf{E} = 0$. Then if the waveguide is in the form of a single hollow conducting pipe, i.e. in waveguides whose walls are electrically connected, the electric field will vanish in its interior. A TEM mode therefore cannot exist inside a single hollow cylinder of infinite conductivity. The propagation of purely transverse electromagnetic waves is however still possible between two or more unconnected conductors, or outside a single conductor.

It may be noted in passing also that, in general, for the propagation of transverse electromagnetic waves with a (z, t)-dependence of the type $\exp(\gamma z - i\omega t)$ we require h to vanish and consequently the propagation constant γ to be purely imaginary. This means that such propagation is possible only under loss-free conditions. For example, in the propagation between plane conductors some longitudinal component will be present if the conductor or dielectric is less than perfect.

As purely transverse electromagnetic waves cannot be transmitted through a hollow waveguide, at least one of the field vectors must have a longitudinal component. The simplest types of transmitted wave are therefore the TM or TE waves. Both types of wave may be transmitted simultaneously in a guide provided the conditions for free transmission are fulfilled. If the system is very nearly ideal there is actually very little coupling between the different types and each may be treated independently. From the opposite point of view, in considering the transmission of a wave in an ideal guiding system we may regard it as consisting of a number, which may be infinite, of these simple types with the appropriate amplitudes and phases.

Consider first the propagation of a TM wave in the rectangular waveguide. The z-component of Eq. (7.26) may be separated in the variables by writing the solution as

$$E_z = X(x)Y(y)e^{(\gamma z - i\omega t)} \quad . \tag{7.41}$$

The equation for E_z then becomes

$$\frac{1}{X}\frac{d^2X}{dx^2} + \frac{1}{Y}\frac{d^2Y}{dy^2} + h^2 = 0 \quad .$$

Since a variation of x affects only the first term of the left hand side and a variation of y affects only the second term, the above is equivalent to two equations

$$\frac{d^2X}{dx^2} + k_1^2 X = 0 \quad , \tag{7.42}$$

$$\frac{d^2Y}{dy^2} + k_2^2 Y = 0 \quad , \tag{7.43}$$

where k_1 and k_2 are real constants related by

$$k_1^2 + k_2^2 = h^2 \quad . \tag{7.44}$$

The solutions of Eqs. (7.42) and (7.43) may be written as

$$X = C_1\cos(k_1x) + C_2\sin(k_1x) \quad ,$$

$$Y = C_3\cos(k_2y) + C_4\sin(k_2y) \quad ,$$

where $C_1,\ldots,C_4$ are constants of integration to be determined from the boundary and initial conditions. The condition that $E_z = 0$ at $x = 0$ for all values of z and t requires that

$$C_1\{C_3\cos(k_2y) + C_4\sin(k_2y)\} = 0 \quad .$$

As this is to be true for all values of y also, we must have $C_1 = 0$. The condition that $E_z = 0$ at $y = 0$ then requires that

$$C_2C_3\sin(k_1x) = 0 \quad .$$

Now C_2 cannot vanish without making X and hence E_z vanish identically. Hence $C_3 = 0$. The condition that $E_z = 0$ at $x = a$ gives

$$C_2C_4\sin(k_1a)\sin(k_2y) = 0$$

for all values of y. This requires that

$$k_1a = m\pi \quad ,$$

where $m = 1,2,3,...$ Similarly the condition that $E_z = 0$ at $y = b$ requires

$$k_2b = n\pi \quad ,$$

where $n = 1,2,3,...$ Then writing $C_{m,n}$ for C_2C_4, we find

$$E_z = C_{m,n}\sin\left(\frac{m\pi}{a}x\right)\sin\left(\frac{n\pi}{b}y\right)e^{(\gamma z - i\omega t)} \quad .$$

Note that negative values of the integers m and n would not give rise to anything new as $C_{m,n}$ is arbitrary, to be determined by the intensity and phase of the wave to be transmitted. Furthermore, m and n cannot be zero without making E_z vanish identically also.

The constant h for a given set of the integers m and n is given by Eq. (7.44)

$$h_{m,n}^2 = \left\{\left(\frac{m}{a}\right)^2 + \left(\frac{n}{b}\right)^2\right\}\pi^2 \quad . \tag{7.45}$$

The corresponding propagation constant is then

$$\gamma = \sqrt{h_{m,n}^2 - \mu\varepsilon\omega^2} \quad . \tag{7.46}$$

The condition for free transmission is that γ should be purely imaginary. This requires the angular frequency ω to be such that

$$\omega > vh_{m,n} \equiv \omega_{m,n} \quad , \tag{7.47}$$

where $v = (\mu\varepsilon)^{-\frac{1}{2}}$ is the phase velocity of the wave in the dielectric of the guiding system if it were of infinite extension. $\omega_{m,n}$ is the lowest angular frequency of the wave that may be transmitted by the guide without attenuation and is called the *cutoff* or *critical* (*angular*) *frequency* for the given set of integers m and n.

The integers m and n denote the mode of a wave. If we follow the convention that the x-axis is always chosen to be parallel to the larger transverse dimension of a rectangular waveguide, then the type of a wave is unambiguously specified by symbols like $TM_{m,n}$. The lowest frequency TM waves that can be transmitted are in the ascending order: $TM_{1,1}$, $TM_{2,1}$, $TM_{1,2}$,... The lowest mode $TM_{1,1}$ is called the *dominant mode*. For the propagation of TM waves of a given frequency the only modes that are allowed are those for which the inequality (7.47) is satisfied, i.e. the critical frequency for an allowed mode must be less than the wave frequency.

For the modes that are transmitted the propagation constant γ is purely imaginary. Writing $\gamma = ik'$, where k' is given by

$$k' = \frac{1}{v}\sqrt{\omega^2 - \omega_{m,n}^2} \tag{7.48}$$

the exponent in the expression for E_z can be written as $i(k'z - \omega t)$. It is now obvious that k' is to be interpreted as the magnitude of the propagation vector of the wave inside the guide, which may be called the *guide propagation vector*. It follows that the phase velocity inside the guiding system is

$$v' = \frac{\omega}{k'} = \frac{v}{\sqrt{1 - \left(\frac{\omega_{m,n}}{\omega}\right)^2}} \quad . \tag{7.49}$$

This relation shows that v' is always greater than v, the phase velocity in the unlimited dielectric. For convenience the former will be called simply the guide phase velocity and the latter the phase velocity in the dielectric of the guide. Since frequency remains unchanged, the corresponding wavelengths λ' and λ follow exactly the same relationship. The *cutoff wavelength* $\lambda_{m,n} = 2\pi v/\omega_{m,n}$ is given by

$$\frac{1}{\lambda_{m,n}} = \sqrt{\frac{1}{2}\left\{\left(\frac{m}{a}\right)^2 + \left(\frac{n}{b}\right)^2\right\}}$$

and is of the same or smaller magnitude as the average transverse dimensions of the guide. Thus, only microwaves are transmitted by waveguides of centimeter dimensions. For waves of larger wavelengths the waveguide required would be too large for practical use.

The group velocity with which energy is transmitted is

$$v_g = \frac{d\omega}{dk'} = v\sqrt{1 - \left(\frac{\omega_{m,n}}{\omega}\right)^2} \tag{7.50}$$

and is always less than v. In particular if air is used as the transmission medium, then $v_g < c$, consistent with special relativity.

By means of Eqs. (7.37)-(7.40) the transverse components of the field vectors for the $TM_{m,n}$ mode may be obtained from the expression for E_z. Thus the field components of the $TM_{m,n}$ mode are

$$E_z = C_{m,n}\sin\left(\frac{m\pi}{a}x\right)\sin\left(\frac{n\pi}{b}y\right)e^{i(k'z-\omega t)} \quad ,$$

$$E_x = iC_{m,n}\frac{k'}{h^2_{m,n}}\frac{m\pi}{a}\cos\left(\frac{m\pi}{a}x\right)\sin\left(\frac{n\pi}{b}y\right)e^{i(k'z-\omega t)} \quad ,$$

$$E_y = iC_{m,n}\frac{k'}{h^2_{m,n}}\frac{n\pi}{b}\sin\left(\frac{m\pi}{a}x\right)\cos\left(\frac{n\pi}{b}y\right)e^{i(k'z-\omega t)} \quad ,$$

$$H_z = 0 \quad ,$$

$$H_x = -\frac{\varepsilon w}{k'} E_y \quad ,$$

$$H_y = \frac{\varepsilon w}{k'} E_x \quad . \tag{7.51}$$

The transmission of TE waves may be similarly treated. In this case $E_z = 0$ and the wave equation (7.27) is to be solved for H_z. To express the boundary condition $E_t = 0$ at the conducting walls, we note that, as $E_z = 0$, Eqs. (7.37) and (7.39) lead to the conditions

$$\frac{\partial H_z}{\partial x} = 0 \ , \quad \text{when} \quad E_y = 0 \quad \text{for} \quad x = 0 \quad \text{and} \quad x = a \quad ,$$

$$\frac{\partial H_z}{\partial y} = 0 \ , \quad \text{when} \quad E_x = 0 \quad \text{for} \quad y = 0 \quad \text{and} \quad y = b \quad .$$

Note that these same conditions could have been obtained from Eq. (7.24), which requires that $H_x = 0$ for $x = 0$, $x = a$, and that $H_y = 0$ for $y = 0$, $y = b$, and Eqs. (7.38) and (7.40).

The same procedure then leads to the following configuration for the electromagnetic field of the $TE_{m,n}$ wave in the dielectric of the guide:

$$H_z = D_{m,n} \cos(\frac{m\pi}{a} x) \cos(\frac{n\pi}{b} y) e^{i(k'z - \omega t)} \quad ,$$

$$H_x = -iD_{m,n} \frac{k'}{h^2_{m,n}} \frac{m\pi}{a} \sin(\frac{m\pi}{a} x) \cos(\frac{n\pi}{b} y) e^{i(k'z - \omega t)} \quad ,$$

$$H_y = -iD_{m,n} \frac{k'}{h^2_{m,n}} \frac{n\pi}{b} \cos(\frac{m\pi}{a} x) \sin(\frac{n\pi}{b} y) e^{i(k'z - \omega t)} \quad ,$$

$$E_z = 0 \quad ,$$

$$E_x = \frac{\mu\omega}{k'} H_y \quad ,$$

$$E_y = -\frac{\mu\omega}{k'} H_x \quad , \tag{7.52}$$

where the amplitude $D_{m,n}$ is related to the intensity and phase of the wave being transmitted. The parameters m and n are positive integers as before. However, for the TE modes one but not both of these may take the value zero. The lowest order TE mode in a rectangular guide is therefore $TE_{1,0}$. The condition for free transmission and the expressions for the phase and group velocities are identical with those for the TM modes of the same order.

7.6 Transmission by a Circular Waveguide

Consider a waveguide consisting of a perfectly conducting tube of uniform circular cross section of internal radius a enclosing a perfect dielectric of permittivity ε and permeability μ. It is convenient to use cylindrical coordinates (ρ, ϕ, z) with the z-axis coinciding with the axis of the tube as shown in Fig. 7.5 The wave equation now takes the form

$$\left(\frac{\partial^2}{\partial\rho^2} + \frac{1}{\rho}\frac{\partial}{\partial\rho} + \frac{1}{\rho^2}\frac{\partial^2}{\partial\phi^2} + h^2\right)\mathbf{E} = 0 \quad , \tag{7.53}$$

or

$$\left(\frac{\partial^2}{\partial\rho^2} + \frac{1}{\rho}\frac{\partial}{\partial\rho} + \frac{1}{\rho^2}\frac{\partial^2}{\partial\phi^2} + h^2\right)\mathbf{H} = 0 \quad . \tag{7.54}$$

As

$$\nabla \times \mathbf{E} = \begin{vmatrix} \dfrac{\mathbf{i}_\rho}{\rho} & \mathbf{i}_\phi & \dfrac{\mathbf{i}_z}{\rho} \\ \dfrac{\partial}{\partial\rho} & \dfrac{\partial}{\partial\phi} & \dfrac{\partial}{\partial z} \\ E_\rho & \rho E_\phi & E_z \end{vmatrix}$$

in cylindrical coordinates, Maxwell's Eqs. (7.29) and (7.30) have components

$$\frac{1}{\rho}\frac{\partial E_z}{\partial\phi} - \gamma E_\phi = i\mu\omega H_\rho \quad , \tag{7.55}$$

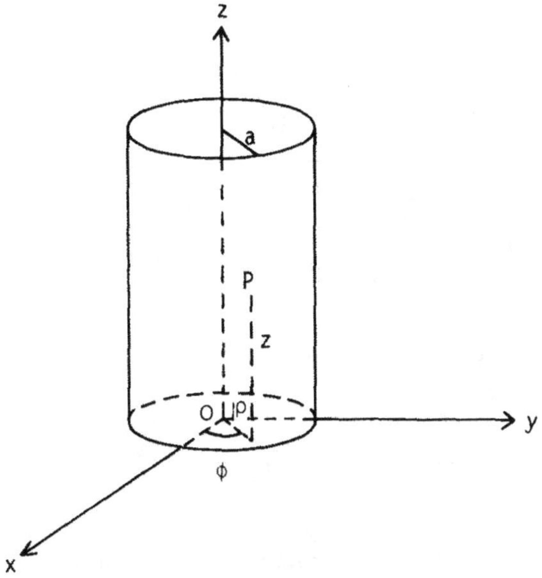

Fig. 7.5 A circular waveguide consisting of a perfectly conducting tube of uniform circular cross section of internal radius a.

$$\gamma E_\rho - \frac{\partial E_z}{\partial \rho} = i\mu\omega H_\phi \quad , \tag{7.56}$$

$$\frac{1}{\rho}\frac{\partial}{\partial \rho}(\rho E_\phi) - \frac{\partial E_\rho}{\rho \partial \phi} = i\mu\omega H_z \quad , \tag{7.57}$$

$$\frac{1}{\rho}\frac{\partial H_z}{\partial \phi} - \gamma H_\phi = - i\varepsilon\omega E_\rho \quad , \tag{7.58}$$

$$\gamma H_\rho - \frac{\partial H_z}{\partial \rho} = - i\varepsilon\omega E_\phi \quad , \tag{7.59}$$

$$\frac{1}{\rho}\frac{\partial}{\partial \rho}(\rho H_\phi) - \frac{1}{\rho}\frac{\partial H_\rho}{\partial \phi} = - i\varepsilon\omega E_z \quad . \tag{7.60}$$

Eliminating first H_ρ and then E_ϕ from Eqs. (7.55) and (7.59) we find the guide equations

$$E_\phi = \frac{1}{h^2}\left(\frac{\gamma}{\rho}\frac{\partial E_z}{\partial \phi} - i\mu\omega\frac{\partial H_z}{\partial \rho}\right) \quad , \tag{7.61}$$

$$H_\rho = \frac{1}{h^2}\left(\gamma\frac{\partial H_z}{\partial \rho} - \frac{i\varepsilon\omega}{\rho}\frac{\partial E_z}{\partial \phi}\right) \quad . \tag{7.62}$$

Similarly Eqs. (7.56) and (7.58) give

$$E_\rho = \frac{1}{h^2}\left(\gamma\frac{\partial E_z}{\partial \rho} + \frac{i\mu\omega}{\rho}\frac{\partial H_z}{\partial \phi}\right) \quad , \tag{7.63}$$

$$H_\phi = \frac{1}{h^2}\left(\frac{\gamma}{\rho}\frac{\partial H_z}{\partial \phi} + i\varepsilon\omega\frac{\partial E_z}{\partial \rho}\right) \quad . \tag{7.64}$$

As shown before, for transmission to occur, **E** and **H** cannot both be transverse.

Consider first the TM waves for which $H_z = 0$. Let

$$E_z = P(\rho)Q(\phi)e^{(\gamma z - i\omega t)} \quad .$$

Substitution in Eq. (7.53) gives

$$\frac{1}{P}\frac{d^2P}{d\rho^2} + \frac{1}{\rho P}\frac{dP}{d\rho} + \frac{1}{\rho^2 Q}\frac{d^2Q}{d\phi^2} + h^2 = 0 \quad .$$

This can be separated into two ordinary differential equations:

$$\frac{d^2Q}{d\phi^2} + m^2Q = 0 \quad , \tag{7.65}$$

$$\frac{1}{h^2}\frac{d^2P}{d\rho^2} + \frac{1}{h^2\rho}\frac{dP}{d\rho} + \left(1 - \frac{m^2}{h^2\rho^2}\right)P = 0 \quad , \tag{7.66}$$

where m is a constant.

The general solution of Eq. (7.65) is

$$Q(\phi) = A_m \cos m\phi + B_m \sin m\phi \quad ,$$

where A_m and B_m are arbitrary constants. For E_z and hence Q to be single-valued, in other words, for

$$Q(\phi) = Q(\phi + 2\pi) \quad ,$$

we require m to be either zero or a positive integer. Note that negative integral values of m will not give rise to anything new since A_m and B_m are arbitrary. $Q(\phi)$ can also be written in terms of a different set of arbitrary constants D_m and δ_m in the form

$$Q(\phi) = D_m \cos(m\phi + \delta_m) \quad .$$

For a given m the phase constant δ_m can be made to vanish by suitably redefining the ϕ-coordinate. In general however δ_m will not vanish for all m for a given set of coordinates.

Equation (7.66) can be written as a Bessel equation for a positive real variable $h\rho$ of integral order m and has the general solution

$$P(h\rho) = A'_m \mathcal{J}_m(h\rho) + B'_m \mathcal{Y}_m(h\rho) \quad ,$$

where $\mathcal{J}_m$ and $\mathcal{Y}_m$ are Bessel's functions of the first and the second kind respectively, A'_m and B'_m are arbitrary constants. Both functions are real for real positive arguments. For $h\rho \to 0$,

$$\mathcal{Y}_m(h\rho) \sim -\frac{(m-1)!}{\pi}\left(\frac{2}{h\rho}\right)^m$$

and becomes infinite for $h\rho = 0$. For a hollow waveguide the intensity must be finite at the axis and we must therefore put $B'_m = 0$. The physically acceptable solution of the wave equation is therefore

$$E_z = C_m \mathcal{J}_m(h\rho)\cos(m\phi + \delta_m)e^{(\gamma z - i\omega t)} \quad , \tag{7.67}$$

where m is zero or a positive integer and $C_m = D_m A'_m$ is a constant related to the intensity and phase of the transmitted wave.

The boundary condition that $E_t = 0$ at the surface of the conducting walls requires that $E_z = 0$ at $\rho = a$, whence

$$\mathcal{J}_m(ha) = 0 \quad . \tag{7.68}$$

The Bessel function $\mathcal{J}_m(\xi)$ of a real positive variable ξ is plotted in Fig. 7.6 for small values of ξ for $m = 0$, 1 and 2. For each given order, Eq. (7.68) has an infinite number of real positive roots, corresponding to an infinite number of possible TM-waves. However, for free transmission we require γ to be purely imaginary or

$$\gamma^2 = h^2 - \mu\varepsilon\omega^2 < 0 \quad ,$$

i.e.

$$h < \frac{\omega}{v} \quad . \tag{7.69}$$

This condition restricts h to those given by the smallest nonzero roots. Zero roots would make $h = 0$, which according to the argument presented in Sec. 7.5 would require all field components to vanish inside a hollow waveguide. If we number the positive roots of $\mathcal{J}_m(ha) = 0$ in ascending order from the smallest one and denote each's place by an integer n, the corresponding h may be specified by the suffixes m and n. The first few roots for the lowest order Bessel functions, $h_{m,n}a$, are given in Table 7.1. Others may be found in standard mathematical tables. For a larger order m, the n-th root is approximately $(n + \frac{m}{2} - \frac{1}{4})\pi$. This approximation is better than 10% for $m \geq 2$. The corresponding freely transmitted modes may be similarly specified as $TM_{m,n}$: $TM_{0,1}$, $TM_{1,1}$, $TM_{2,1}$,... in the ascending order of the cutoff frequency.

The magnitude of the propagation vector of the $TM_{m,n}$ wave is

$$k' = \sqrt{\mu\varepsilon\omega^2 - h_{m,n}^2} = \frac{1}{v}\sqrt{\omega^2 - \omega_{m,n}^2} \quad , \tag{7.70}$$

Table 7.1 Lowest nonzero roots of $\mathcal{J}_m(\xi) = 0$ and $\mathcal{J}'_m(\xi) = 0$.

	ξ for $\mathcal{J}_m(\xi) = 0$			ξ for $\mathcal{J}'_m(\xi) = 0$		
n =	1	2	3	1	2	3
m = 0	2.4048	5.5201	8.6537	3.8317	7.0156	10.1735
1	3.8317	7.0126	10.1735	1.8412	5.3314	8.5363
2	5.1356	8.4172	11.6198	3.0542	6.7061	9.9695
3	6.3802	9.7610	13.0152	4.2012	8.0152	11.3459

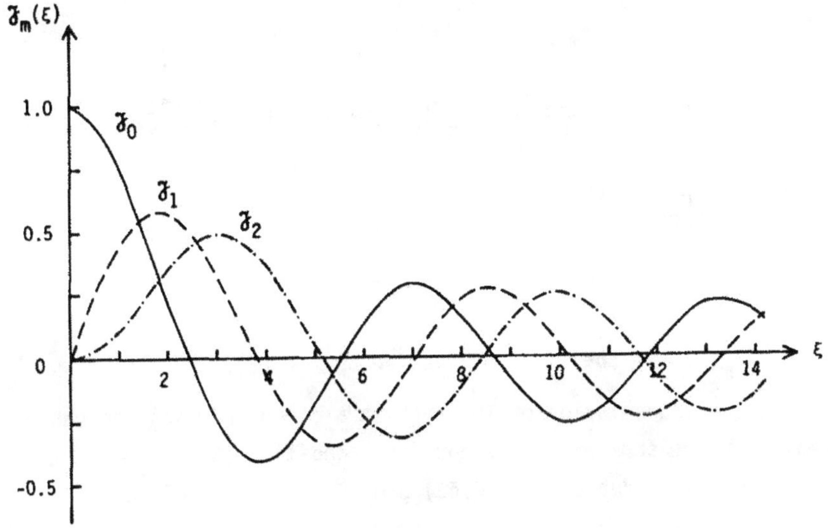

Fig. 7.6 Bessel functions of a real variable ξ: $\mathcal{J}_0(\xi)$, $\mathcal{J}_1(\xi)$ and $\mathcal{J}_2(\xi)$. Each function is oscillatory with an attenuating amplitude. For any order m there is an infinite number of positive roots for the equation $\mathcal{J}_m(\xi) = 0$.

where $\omega_{m,n} = vh_{m,n}$ is the cutoff angular frequency below which the transmission of the $TM_{m,n}$ mode ceases. Thus, expressed in terms of the cutoff frequency $\omega_{m,n}$ and the phase velocity v in the unrestricted dielectric, k' and hence the expressions for the phase and group velocities in the circular guide are the same as for waves in a rectangular guide.

The remaining components of the field vectors for the $TM_{m,n}$ mode are readily calculated by means of Eqs. (7.61)-(7.64). Thus the field components for this mode are

$$E_z = C_{m,n} \mathcal{J}_m(h_{m,n}\rho)\cos(m\phi + \delta_m)e^{i(k'z - \omega t)} \quad ,$$

$$E_\rho = i \frac{C_{m,n}k'}{h_{m,n}} \mathcal{J}'_m(h_{m,n}\rho)\cos(m\phi + \delta_m)e^{i(k'z - \omega t)} \quad ,$$

$$E_\phi = - i \frac{C_{m,n}k'm}{h_{m,n}\rho} \mathcal{J}_m(h_{m,n}\rho)\sin(m\phi + \delta_m)e^{i(k'z - \omega t)} \quad ,$$

$$H_\rho = - \frac{\varepsilon\omega}{k'} E_\phi \quad ,$$

$$H_\phi = \frac{\varepsilon\omega}{k'} E_\rho \quad ,$$

where $\mathcal{J}'_m(\xi)$ is the first derivative of $\mathcal{J}_m(\xi)$.

The transmission of TE waves in a circular waveguide can be treated in the same way. The boundary condition that $E_\phi = 0$ at $\rho = a$ now gives, by virtue of Eq. (7.61) and $E_z = 0$, the following

$$\frac{\partial H_z}{\partial \rho} = 0 \quad , \qquad \text{at} \qquad \rho = a \quad .$$

The solution of the wave equation for H_z may be expressed as

$$H_z = D_m \mathcal{J}_m(h\rho)\cos(m\phi + \delta_m)e^{(\gamma z - i\omega t)} \quad .$$

The boundary condition then requires h to be given by the positive roots of

$$\mathcal{J}'_m(ha) = 0 \quad . \tag{7.71}$$

The smallest root for $m=0$ is zero. As this is of no physical interest we shall start numbering the positive roots from the smallest non-zero root and call the number n. The smallest roots of Eq. (7.71) for several orders are given in Table 7.1. Others may be found from standard tables. For large orders, the n-th root is approximately $(n+\frac{m}{2}-\frac{3}{4})\pi$. A TE mode is then specified by the integers m and n.

The components of the electromagnetic field for the $TE_{m,n}$ mode are

$$H_z = D_{m,n}\mathcal{J}_m(h_{m,n}\rho)\cos(m\phi+\delta_m)e^{i(k'z-\omega t)}$$

$$H_\rho = i\frac{D_{m,n}k'}{h_{m,n}}\mathcal{J}'_m(h_{m,n}\rho)\cos(m\phi+\delta_m)e^{i(k'z-\omega t)}$$

$$H_\phi = -i\frac{D_{m,n}k'm}{h^2_{m,n}\rho}\mathcal{J}_m(h_{m,n}\rho)\sin(m\phi+\delta_m)e^{i(k'z-\omega t)}$$

$$E_z = 0 \quad ,$$

$$E_\rho = \frac{\mu\omega}{k'}H_\phi \quad ,$$

$$E_\phi = -\frac{\mu\omega}{k'}H_\rho \quad ,$$

where k' is the magnitude of the guide propagation vector and is given by the same equation as for the $TM_{m,n}$ mode.

In the above we have analysed the propagation of electromagnetic waves in simple guiding systems, but avoided the important questions of excitation and reception of the guided waves. Briefly speaking, three methods are commonly employed for the excitation of a particular wave in a guide. At one end of the waveguide a probe or a straight-wire antenna may be introduced oriented parallel to the electric field, or a loop may be inserted oriented in a plane normal to the magnetic field of the desired mode, in each case near the place of maximum intensity. The guiding system may also be coupled to the exciting field, such as the relatively large magnetic field that exists in the vicinity of the short-circuited terminal of a coaxial line, through an iris or slit in such a way that there is some common field component between the desired mode and the exciting source. In the third method, currents are introduced from radio frequency sources in such a manner that currents flow in the guide walls in directions which will induce the desired fields inside the guide. In general, a multitude of modes will be excited by the source and more than one can be transmitted if the guide is large enough. Frequently, however, a guide is designed so that only one mode is above the cutoff. This is usually desirable since it is difficult to design a system that has a high efficiency simultaneously for several modes. For instance, for frequencies between cutoff and twice the cutoff or (a/b) times cutoff, whichever is the smaller, TE_{10} is the only propagating mode for a rectangular waveguide. The receiving process is exactly the reverse and the same methods may be applied.

7.7 Power Transfer and Attenuation in Waveguides

The average rate of energy flow or the power transfer in a waveguide is the integral of the time-averaged Poynting vector as given by Eq. (4.68) over the area A of a cross section:

$$\langle P \rangle = \frac{1}{2}\int_A \mathrm{Re}(\mathbf{E}\times\mathbf{H}^*)\cdot d\mathbf{S} = \frac{1}{2}\int_A (\mathbf{E}\times\mathbf{H}^*)_z dS \quad . \qquad (7.72)$$

The transverse components of $\mathbf{E}\times\mathbf{H}^*$ are purely imaginary and do not contribute to the integral. Using Maxwell's Eqs. (7.31) through (7.36) or (7.55) through (7.60) we may write the above as

$$\langle P \rangle = \frac{1}{2} Z \int_A \mathbf{H}_\perp \cdot \mathbf{H}_\perp^* \, dS \quad , \qquad (7.73)$$

where $\mathbf{H}_\perp$ is the transverse component of $\mathbf{H}$ and Z is the *guide impedance* having the values

$$Z_{TM} = \frac{k'}{\varepsilon\omega} = \sqrt{\frac{\mu}{\varepsilon}}\sqrt{1 - \left(\frac{\omega_{m,n}}{\omega}\right)^2} \tag{7.74}$$

for TM waves and

$$Z_{TE} = \frac{\mu\omega}{k'} = \sqrt{\frac{\mu}{\varepsilon}}\frac{1}{\sqrt{1 - \left(\frac{\omega_{m,n}}{\omega}\right)^2}} \tag{7.75}$$

for TE waves. Note that the guide impedance is imaginary or purely reactive for frequencies below the cutoff so that no average power transfer can take place for such frequencies. For large frequencies the impedances approach the intrinsic impedance of the dielectric.

So far we have assumed the walls of the guide to be perfectly conducting and the medium of propagation to be a perfect dielectric so that no loss is incurred in transmission provided the frequency is above cutoff. In a real situation these ideal conditions cannot be realized and ohmic losses occur. Provided however that the losses are small, as they should be if the guiding system is of any practical interest at all, the condition for propagation, the phase and group velocities, and the field configuration are to a good approximation the same as those obtained for the ideal waveguide.

Consider first the attenuation due to the finite resistance of the conducting walls. The induced linear current density $\mathbf{I}_\ell$ is given approximately by the boundary condition (7.22)

$$\mathbf{I}_\ell = \mathbf{n} \times \mathbf{H} \quad .$$

Since the reactive part of an impedance plays no role in energy dissipation, only the resistive part needs to be considered. The resistance per unit square of the conducting wall is given by Eq. (7.20)

$$R_s = \frac{1}{\sigma\delta}\sqrt{\frac{\mu\omega}{2\sigma}} \quad .$$

The average Joule heat loss per unit time per unit length of the guide is therefore

$$-\frac{d\langle P\rangle}{dz} = R_s \oint_C \langle (\mathrm{Re}\,\mathbf{I}_\ell)^2 \rangle \, d\ell \quad ,$$

where the averaging is to be taken over one cycle of oscillation and the integration carried out over the periphery C of a cross section. Although the expression for R_s has been obtained for a semi-infinite plane conductor it may be applied to a conductor of any shape if the radius of curvature at the surface is everywhere much greater than the skin depth. In the case of a rectangular waveguide, the sharp corners constitute only a small fraction of the total periphery so that any error introduced by applying plane conductor approximation is not serious.

We have

$$\begin{aligned}\langle (\mathrm{Re}\,\mathbf{I}_\ell)^2 \rangle &= \langle \frac{1}{4} (\mathbf{I}_\ell + \mathbf{I}_\ell^*)^2 \rangle \\ &= \frac{1}{4} (\langle I_\ell^2 \rangle + \langle I_\ell^{*2} \rangle + 2 \langle \mathbf{I}_\ell \cdot \mathbf{I}_\ell^* \rangle) \\ &= \frac{1}{2} \mathbf{I}_\ell \cdot \mathbf{I}_\ell^* \quad ,\end{aligned}$$

as I_ℓ^2 and I_ℓ^{*2}, which contain the time factors $\exp(\mp 2i\omega t)$, vanish on averaging. Furthermore, Eq. (7.24) shows that $\mathbf{H}$ is tangential at the surface of a conductor so that $\mathbf{I}_\ell \cdot \mathbf{I}_\ell^* = \mathbf{H} \cdot \mathbf{H}^*$. The power dissipated per unit length of the guide is therefore

$$-\frac{d\langle P\rangle}{dz} = \frac{1}{2\sigma\delta} \oint_C \mathbf{H} \cdot \mathbf{H}^* d\ell \quad . \tag{7.76}$$

At any point along the length of the waveguide specified by z, the power transfer is given by Eq. (7.73) provided dissipation is small. Then as the x and y coordinates are eliminated by integration, both $\langle P\rangle$ and $-\frac{d\langle P\rangle}{dz}$ will be proportional to the square of the amplitude of H_z and hence to each other. Thus we may write

$$-\frac{d\langle P\rangle}{dz} = 2\alpha_c \langle P\rangle \quad ,$$

or

$$\langle P(z)\rangle = \langle P(0)\rangle e^{-2\alpha_c z} \quad , \tag{7.77}$$

where α_c is a constant depending on the frequency and mode of the wave and the material and geometry of the guide and is called the *attenuation constant* for losses due to imperfectly conducting walls.

As mentioned before, for small losses the field vector to be used in Eq. (7.76) may be approximated by that for an ideal waveguide. Then as the normal component of $\mathbf{H}$ vanishes on the conducting walls, the power dissipated per unit length may be written for a rectangular guide as follows

$$-\frac{d\langle P\rangle}{dz} = \frac{1}{\sigma\delta}\int_0^a (H_x H_x^* + H_z H_z^*)_{y=0}dx + \frac{1}{\sigma\delta}\int_0^b (H_y H_y^* + H_z H_z^*)_{x=0}dy$$

for direct use in a computation.

Consider next the attenuation due to the finite conductivity of the dielectric. We have seen in Sec. 5.4 that for a medium with a conductivity σ the equivalent permittivity to be used in the propagation wave equation is the complex quantity $\varepsilon(1+\frac{i\sigma}{\omega\varepsilon})$. The propagation constant is then complex and is given by

$$\gamma^2 = h_{m,n}^2 - \mu\varepsilon\omega^2 - i\mu\sigma\omega \quad . \tag{7.78}$$

As we are only interested in a dielectric with a small conductivity, the imaginary part of γ^2 will have a magnitude much smaller than that of the real part, and the latter is negative for essentially loss-free transmission. We may therefore write

$$\gamma = i(k' + i\alpha_d) \quad . \tag{7.79}$$

A comparison with Eq. (7.78) gives

$$k'^2 - \alpha_d^2 = \mu\varepsilon\omega^2 - h_{m,n}^2$$

and

$$2k'\alpha_d = \mu\sigma\omega \quad .$$

Under the condition of small conductivity, $\alpha_d \ll k'$ and we have the approximate solution

$$k' \simeq \sqrt{\mu\varepsilon\omega^2 - h_{m,n}^2} = \frac{\omega}{v}\sqrt{1 - \left(\frac{\omega_{m,n}}{\omega}\right)^2} \quad ,$$

which is the same as for a perfect waveguide, and

$$\alpha_d = \frac{\mu\sigma\omega}{2k} \simeq \frac{\sigma}{2}\sqrt{\frac{\mu}{\varepsilon}}\,\frac{1}{\sqrt{1 - \left(\frac{\omega_{m,n}}{\omega}\right)^2}} \quad . \tag{7.80}$$

The presence of a real part in γ gives rise to an attenuation factor $\exp(-\alpha_d z)$ in the amplitude of the transmitted wave. α_d is therefore the attenuation constant for the dielectric loss. Note that α_d is greatest near the cutoff frequency and decreases to a minimum of $\frac{\sigma}{2}\sqrt{\frac{\mu}{\varepsilon}}$ at large frequencies.

In general both types of attenuation will occur in a real waveguide. If these are comparable in magnitude the total attenuation constant is their sum $\alpha_c + \alpha_d$.

7.8 Waveguides as Cavity Resonators

If a finite length of a waveguide is closed at both ends with plane conductors perpendicular to the axis, the waves incident on the end surfaces will be reflected back in accordance with the boundary conditions Eqs. (7.21) and (7.24). The superposition of the forward and backward waves may cause a system of standing waves to be set up in the space or cavity bounded by the conducting walls, similar to the standing waves set up on a vibrating string. If the dielectric of the cavity and the conductor of the walls are nearly perfect, an oscillating electromagnetic field may be maintained in the cavity with very little loss. Such a device is called a *cavity resonator* or a *rhumbatron* and may be used in place of the conventional LC-resonant circuit.

At high frequencies, e.g. microwave frequencies, a conventional circuit with dimensions of the same or larger order of magnitude as the wavelength will lose energy by radiation. Furthermore the skin effect will greatly increase the effective resistance of an ordinary conducting wire. These effects make ordinary circuit elements inefficient at high frequencies. To prevent the radiation leakage we might surround the circuit region with a good conductor. To minimize the skin effect the current paths may be extended to as large an area as possible. Both these requirements are satisfied by a cavity resonator. Although power is still being dissipated in a real cavity, compared to that in a conventional LC-circuit the loss may be reduced by a factor of as much as twenty.

A cavity resonator may be excited in the same way as the excitation of a waveguide described earlier. If the frequency of the impressed signal differs appreciably from any of the resonant frequencies of the resonator, the electromagnetic field in the resonator will be extremely small. The field increases as a resonant frequency is approached. When the exciting frequency is equal to a resonant frequency, pronounced electromagnetic oscillation takes place setting up strong standing electromagnetic waves in the resonator. The various modes and frequencies that may be generated in a cavity depends on the geometry and the method of excitation. Exact analysis of the electromagnetic field configuration in a cavity is possible only for certain simpler geometrical types. As the first example we shall consider a rectangular cavity of width a, height b and length ℓ, and choose a Cartesian system as shown in Fig. 7.4 with the end surfaces at $z=0$ and $z=\ell$.

Suppose the wave generated is sinusoidal in time with angular frequency ω, then the wave equations for **E** and **H** are

$$\nabla^2\mathbf{E} + k^2\mathbf{E} = 0 \quad , \tag{7.81}$$

and

$$\nabla^2\mathbf{H} + k^2\mathbf{H} = 0 \quad . \tag{7.82}$$

where $k^2 = \mu\varepsilon\omega^2$.

Consider the x-component of Eq. (7.81). By writing the solution as $E_x = X_1(x)Y_1(y)Z_1(z)e^{-i\omega t}$, it may be separated into three ordinary differential equations:

$$\frac{d^2X_1}{dx^2} + k_1^2X_1 = 0 \quad ,$$

$$\frac{d^2Y_1}{dy^2} + k_2^2Y_1 = 0 \quad ,$$

$$\frac{d^2Z_1}{dz^2} + k_3^2Z_1 = 0 \quad ,$$

where

$$k_1^2 + k_2^2 + k_2^2 = k^2 = \mu\varepsilon\omega^2 \quad . \tag{7.83}$$

The first equation for example has the general solution $X_1(x) = A_1\cos(k_1x) + B_1\sin(k_1x)$, where A_1 and B_1 are constants. The boundary condition to be applied is $E_x = 0$ for $y = 0, b$, and for $z = 0, \ell$, which requires the solution for E_x to be

$$E_x = X_1(x)\sin(k_2y)\sin(k_3z)e^{-i\omega t} \quad ,$$

where $k_2b = n\pi$, $k_3\ell = p\pi$, n and p being positive integers. Negative values will only introduce a change of sign.

Similarly, considering the remaining component equations of Eq. (7.81) and taking account of the boundary conditions that $E_y = 0$ for $x = 0, a$ and for $z = 0, \ell$, and that $E_z = 0$ for $x = 0, a$, and for $y = 0, b,$ we find that

$$E_y = Y_2(y)\sin(k_1'x)\sin(k_3'z)e^{-i\omega t}$$

and

$$E_z = Z_3(z)\sin(k_1''x)\sin(k_2''y)e^{-i\omega t} \quad ,$$

where $k_1'a = m\pi$, $k_1''a = m'\pi$, $k_2''b = n'\pi$, $k_3'\ell = p'\pi$, the parameters m, m', n', p' being positive integers.

The components of $\mathbf{E}$ are not independent but are related by $\nabla \cdot \mathbf{E} = 0$, which holds since no charge is present. The divergence condition requires that

$$\frac{dX_1}{dx}\sin(k_2y)\sin(k_3z) + \frac{dY_2}{dy}\sin(k_1'x)\sin(k_3'z) + \frac{dZ_3}{dz}\sin(k_1''x)\sin(k_2''y)$$
$$= 0 \quad .$$

This equation may be satisfied if

$$k_1'' = k_1' = k_1 = \frac{m\pi}{a} \quad , \quad k_2'' = k_2 = \frac{n\pi}{b} \quad ,$$

$$k_3' = k_3 = \frac{p\pi}{\ell} \tag{7.84}$$

and

$$X_1 = E_1\cos(k_1x) \quad , \quad Y_2 = E_2\cos(k_2y) \quad ,$$

$$Z_3 = E_3\cos(k_3z) \quad ,$$

where the amplitudes E_1, E_2 and E_3 are related by

$$k_1E_1 + k_2E_2 + k_3E_3 = 0 \quad . \tag{7.85}$$

Equation (7.83) then gives

$$\omega^2 = \frac{\pi^2}{\mu\varepsilon}\left(\frac{m^2}{a^2} + \frac{n^2}{b^2} + \frac{p^2}{\ell^2}\right) \quad , \tag{7.86}$$

which specifies the resonance frequencies of the cavity. Note that at most one of the integers m, n, p may take the value zero without making all the components vanish identically. If one of the integers vanishes, the electric field **E** is parallel to one axis of symmetry. Each set of the integers m, n, p defines a mode of oscillation with resonance frequency $\omega_{m,n,p}$ given by Eq. (7.86). The total field in the cavity is in general a sum of all the allowed modes, with relative amplitudes depending on the method of excitation.

The components of the magnetic vector for an m, n, p-mode may be obtained from Maxwell's equation $i\mu\omega\mathbf{H}=\nabla\times\mathbf{E}$. Thus the components of the electromagnetic field of the m, n, p-mode are given by

$$E_x = E_1\cos(k_1x)\sin(k_2y)\sin(k_3z)e^{-i\omega t} \quad ,$$

$$E_y = E_2\sin(k_1x)\cos(k_2y)\sin(k_3z)e^{-i\omega t} \quad ,$$

$$E_z = E_3\sin(k_1x)\sin(k_2y)\cos(k_3z)e^{-i\omega t} \quad ,$$

$$H_x = -\frac{i}{\mu\omega}(k_2E_3 - k_3E_2)\sin(k_1x)\cos(k_2y)\cos(k_3z)e^{-i\omega t} \quad ,$$

$$H_y = -\frac{i}{\mu\omega}(k_3E_1 - k_1E_3)\cos(k_1x)\sin(k_2y)\cos(k_3z)e^{-i\omega t} \quad ,$$

$$H_z = -\frac{i}{\mu\omega}(k_1E_2 - k_2E_1)\cos(k_1x)\cos(k_2y)\sin(k_3z)e^{-i\omega t} \quad ,$$

(7.87)

where k_1, k_2, k_3 are given in Eq. (7.84), and ω is given by Eq. (7.86).

Along any axis of symmetry of the cavity, the electric and magnetic fields cannot both be transverse. For example, if in Eq. (7.87) both E_z and H_z vanish, then, as $E_3 = 0$ and $E_2 \propto E_1$, Eq. (7.85) will require that E_1, and hence **E** vanishes also. Hence along any one axis, the standing waves may be decomposed into TE or TM waves or both, but not TEM waves. It should be noted however that such a description of the wave pattern is not unique, for a TE wave along one axis of symmetry may be considered TM along another. The nomenclature becomes unique if we take the coordinate axes as shown in Fig. 7.4 with the dimensions $\ell > a > b$ and the convention that TE and TM refer to the z-axis always. The various modes may then be denoted uniquely $TE_{m,n,p}$ or $TM_{m,n,p}$. Then as Eq. (7.86) applies to both modes it is obvious that TE and TM modes of the same order m, n, p have the same frequency. Modes of the same resonance frequency but different field configurations are said to *degenerate*. Degeneracy also occurs if two or all of the dimensions of a rectangular cavity are identical.

As noted previously the standing wave patterns may also be obtained by the superposition of the fields of the incident and reflected travelling waves along an axis of symmetry. For example, the $TM_{m,n,p}$ mode may be found by adding the components of the oppositely travelling $TM_{m,n}$ waves which have propagation vectors $\mathbf{k}'$ and $-\mathbf{k}'$, with the amplitudes $C^+_{m,n}$ and $C^-_{m,n}$ chosen to make $E_t = 0$ and $H_n = 0$ at the conducting walls. We shall consider in this manner the oscillations in an ideal right circular cylinder of radius a and length ℓ. Using the cylindrical coordinates shown in Fig. 7.5, we see that the boundary conditions at the end surfaces $z = 0$ and $z = \ell$ are $E_\rho = E_\phi = 0$ and $H_z = 0$.

From the analysis of a cylindrical waveguide, it is seen that for the $TM_{m,n}$ mode the total electric field has the component

$$E_\rho = \frac{i}{h_{m,n}} \mathcal{J}'_m(h_{m,n}\rho)\cos(m\phi + \delta_m)(k'C^+_{m,n}e^{ik'z} - k'C^-_{m,n}e^{-ik'z})e^{-i\omega t} .$$

The condition that $E_\rho = 0$ for $z = 0$ requires that $C^+_{m,n} = C^-_{m,n} \equiv C_{m,n}$, say, giving

$$E_\rho = -2\frac{C_{m,n}k'}{h_{m,n}} \mathcal{J}'_m(h_{m,n}\rho)\cos(m\phi + \delta_m)\sin(k'z)e^{-i\omega t} .$$

The remaining components are obtained in the same way to be

$$E_z = 2C_{m,n}\mathcal{J}_m(h_{m,n}\rho)\cos(m\phi+\delta_m)\cos(k'z)e^{-i\omega t} \quad ,$$

$$E_\phi = 2\frac{C_{m,n}k'm}{h^2_{m,n}\rho}\mathcal{J}_m(h_{m,n}\rho)\sin(m\phi+\delta_m)\sin(k'z)e^{-i\omega t} \quad ,$$

$$H_z = 0 \quad ,$$

$$H_\rho = i2C_{m,n}\frac{\varepsilon\omega m}{h^2_{m,n}\rho}\mathcal{J}_m(h_{m,n}\rho)\sin(m\phi+\delta_m)\cos(k'z)e^{-i\omega t} \quad ,$$

$$H_\phi = i2C_{m,n}\frac{\varepsilon\omega}{h_{m,n}}\mathcal{J}'_m(h_{m,n}\rho)\cos(m\phi+\delta_m)\cos(k'z)e^{-i\omega t} \quad . \qquad (7.88)$$

The resonance frequency is given by the condition $E_\rho = 0$ for $z = \ell$. This requires that

$$k'\ell = p\pi \quad ,$$

where $p = 0,1,2,\ldots$ As the magnitude of the propagation vector is given by Eq. (7.70), the resonance frequency $\omega_{m,n,p}$ may be found from

$$\mu\varepsilon\omega^2_{m,n,p} = h^2_{m,n} + \left(\frac{p\pi}{\ell}\right)^2 \quad , \qquad (7.89)$$

where $h_{m,n}$ is given by the root of Eq. (7.68).

The condition that $\nabla\cdot\mathbf{E} = 0$ is also satisfied, since on substitution of (7.88) it may be reduced to

$$\mathcal{J}''_m(h_{m,n}\rho) + \frac{1}{h_{m,n}\rho}\mathcal{J}'_m(h_{m,n}\rho) + \left(1-\frac{m^2}{h^2_{m,n}\rho^2}\right)\mathcal{J}_m(h_{m,n}\rho) = 0$$

which is simply *Bessel's equation* of order m.

The expressions for the field components of the TE mode oscillations can be similarly obtained. The resonance frequencies $\omega_{m,n,p}$ are given by the same Eq. (7.89) with $h_{m,n}$ given by the roots of Eq. (7.71).

7.9 Q of a Cavity Resonator

The energy stored in a cavity resonator is the volume integral of the sum of the energy densities:

$$U_e = \frac{\varepsilon}{2}(\mathrm{Re}\mathbf{E})^2 = \frac{\varepsilon}{8}(E^2 + 2\mathbf{E}\cdot\mathbf{E}^* + E^{*2}) \quad ,$$

$$U_m = \frac{\mu}{2}(\mathrm{Re}\mathbf{H})^2 = \frac{\mu}{8}(H^2 + 2\mathbf{H}\cdot\mathbf{H}^* + H^{*2}) \quad .$$

As **E** and **H** are related by Maxwell's equations (7.29) and (7.30), we may transform the volume integral as follows:

$$\begin{aligned}\int_V (\varepsilon E^2 + \mu H^2)dV &= \frac{i}{\omega}\int_V (\mathbf{E}\cdot\nabla\times\mathbf{H} - \mathbf{H}\cdot\nabla\times\mathbf{E})dV \\ &= \frac{i}{\omega}\int_V \nabla\cdot(\mathbf{H}\times\mathbf{E})dV \\ &= \frac{i}{\omega}\oint_S \mathbf{H}\times\mathbf{E}\cdot d\mathbf{S} \\ &= \frac{i}{\omega}\oint_S \mathbf{H}\cdot\mathbf{E}\times d\mathbf{S} \quad ,\end{aligned}$$

where V and S are respectively the volume and surface of the cavity. Then as **E** is normal to the conducting boundary the above integral vanishes. The same conclusion holds for the integral of $\varepsilon E^{*2} + \mu H^{*2}$. The total amount of energy stored in the cavity is therefore

$$W = \int_V \left(\frac{1}{4}\varepsilon\mathbf{E}\cdot\mathbf{E}^* + \frac{1}{4}\mu\mathbf{H}\cdot\mathbf{H}^*\right)dV \quad .$$

The integral is now free of any time factor, showing that the energy stored in an ideal cavity is a constant. Furthermore, the two terms of the integrand represent the time-averaged electric and magnetic energy densities as might be expected.

The energy stored in a cavity is on the average equally shared between the electric and the magnetic field. This can be seen by considering the difference

$$<W_e> - <W_m> = \frac{1}{4}\int_V \varepsilon\mathbf{E}\cdot\mathbf{E}^*dV - \frac{1}{4}\int_V \mu\mathbf{H}\cdot\mathbf{H}^*dV$$

$$= \frac{i}{4\omega}\int_V (\mathbf{E}^*\cdot\nabla\times\mathbf{H} - \mathbf{H}\cdot\nabla\times\mathbf{E}^*)dV$$

$$= \frac{i}{4\omega}\int_V \nabla\cdot(\mathbf{H}\times\mathbf{E}^*)dV$$

$$= \frac{i}{4\omega}\oint_S \mathbf{H}\times\mathbf{E}^*\cdot d\mathbf{S} \quad ,$$

which vanishes for the reason mentioned above. We may therefore write

$$W = \frac{\varepsilon}{2}\int_V \mathbf{E}\cdot\mathbf{E}^*dV = \frac{\mu}{2}\int_V \mathbf{H}\cdot\mathbf{H}^*dV \quad . \qquad (7.90)$$

In a real cavity resonator, energy is dissipated in the conducting walls as well as in the dielectric of the cavity. This power loss has the effect of smearing out the sharp resonance frequency expected for an ideal resonator. As a result, a real cavity may respond appreciably to excitations of frequencies within a narrow but finite range about the resonance frequency. A measure of the selectivity of a resonant circuit is given by the *quality factor* or Q of the circuit defined as

$$Q = \frac{2\pi \times \text{total energy stored}}{\text{energy dissipated per cycle}} = \frac{\omega_0 W}{<P>} \quad , \qquad (7.91)$$

where ω_0 is the resonance angular frequency under loss-free conditions and $<P>$ the average power loss. If the power loss is due mainly to dissipation in the conducting walls, $<P>$ may be found by an extension of Eq. (7.76):

$$P = \frac{1}{2\sigma\delta}\oint_S \mathbf{H}\cdot\mathbf{H}^*dS \quad , \qquad (7.92)$$

integrating over the entire surface of the cavity. If ohmic loss in the dielectric is appreciable, it must also be included in $\langle P \rangle$.

Since by energy conservation

$$-\frac{dW}{dt} = \langle P \rangle = \frac{\omega_0}{Q} W \quad , \tag{7.93}$$

and Q is approximately a constant if energy dissipation is small, W decays exponentially and the field amplitudes have an attenuation constant

$$\alpha = \frac{\omega_0}{2Q} \quad . \tag{7.94}$$

As for a conventional resonant circuit, Q measures the sharpness of response. If ω is the angular frequency at which the amplitude response is reduced from the resonant amplitude by a factor of $2^{1/2}$, it may be shown that

$$\frac{1}{Q} \simeq \frac{\omega_0 - \omega}{\omega_0} \quad . \tag{7.95}$$

7.10 Ideal Two-Conductor Transmission Lines

Apart from the waveguides considered above other guiding systems for electromagnetic waves are also possible. A common method of power or signal transmission is by means of two parallel long straight cylindrical conductors, such as parallel wires embedded in a dielectric medium, or coaxial lines, or a long cylindrical conductor with an insulating cover laid in sea-water, the latter acting as the second conductor. We shall consider a uniform two-conductor transmission line made of, in the first instance, perfect conductors and perfect dielectric, of which the length of the conductors is much greater than their separation and radii.

As in the case of waveguides, we shall consider the propagation of a wave with a (z,t)-dependence given by $e^{(\gamma z - i\omega t)}$, the z-axis being taken parallel to the two conductors. The wave equations (7.26) and

(7.27), as well as Maxwell's equations, (7.31) to (7.36), still apply. The objection to the transmission of a TEM wave in a hollow tube waveguide as presented in Sec. 7.5 is however no longer valid as the two conductors may be charged to different potentials. Although the transmission of other modes are also possible we shall confine ourselves to that of TEM waves only.

Let ε_1 and μ_1 be respectively the permittivity and permeability of the dielectric of the system. As $E_z = H_z = 0$, Eqs. (7.37) to (7.40) require that, if free transmission is to take place,

$$h^2 = \gamma^2 + \mu_1\varepsilon_1\omega^2 = 0$$

or

$$\gamma = i\omega\sqrt{\mu_1\varepsilon_1} = ik \quad ,$$

showing that the wave must travel with the velocity $v = (\mu_1\varepsilon_1)^{-\frac{1}{2}}$. Thus the wave velocity is not modified by the presence of the conductors. In the following derivations we shall assume the wave to be propagating in the positive z-direction. Relations for a wave travelling in the opposite direction may be obtained by changing the sign of v, or equivalently, that of γ or k.

The transverse components of the electric intensity may be obtained from the wave equation, Eq. (7.26), which, for $h^2 = 0$, becomes

$$\left(\frac{\partial^2}{\partial x^2} + \frac{\partial^2}{\partial y^2}\right)\begin{matrix} E_x \\ E_y \end{matrix} = (\nabla_t^2)\begin{matrix} E_x \\ E_y \end{matrix} = 0 \quad , \tag{7.96}$$

where ∇_t is the two dimensional del. Expressing $\mathbf{E}$ in terms of a scalar function $V(x, y, z, t)$ as

$$E_x = -\frac{\partial V}{\partial x} \quad , \quad E_y = -\frac{\partial V}{\partial y} \quad , \tag{7.97}$$

we may expect V to have the form $V_0(x,y)e^{(\gamma z - i\omega t)}$, with V_0

satisfying the two dimensional Laplace's equation

$$\nabla_t^2 V_o = 0 \quad . \tag{7.98}$$

The boundary condition that $E_t = 0$ on the surfaces of the conductors requires that V_o be a constant on each conductor surface. H_t and E_n, however, need not vanish on these surfaces. The transverse components of the magnetic intensity may be obtained then from Eqs. (7.31) and (7.32), which in the vector form become

$$\mathbf{H} = \frac{1}{\mu_1 \omega} \mathbf{k} \times \mathbf{E} \quad , \tag{7.99}$$

where $\mathbf{k}$ is the propagation vector having a magnitude k and a direction along that of wave propagation. Actually it can be shown quite readily that $\text{curl} = \mathbf{k} \times$ for transverse waves.

From the above discussion we see that the configuration of the electric field in a transverse plane at any given instant is the same as that for two parallel conductors charged to potentials equal to the given values of V. The TEM waves which are being transmitted by the system are the same as the plane electromagnetic waves propagating in an infinite dielectric considered in Sec. 2.8, the only difference being that in the present case the fields $\mathbf{E}$ and $\mathbf{H}$ in a transverse plane are functions of position, i.e., the waves are no longer uniform.

From Gauss' flux theorem, Eq. (2.6), we can obtain the charge per unit length or the *linear charge density* T_1 of conductor 1:

$$T_1 = \frac{1}{\Delta z} \int_{S_1} D_n dS = -\varepsilon_1 \oint_{C_1} \frac{\partial V}{\partial n} d\ell \quad , \tag{7.100}$$

where S_1 is the curved surface, the normal to which is designated $\mathbf{n}$, of a section of length Δz of conductor 1 and C_1 is the periphery of a cross section of the conductor, the end surfaces not contributing as $\mathbf{D} = 0$ inside a conductor. Denoting a cylindrical surface of infinite radius and length Δz enclosing and parallel to the conductors by S_∞, we can express the total charge carried by unit length of the twin circuit as

$$T_1 + T_2 = \frac{1}{\partial z}\int_{S_\infty} D_n dS = 0$$

since **D** vanishes at infinity[b]. It follows that $T_2 = -T_1$, i.e. the conductors carry equal and opposite charges.

The current carried by conductor 1 is by Eq. (7.22)

$$I_1 = \oint_{C_1} |\mathbf{n}\times\mathbf{H}|d\ell = \frac{1}{\mu_1\omega}\oint_{C_1} |\mathbf{n}\times(\mathbf{k}\times\mathbf{E})|d\ell$$

$$= \frac{k}{\mu_1\omega}\oint_{C_1} E_n d\ell = -\frac{k}{\mu_1\omega}\oint_{C_1}\frac{\partial V}{\partial n}d\ell = vT_1 \quad , \qquad (7.101)$$

making use of the fact that $E_t = 0$ on the conductor surface. It follows immediately that $I_2 = -I_1$, i.e. the current which flows in one conductor must return through the other.

Let $\Phi = V_1 - V_2$, where V_1 and V_2 are the values of V on conductors 1 and 2 respectively. As the conductors carry equal and opposite charges and the equations and boundary conditions for **E** are the same as those in electrostatics, we may assume T_1 to be proportional to Φ and define the capacitance per unit length C by

$$T_1 = C\Phi \quad . \qquad (7.102)$$

The magnetic energy per unit length of the line is

$$W_m = \frac{\mu_1}{2\Delta z}\int_{V_\infty} H^2 dV = \frac{1}{2\Delta z}\frac{k^2}{\mu_1\omega^2}\int_{V_\infty} E^2 dV = \frac{\varepsilon_1}{2\Delta z}\int_{V_\infty}(\nabla_t V)^2 dV \quad ,$$

---

[b] For an infinitely long charged line, $D \sim r^{-1}$, where r is the distance from the line. Note also that S_∞ is a constant independent of r.

where V_∞ is the volume of a cylinder of length Δz which is bounded externally by S_∞, internally by S_1 and S_2, and at the two ends by parallel planes perpendicular to the conductors. As

$$(\nabla_t V)^2 = \nabla_t \cdot (V\nabla_t V) - V\nabla_t^2 V = \nabla_t \cdot (V\nabla_t V) \quad ,$$

the integral becomes

$$W_m = \frac{\varepsilon_1}{2\Delta z} \int_{S_\infty + S_1 + S_2} V\nabla_t V \cdot d\mathbf{S}$$

using the corollary of Gauss' theorem (Eq. (A.12) of Appendix). Note that as $E_z = 0$ the end surfaces do not contribute to the integral. Since $\nabla_t V = 0$ at infinity and the normals to S_1 and S_2, the inner surfaces of V_∞, are directed into the conductors, we have

$$\begin{aligned} W_m &= -\frac{\varepsilon_1 V_1}{2\Delta z} \int_{S_1} \frac{\partial V}{\partial n} dS - \frac{\varepsilon_1 V_2}{2\Delta z} \int_{S_2} \frac{\partial V}{\partial n} dS \\ &= \frac{1}{2} (T_1 V_1 + T_2 V_2) = \frac{1}{2} T_1 \Phi \quad . \end{aligned}$$

If we then define, by analogy to Eq. (4.20), an *"external inductance"* per unit length L_e for the line by

$$\frac{1}{2} L_e I_1^2 = \frac{1}{2} T_1 \Phi \quad ,$$

then by virtue of Eqs. (7.101) and (7.102) we have

$$L_e C = v^{-2} = \mu_1 \varepsilon_1 \tag{7.103}$$

and

$$\Phi = vL_e I_1 = \sqrt{\frac{L_e}{C}}\, I_1 \quad . \tag{7.104}$$

The quantity $\sqrt{\frac{L_e}{C}}$ which has the dimensions of a resistance is known as the *wave resistance* of the system.

Equation (7.104) gives

$$ik\Phi = i\omega L_e I_1 \quad ,$$

or, using the identities $\frac{\partial}{\partial z} = ik$ and $\frac{\partial}{\partial t} = -i\omega$,

$$\frac{\partial \Phi}{\partial z} + L_e \frac{\partial I_1}{\partial t} = 0 \quad . \tag{7.105}$$

Equations (7.101) and (7.102) give

$$ikI_1 = i\omega C\Phi \quad ,$$

or

$$\frac{\partial I_1}{\partial z} + C \frac{\partial \Phi}{\partial t} = 0 \quad . \tag{7.106}$$

These equations may be combined to give the wave equation

$$\frac{\partial^2 \Phi}{\partial z^2} - L_e C \frac{\partial^2 \Phi}{\partial t^2} = 0 \quad , \tag{7.107}$$

which also holds for I_1. Although we have derived these equations for harmonic waves, they may be shown by a Fourier expansion to hold generally for arbitrary (z, t)-dependences of the type $f(kz \pm \omega t)$, where f is an arbitrary function of the argument indicated.

For the interpretation of Eqs. (7.105) and (7.106), consider a section AB of a circuit shown in Fig. 7.7. The increment of the voltage V between the two parallel wires across the inductance $L_e \Delta z$ is

$$\Delta V = - L_e \Delta z \frac{\partial I_1}{\partial t} \quad ,$$

or

$$\frac{\Delta V}{\Delta z} + L_e \frac{\partial I_1}{\partial t} = 0 \quad ,$$

which is identical with Eq. (7.105) in the limit $\Delta z \to 0$. The time variation of the charge $CV\Delta z$ carried by the condenser produces an increment of the current I_1 of

$$\Delta I_1 = - C\Delta z \frac{\partial V}{\partial t} \quad ,$$

$$\frac{\Delta I_1}{\Delta z} + C \frac{\partial V}{\partial t} = 0 \quad .$$

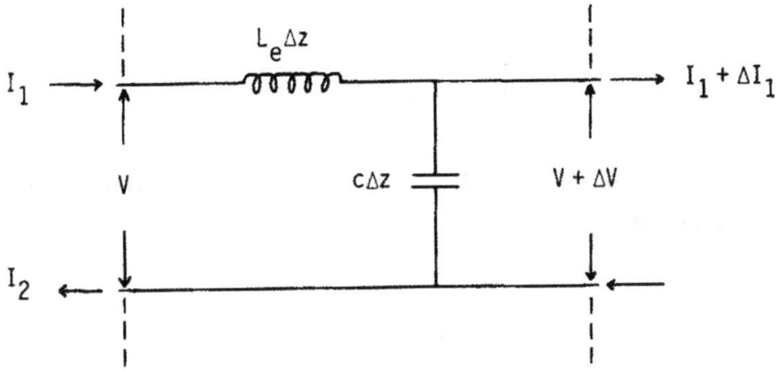

Fig. 7.7 A section of a circuit consisting of two parallel wires, whose impedances may be neglected, with lumped circuit-elements inserted as shown.

This is identical with Eq. (7.106) in the limit $\Delta z \to 0$.

As we are able to reproduce exactly the same equations (7.105) and (7.106) with the physical circuit we may interpret Φ as the instantaneous voltage difference between the two conductors of a transmission line at the point z. Equation (7.105) is then simply a special case of Kirchhoff's equation (4.12), while Eq. (7.106) represents charge conservation. Furthermore, we may alter the parameters L_e and C of a two-conductor transmission line by inserting suitable inductive and capacitive elements at sufficiently small intervals.

To consider the effect of boundary conditions imposed at the ends of the lines, assume that an alternating voltage $\Phi_0 e^{-i\omega t}$ is applied between the wires at the end $z = 0$, and that an ohmic resistance R is connected between them at the other end, where $z = \ell$. The general solution of Eq. (7.107) is the sum of a wave going in the $+z$ direction and one going in the $-z$ direction:

$$\Phi = \Phi_+ e^{i(kz - \omega t)} + \Phi_- e^{-i(kz + \omega t)} \quad .$$

Correspondingly we also have

$$I_1 = I_+ e^{i(kz - \omega t)} + I_- e^{-i(kz + \omega t)} \quad .$$

The relation between Φ and I_1 depends on the direction of propagation. Equation (7.104) gives

$$\Phi_+ = \sqrt{\frac{L_e}{C}}\, I_+ \quad , \quad \Phi_- = -\sqrt{\frac{L_e}{C}}\, I_- \quad .$$

Then as $\Phi = RI_1$ at $z = \ell$ we have

$$\Phi_+ e^{ik\ell} + \Phi_- e^{-ik\ell} = R\sqrt{\frac{C}{L_e}}\,(\Phi_+ e^{ik\ell} - \Phi_- e^{-ik\ell}) \quad . \qquad (7.108)$$

At $z = 0$, we require that

$$\Phi_+ + \Phi_- = \Phi_0 \quad . \qquad (7.109)$$

These equations determine Φ_+ and Φ_- for various values of R.

Consider two special cases:

(a). $R = \infty$, i.e. the circuit is open at one end. For the left-hand side of (7.108) to be finite we must have

$$\frac{\Phi_-}{\Phi_+} = e^{2ik\ell} \quad .$$

This shows that the reflected and the outgoing waves will have the same amplitude and a phase difference of $2k\ell$. Equations (7.108) and (7.109) then give

$$\Phi = \Phi_0 \frac{\cos k(\ell - z)}{\cos k\ell}\, e^{-i\omega t}$$

and

$$I_1 = i\sqrt{\frac{C}{L_e}}\,\Phi_0 \frac{\sin k(\ell - z)}{\cos k\ell}\, e^{-i\omega t} \quad .$$

Thus $\Phi = 0$ on transverse planes for which $k(\ell - z) = (2n + 1)\frac{\pi}{2}$, where n is an integer, i.e. for $\ell - z = \frac{1}{4}\lambda, \frac{3}{4}\lambda, \frac{5}{4}\lambda, \ldots$ at all times and similarly $I_1 = 0$ for $\ell - z = 0, \frac{1}{2}\lambda, \lambda, \frac{3}{2}\lambda, \ldots,$ where $\lambda = 2\pi/k$ is the wavelength. In this case the incident and reflected waves superpose to give rise to standing waves.

(b). $R = \sqrt{\frac{L_e}{C}}$, i.e. the applied resistance matches the wave resistance. In this case $\Phi_- = 0$, showing that the outgoing wave is completely absorbed by the resistance and reflection does not occur. Φ and I_1 then have the same phase at all points of the transmission line.

7.11 Dissipative Transmission Lines

In a realistic situation the conductors and dielectric that form a guiding system will have finite conductivities. Joule heating will then take place causing the waves to attenuate during transmission. The damping as well as the wave velocity will in general be frequency-dependent, giving rise to distortion when a signal is transmitted. Actually, when attenuation takes place, purely transverse waves are no longer possible (Sec. 7.5). However, we shall assume that any departure of conditions from ideal is so small that to the first approximation the previous analysis still applies. Only small corrections need to be introduced.

Let the permittivities, permeabilities and conductivities of the dielectric and conductors be $\varepsilon_1, \mu_1, \sigma_1$ and $\varepsilon_2, \mu_2, \sigma_2$ respectively. We consider first the modifications required by the finite conductivity of the dielectric. Putting $h = 0$ in Eq. (7.78) we see that the results obtained for the ideal system will have to be modified according to the rule

$$k \to k\sqrt{1 + \frac{i\sigma_1}{\omega\varepsilon_1}} \quad ,$$

where on the right-hand side $k = \omega\sqrt{\mu_1\varepsilon_1} = \omega/v$ as before. As the expression of I_1 in terms of T_1 and that of W_m in terms of the integral over E^2 both involve k, Eq. (7.101) becomes

$$I_1 = vT_1 \sqrt{1 + \frac{i\sigma_1}{\omega\varepsilon_1}} \quad , \tag{7.110}$$

and Eq. (7.104) becomes

$$\Phi = vL_e I_1 \left(1+\frac{i\sigma_1}{\omega\varepsilon_1}\right)^{-\frac{1}{2}} \quad , \tag{7.111}$$

so that

$$L_e C = \mu_1 \varepsilon_1$$

as before. Note that as the definition of C by Eq. (7.102) is not affected, L_e remains the same as in the case of a perfect dielectric.

The average Joule heat dissipation in the dielectric per unit volume per unit time is

$$< \mathrm{Re}\mathbf{J} \cdot \mathrm{Re}\mathbf{E} > = \frac{1}{2}\sigma_1 \mathbf{E}\cdot\mathbf{E}^* \quad .$$

Assuming that the longitudinal component of the electric field in the dielectric is negligible compared with the transverse components and following the procedure for obtaining W_m, we find that the rate of Joule dissipation per unit length of the line is

$$\frac{\sigma_1}{2\Delta z}\int_{V_\infty} \mathbf{E}\cdot\mathbf{E}^* dV = \frac{\sigma_1}{2\varepsilon_1} T_1 \Phi^* \quad .$$

As T_1 is proportional to Φ the above is proportional to $\Phi\Phi^*$. We introduce the *shunt conductance* per unit length, G, by writing the above result as $\frac{1}{2}G\Phi\Phi^*$ and obtain for G

$$\frac{\sigma_1}{\varepsilon_1} = \frac{G}{C} \quad . \tag{7.112}$$

From Eqs. (7.110) and (7.111) we obtain

$$\frac{\partial I_1}{\partial z} = ik\left(1+\frac{iG}{\omega C}\right)^{\frac{1}{2}} I_1 = ik\left(1+\frac{iG}{\omega C}\right)vC\Phi = i\omega C\Phi - G\Phi = -C\frac{\partial\Phi}{\partial t} - G\Phi \quad ,$$

or

$$\frac{\partial I_1}{\partial z} + C\,\frac{\partial \Phi}{\partial t} + G\Phi = 0 \quad , \tag{7.113}$$

which is the modified form of Eq. (7.106). Equation (7.105) on the other hand requires no change. Thus the effect of a finite conductivity of the dielectric is to introduce a leak current $G\Phi$ per unit length of the transmission line which flows between the two conductors.

Consider next the finite resistance of the conductors. As we have seen in Sec. 7.1, this will allow the fields to penetrate into the conductors to the extent of the skin depth δ. Joule heating must then develop in the conductors giving rise to additional damping of the transmitted waves. This effect may be taken into account by introducing an internal impedance which has a resistive part and an inductive part. Let the internal resistance and inductance per unit length of the line be R_i and L_i respectively. In general, these quantities are functions of frequency and depend on the radii of the conductors and perhaps on their separation also. However, for the two-conductor system under consideration they may be taken to be approximately twice the resistance and internal inductance per unit length per conductor calculated in Sec. 7.3.

Equation (7.105) must now be modified to take account of the potential drop across the additional impedance to

$$\frac{\partial \Phi}{\partial z} + (L_e + L_i)\,\frac{\partial I_1}{\partial t} + R_i I_1 = 0 \quad . \tag{7.114}$$

Equation (7.106) or (7.113), as the case may be, however, still applies. Eliminating Φ from Eqs. (7.113) and (7.114) we obtain the "telegraph equation" for a line where both losses are present:

$$\frac{\partial^2 I_1}{\partial z^2} - C(L_e + L_i)\,\frac{\partial^2 I_1}{\partial t^2} - \{CR_i + G(L_e + L_i)\}\,\frac{\partial I_1}{\partial t} - GR_i I_1 = 0 \quad . \tag{7.115}$$

The same equation holds for Φ also. Equation (7.115) describes the behaviour of a transmission line system. For a given line the parameters C, L_e, L_i, R_i, and G may be calculated and its behaviour determined by means of the telegraph equation. Note that if $R_i = 0$, $L_i = 0$ and $G = 0$, Eq. (7.115) reduces to the wave equation (7.107).

To find the propagation constant, substitute $I_1 = I_o e^{(\gamma z - i\omega t)}$ in Eq. (7.115). This gives

$$\gamma^2 = (G - i\omega C)(R_i - i\omega L) \quad , \tag{7.116}$$

where

$$L = L_e + L_i \quad .$$

As γ is now complex, let $\gamma = i(k' + i\alpha)$, where k' and α are real numbers, as in the case of a waveguide. Equation (7.116) then gives

$$\alpha^2 - k'^2 = GR_i - \omega^2 CL \quad ,$$

and

$$2\alpha k' = (CR_i + GL)\omega \quad . \tag{7.117}$$

The attenuation constant α is seen to be frequency-dependent, as it varies from $\sqrt{GR_i}$ for low frequencies to $\frac{1}{2}(CR_i + GL)(CL)^{-\frac{1}{2}}$ for high frequencies. k', and hence the phase velocity, also varies with frequency. Thus a signal consisting of a spread of frequencies will in general be distorted on transmission.

The two equations (7.117) may be combined to give

$$\alpha^2 - GR_i = \left\{\left(\frac{CR_i + GL}{2\alpha}\right)^2 - CL\right\}\omega^2 \quad . \tag{7.118}$$

This equation may be satisfied while at the same time α^2 made independent of ω if

$$\frac{R_i}{L} = \frac{G}{C}$$

and

$$\alpha^2 = GR_i \quad .$$

Under such conditions, the wave velocity

$$\frac{\omega}{k'} = \frac{1}{\sqrt{CL}}$$

is also independent of frequency. Signals will then be transmitted free of distortion. This conclusion is however strictly true only for lines with constant parameters. For the two-conductor system under consideration R_i and L_i, which arise from the skin effect, are themselves dependent on the frequency to a small degree, i.e. to the power of one half. Transmission will be distortion-free under the above conditions if the frequency spread is small.

Equations (7.114) and (7.116) give a relation between Φ and I_1:

$$\Phi = \pm I_1 \sqrt{\frac{R_i - i\omega L}{G - i\omega C}} ,$$

where the uncertainty in sign comes from the square-rooting of γ^2. To be consistent with Eq. (7.104) for the ideal case where $R_i, G, L_i \to 0$, we have to choose the positive sign for a wave propagating in the positive z-direction. Defining the *characteristic* or *wave impedance* Z_o by

$$Z_o = \sqrt{\frac{R_1 - i\omega L}{G - i\omega C}} ,$$

we have $\Phi_+ = Z_o I_+$ for a wave propagating in the positive z-direction and $\Phi_- = -Z_o I_-$ for one propagating in the negative z-direction.

Suppose now the line is terminated by a load impedance Z_ℓ. The boundary condition is then $\Phi = Z_\ell I_1$ at $z = \ell$. This gives

$$\begin{aligned}\Phi_+ e^{\gamma\ell} + \Phi_- e^{-\gamma\ell} &= Z_\ell (I_+ e^{\gamma\ell} + I_- e^{-\gamma\ell}) \\ &= \frac{Z_\ell}{Z_o} (\Phi_+ e^{\gamma\ell} - \Phi_- e^{-\gamma\ell}) ,\end{aligned}$$

or

$$\frac{\Phi_-}{\Phi_+} = \frac{Z_\ell - Z_o}{Z_\ell + Z_o} e^{2\gamma\ell} \tag{7.119}$$

being the amplitude reflection coefficient for an incident wave $\Phi_+ e^{(\gamma z - i\omega t)}$. In particular, if the parameters are such that $Z_\ell = Z_o$, no reflection will occur. The circuit is then said to be matched. Obviously energy transmission is most efficient under such conditions.

Suppose at $z = 0$, $\Phi = \Phi_o e^{-i\omega t}$ and $I_1 = I_o e^{-i\omega t}$, the impedance observed at the transmitting end is

$$\frac{\Phi_o}{I_o} = \frac{\Phi_+ + \Phi_-}{\Phi_+ - \Phi_-} Z_o = \frac{Z_\ell - Z_o \tanh(\gamma\ell)}{Z_o - Z_\ell \tanh(\gamma\ell)} Z_o \quad .$$

By suitably choosing Z_ℓ and ℓ the transmitting impedance of a two-conductor transmission line can be made to have any desired value, a fact that makes such lines a very useful matching device.

In the above discussion we have considered two-conductor lines where the resistance, shunt conductance, inductance and capacitance are distributed uniformly. This is of course the case of the parallel-wire transmission line. It is however possible to insert suitable lumped circuit elements at appropriate intervals along the line and apply the same equations at least as a first approximation. By the same method the line parameters, R_i, G, L or C may be changed to any desired value. One such artificial transmission line is shown schematically in Fig. 7.8.

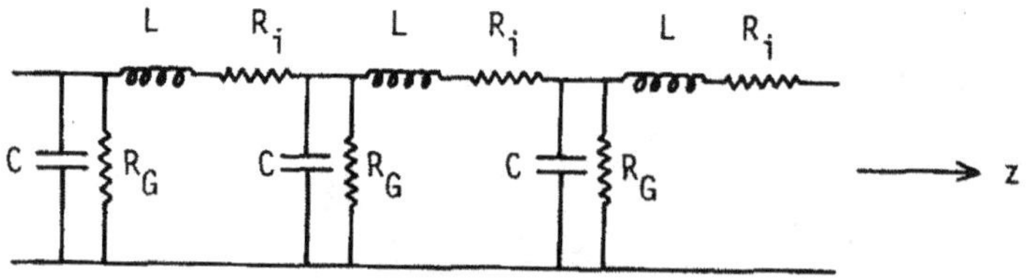

Fig. 7.8 Schematic representation of a lumped-circuit transmission line which may be treated as a continuous line. R_G is the equivalent resistance due to shunt conductance G.

7.12 Waves Guided by Dielectrics

In the treatment of waves guided by conductors, we have seen that waves propagating in a dielectric can be prevented from lateral spreading by conducting walls by the mechanism of reflection. The same can be achieved by suitable dielectric walls also. Such is the case of *optical fibres*. A type of optical fibre consists of a cylindrical core of a dielectric of refractive index n_1 surrounded by a dielectric cladding of refractive index n_2 in the form of a concentric cylindrical tube. If $n_1 > n_2$, electromagnetic waves that propagate in the core can be prevented from radial spreading by total internal reflection at the boundary surface. Some evanescent waves will be present in the cladding, as one might expect for total internal reflection, so that the cladding must be of a sufficient thickness to avoid energy leakage.

Let the radius of the core be a and use the symbols and geometry of Sec. 7.6. The waves may be assumed to propagate along the axis with a (z, t)-dependence of $\exp(\gamma z - i\omega t)$, and the same wave equation and guide equations will again apply. The requirement of finite amplitude along the axis limits the acceptable solution of the radial wave equation for the core to Bessel's functions of the first kind. No such requirement applies to the cladding and the solution there is best expressed in terms of Hankel's functions of the second kind $\mathcal{H}_m^{(2)}$ which are linear combinations of Bessel's functions of the first and second kinds and have the correct asymptotic behaviour. Thus we have the following solutions for the wave equations for E_z and H_z in the core and cladding denoted by subscripts 1 and 2 respectively:

$$E_{1z} = C_1 \mathcal{J}_m(h_1\rho)\cos(m\phi + \delta_m) \quad ,$$

$$H_{1z} = D_1 \mathcal{H}_m^{(2)}(h_1\rho)\cos(m\phi + \delta_m) \quad ,$$

$$E_{2z} = C_2 \mathcal{J}_m(h_2\rho)\cos(m\phi + \delta_m) \quad ,$$

$$H_{2z} = D_2 \mathcal{H}_m^{(2)}(h_2\rho)\cos(m\phi + \delta_m) \quad ,$$

where m is zero or a positive integer for single-valuedness of the solutions and h_1 and h_2 are given by $h_1^2 = \gamma^2 + \mu_1\varepsilon_1\omega^2$ and $h_2^2 = \gamma^2 + \mu_2\varepsilon_2\omega^2$ respectively, the factor $\exp(\gamma z - i\omega t)$ in each of the above expressions having been omitted.

The boundary conditions $E_{1t} = E_{2t}$ and $H_{1t} = H_{2t}$ apply to E_z, E_φ, H_z and H_φ at $\rho = a$. In general, all the six components of the field vectors **E** and **H** will be present and it will not be possible to separate them into the simpler TM and TE modes as in the case of conductor waveguides. In the case of $m = 0$, however, the waves that can be propagated in the fibre automatically separate into TM and TE modes, as can be seen from the following.

With $m = 0$ the boundary conditions give

$$C_1 \mathcal{J}_0(h_1 a) = C_2 \mathcal{H}_0^{(2)}(h_2 a) \quad ,$$

$$\frac{\varepsilon_1}{h_1} C_1 \mathcal{J}_0'(h_1 a) = \frac{\varepsilon_2}{h_2} C_2 \mathcal{H}_0^{(2)'}(h_2 a) \quad ,$$

$$D_1 \mathcal{J}_0(h_1 a) = D_2 \mathcal{H}_0^{(2)}(h_2 a) \quad ,$$

$$\frac{1}{h_1} D_1 \mathcal{J}_0'(h_1 a) = \frac{1}{h_2} D_2 \mathcal{H}_0^{(2)'}(h_2 a) \quad ,$$

use having been made of the guide equations (7.61)-(7.64). The first two equations are homogeneous in C_1 and C_2 while the last two equations are homogeneous in D_1 and D_2. We therefore require

$$\frac{\varepsilon_2 h_1 \mathcal{J}_0(h_1 a)}{\varepsilon_1 h_2 \mathcal{J}_0'(h_1 a)} = \frac{\mathcal{H}_0^{(2)}(h_2 a)}{\mathcal{H}_0^{(2)'}(h_2 a)} \tag{7.120}$$

for non-zero C_1 and C_2; and

$$\frac{h_1 \mathcal{J}_0(h_1 a)}{h_2 \mathcal{J}_0'(h_1 a)} = \frac{\mathcal{H}_0^{(2)}(h_2 a)}{\mathcal{H}_0^{(2)'}(h_2 a)} \tag{7.121}$$

for non-zero D_1 and D_2. Since Eqs. (7.120) and (7.121) are not compatible, we have the following two possibilities:

(a) Equation (7.120) is satisfied and $D_1 = D_2 = 0$, corresponding to TM waves;

(b) Equation (7.121) is satisfied and $C_1 = C_2 = 0$, corresponding to TE waves.

The cutoff condition is determined by the requirements that the waves are to be confined predominantly in the core and to attenuate rapidly in the cladding as the radial distance from the surface of the core increases. For large values of $h_2\rho$, Hankel's function has the asymptotic expression

$$\mathcal{H}_m^{(2)}(h_2\rho) = \sqrt{\frac{2}{\pi h_2\rho}}\, e^{-i(h_2\rho - m\pi/2 - \pi/4)}$$

To satisfy the above requirements, h_2 has to be negative imaginary (hence the use of Hankel's functions $\mathcal{H}_m^{(2)}(h_2\rho)$) and $(h_2\rho)^2$ negative. From the definitions of h_1 and h_2, we find

$$\omega = \frac{1}{a}\sqrt{\frac{(h_1a)^2 - (h_2a)^2}{\mu_1\varepsilon_1 - \mu_2\varepsilon_2}}$$

The cutoff frequency for the particular mode is then obtained by putting $h_2a = 0$ in the above equation:

$$\omega_c = \frac{h_1a}{a\sqrt{\mu_1\varepsilon_1 - \mu_2\varepsilon_2}} \quad .$$

The corresponding propagation constant k is then given by

$$\gamma = ik = \sqrt{h_1^2 - \mu_1\varepsilon_1\omega^2} = i\sqrt{\mu_2\varepsilon_2}\,\omega \quad ,$$

showing that the waves will travel with a phase velocity $(\mu_2\varepsilon_2)^{-\frac{1}{2}}$ which is characteristic of the cladding.

Consider the case of TM and TE waves for which $m = 0$. For

small values of h_2a the left-hand side of Eqs. (7.120) and (7.121) has the asymptotic form

$$\frac{\mathcal{H}_0^{(2)}(h_2a)}{\mathcal{H}_0^{(2)\prime}(h_2a)} \sim -(h_2a)\ln(h_2a) \quad .$$

As h_2a goes to zero faster than $\ln(h_2a)$ grows to minus infinity, both equations are satisfied by

$$\mathcal{J}_0(h_1a) = 0 \quad . \tag{7.122}$$

The roots of Eq. (7.122) determines the cutoff frequencies of the various TM and TE modes. The lowest positive root of Eq. (7.122) is 2.4048 (see Table 7.1) giving a cutoff frequency of

$$f_c = \frac{2.4048c}{2\pi a\sqrt{n_1^2 - n_2^2}} \quad ,$$

where c is the velocity of light in free space and $\mu_1 \approx \mu_2 \approx \mu_0$ has been assumed. This however is not the lowest cutoff frequency. For $m = 1$ the lowest mode has no cutoff frequency. A fibre designed to allow the propagation of waves of frequencies between zero and that given by the above equation can carry that one mode only. Such a fibre is called a *single mode fibre*.

PROBLEMS

1. Equation (7.13) may be written as

$$\frac{d^2J_0}{d\rho^2} + \frac{1}{\rho}\frac{dJ_0}{d\rho} - T^2J_0 = 0 \quad ,$$

where $T^2 = -i\mu\sigma\omega$. The solution finite at $\rho = 0$ is then $J_0 = \mathcal{J}_0(\xi e^{i3\pi/4})$, where $\xi = \sqrt{2}\,\rho/\delta$. The zero order Bessel function may be written as

$$\mathcal{J}_o(\xi e^{i3\pi/4}) = \mathrm{ber}\xi + i\mathrm{bei}\xi$$

in terms of the following real functions:

$$\mathrm{ber}\xi = 1 - \frac{1}{(2!)^2}\left(\frac{\xi^2}{4}\right)^2 + \frac{1}{(4!)^2}\left(\frac{\xi^2}{4}\right)^4 - \ldots \quad ,$$

$$\mathrm{bei}\xi = \frac{1}{(1!)^2}\frac{\xi^2}{4} - \frac{1}{(3!)^2}\left(\frac{\xi^2}{4}\right)^3 + \frac{1}{(5!)^2}\left(\frac{\xi^2}{4}\right)^5 - \ldots \quad .$$

Express J_o in terms of these functions and the surface amplitude J_s, and show that for $\delta << a$

$$\frac{J_o}{J_s} \approx \left(\frac{8\delta^4 + \rho^4}{8\delta^4 + a^4}\right)^{\frac{1}{2}} \quad .$$

2. A plane electromagnetic wave propagates in a dielectric of thickness a sandwiched between two parallel plane conductors, both the dielectric and conductor are perfect. Show by solving the wave equation that the guide propagation vector **k'** and the propagation vector **k** for the unbounded dielectric are related by

$$k'^2 + \left(\frac{m\pi}{a}\right)^2 = k^2 \quad ,$$

where m is an integer. Hence show that the guide wavelength λ' and the impressed wavelength λ are related by

$$\lambda' = \frac{\lambda}{\sqrt{1 - \left(\frac{m\pi}{ka}\right)^2}} \quad .$$

Show also that the group velocity v_g and the guide phase velocity v' are related by

$$v_g v' = v^2 \quad ,$$

where v is the phase velocity in the unbounded dielectric.

3. Investigate the propagation of plane electromagnetic waves between two parallel perfect plane conductors by solving the wave equations (7.26) and (7.27) and applying the guide equations (7.37) to (7.40). Compare the results with those given in the text.

4. Show that the time-averaged energy flux in an ideal waveguide is parallel to the guide axis. Show also that the surface currents are longitudinal for TM modes.

5. The differential equations for a line of force may be obtained from the definition that a displacement $d\ell$ along the line is parallel to the local field. Thus the equations for the electric lines of force are obtained from

$$\frac{dx}{E_x} = \frac{dy}{E_y} = \frac{dz}{E_z} \quad .$$

Show that for the $TE_{1,0}$ wave in the rectangular waveguide shown in Fig. 7.4 the magnetic lines of force in the xz-plane are the family of curves given by

$$\cos(k'z - \omega t)\sin\left(\frac{\pi x}{a}\right) = \text{constant} \quad ,$$

where k' is the magnitude of the guide propagation vector. Sketch the configuration of the electric and magnetic lines of force in the three principal planes of the guide for a given instant of time t.

6. The internal cross section of an ideal waveguide consists of the first quadrant of a circle of radius a. Show that h as defined by Eq. (7.28) is given by the nonzero roots of $\mathcal{J}_{2m}(ha) = 0$ for TM modes and those of $\mathcal{J}'_{2m}(ha) = 0$ for TE modes, where m is zero or a positive integer.

7. A coaxial line consists of a hollow conducting pipe of inner radius b enclosing a cylindrical conductor of radius a, where $b > a$. Assuming the conductors and the dielectric filling the gap between the conductors to be perfect, show that for loss-free transmission the critical angular frequency is vh, where v is the phase velocity in the unbounded dielectric and h is the solution of the equation

$$\frac{\mathcal{J}_m(ha)}{\mathcal{J}_m(hb)} = \frac{\mathcal{Y}_m(ha)}{\mathcal{Y}_m(hb)}$$

for a TM wave, and the solution of

$$\frac{\mathcal{J}_m'(ha)}{\mathcal{J}_m'(hb)} = \frac{\mathcal{Y}_m'(ha)}{\mathcal{Y}_m'(hb)}$$

for a TE wave. Find the field configuration in each case. Can a TEM wave be propagated in the coaxial line?

8. Show that the attenuation constant α_c due to the ohmic loss in the conducting walls of a rectangular waveguide of width a and height b for the $TM_{m,n}$ mode is

$$\alpha_c = 2R_s \sqrt{\frac{\varepsilon}{\mu}} \left(1 - \frac{\omega_{m,n}^2}{\omega^2}\right)^{-\frac{1}{2}} \left(\frac{n^2}{b^3} + \frac{m^2}{a^3}\right)\left(\frac{n^2}{b^2} + \frac{m^2}{a^2}\right)^{-1} ,$$

where R_s is the surface resistivity of the conductor, ε and μ the permittivity and permeability of the dielectric filling the guide. Find α_c for the $TE_{m,n}$ mode.

9. Show that if dielectric loss may be neglected the quality factor of a rectangular cavity resonator of width a, height b and length ℓ for the $TM_{m,n,p}$ resonant mode is

$$Q_{m,n,p} = \frac{\mu\omega}{4R_s} \left\{ \left(\frac{m}{a}\right)^2 + \left(\frac{n}{b}\right)^2 \right\} \left\{ \left(\frac{m}{a}\right)^2 \left(\frac{1}{a} + \frac{1}{\ell}\right) + \left(\frac{n}{b}\right)^2 \left(\frac{1}{b} + \frac{1}{\ell}\right) \right\}^{-1} ,$$

where R_s is the surface resistivity of the conducting walls, μ the permeability of the dielectric filling the cavity and ω the resonance frequency, assuming that $m,n,p \neq 0$.

10. Show that the lowest resonant mode for TM oscillations in the cavity of Prob. (7.9) is $TM_{1,1,0}$ and

$$Q_{1,1,0} = \frac{\mu\omega}{2R_s}\left(\frac{1}{a^2}+\frac{1}{b^2}\right)\left\{\frac{1}{a^2}\left(\frac{2}{a}+\frac{1}{\ell}\right)+\frac{1}{b^2}\left(\frac{2}{b}+\frac{1}{\ell}\right)\right\}^{-1} ,$$

provided dielectric loss may be neglected. What is the lowest TE mode?

11. Considering the $TM_{m,n,p}$ mode oscillation in an ideal rectangular cavity resonator as the superposition of $TM_{m,n}$ waves incident on and reflected from the end surfaces, obtain the field configuration and the resonance frequency. Show that the condition $\nabla \cdot \mathbf{E} = 0$ is satisfied in the dielectric of the cavity.

12. Investigate TE mode oscillations in an ideal cylindrical cavity of radius a and length ℓ and show that the frequency of the lowest mode is

$$f = 0.2930\left(\frac{v}{a}\right)\left\{1 + 2.911\left(\frac{a}{\ell}\right)^2\right\}^{\frac{1}{2}} \text{ Hz} ,$$

where v is the phase velocity in the dielectric of the cavity if unbounded.

13. An ideal cubical cavity resonator has sides of length a. Show that for wavelengths short compared with a the number of resonant modes per unit angular frequency interval including both TE and TM modes is

$$\frac{dN}{d\omega} = \frac{a^3\omega^2}{\pi^2 v^3} ,$$

where v is the phase velocity in the dielectric of the cavity if unbounded.

Hint: Consider the space of rectangular axes m,n,p. Each unit volume in this space corresponds to a set of (m,n,p).

14. Show that the instantaneous power transmitted by a loss-free transmission line is

$$P(z,t) = \left(\frac{C}{L_e}\right)^{\frac{1}{2}} \{\Phi_+^2 \cos^2(kz - \omega t) - \Phi_-^2 \cos^2(kz + \omega t)\} \quad .$$

Hence show that the time-averaged power is equal to

$$< P(z) > = \frac{1}{2} \left(\frac{C}{L_e}\right)^{\frac{1}{2}} (\Phi_+^2 - \Phi_-^2) \quad .$$

15. For a parallel-wire transmission line in air, the shunt conductance G may be assumed zero. Find the expressions for k' and α in this case. Hence show that for heavy damping, i.e. $R_i^2 >> (\omega L)^2$,

$$k' \simeq \alpha \simeq \left(\frac{\omega R_i C}{2}\right)^{\frac{1}{2}} \quad ,$$

and that for light damping, i.e. $R_i^2 << (\omega L)^2$,

$$k' \simeq \omega\left\{1 + \frac{1}{8}\left(\frac{R_i}{\omega L}\right)^2\right\}(LC)^{\frac{1}{2}} \quad , \quad \alpha \simeq \frac{R_i}{2}\left(\frac{C}{L}\right)^{\frac{1}{2}} \quad ,$$

where $L = L_e + L_i$.

16. If damping may be neglected, i.e. $R_i = 0$, $G = 0$, show that the propagation constant γ for a parallel-wire transmission line is purely imaginary, and find the length ℓ of the line for which (i) the transmitting impedance equals the load impedance, (ii) the characteristic impedance is the geometric mean of the transmitting and load impedances.

17. A Lecher wire system consists of two long parallel wires spaced at d in a dielectric medium of permittivity ε_1 and permeability

μ_1, each of radius a and carrying equal and opposite charges and currents. In the first instance, let each wire be of length 2ℓ and obtain the potential at a point on the transverse plane through the mid-points of the wires assuming $d \gg a$. Then letting $\ell \to \infty$ show that the capacitance per unit length of the line is

$$C = \frac{\pi\varepsilon_1}{\ln(\frac{d}{a})} \quad .$$

Hence show that the external inductance per unit length is

$$L_e = \frac{\mu_1}{\pi} \ln(\frac{d}{a}) \quad .$$

18. A telephone line consists of two parallel copper wires each of radius $a = 0.5$ mm spaced at $d = 10a$. If the conductivity of copper is 5.9×10^7 ohm^{-1}m^{-1} and an angular frequency $\omega = 5 \times 10^6$ s^{-1} is used, calculate the resistance and internal inductance per unit length of the line. Calculate also the capacitance and external inductance of the line and indicate whether the damping is heavy or light. Hence calculate the attenuation constant and find the range through which a signal may be transmitted without further amplification.

Chapter VIII

ELECTRODYNAMICS AND SPECIAL RELATIVITY

Maxwell's equations are covariant under the Lorentz transformation and therefore compatible with the principle of relativity. The application of Einstein's special relativity in electrodynamics is however not trivial. It has introduced some fundamental conceptual changes, which include the abolition of the hypothetical ether as the carrier of electromagnetic stresses and waves, and the emphasis on the role played by the velocity of light in free space. It has also served to clarify many ideas and deepened our understanding of Maxwell's electrodynamic theory.

8.1 Covariance of Maxwell's Equations

In the preceding chapters we have seen how Maxwell's equations are capable of correlating the diversive electromagnetic phenomena, of giving a consistent interpretation to the empirical results of physical optics, and of accounting for the propagation of electromagnetic waves in general. The question one would naturally ask is whether these equations do represent general physical laws, in the sense that they are independent of the state of motion of the observer.

Suppose a phenomenon is being observed by two observers in relative motion and each makes a record of the phenomenon with reference to a coordinate frame attached to himself. A happening or an event will be recorded as taking place at a location (x, y, z) and a time t by one observer and at (x', y', z') and t' by the other. From the observation of a series of events associated with the phenomenon which he has made, each observer deduces a law correlating certain physical quantities defined in some way common to both. If the law deduced by the observers is found to have exactly the same form, for example, if the equation of the motion of a certain body is deduced by one observer to be

$$\mathbf{F} = m \frac{d^2 \mathbf{r}}{dt^2} \quad ,$$

and by the other to be

$$\mathbf{F}' = m' \frac{d^2 \mathbf{r}'}{dt'^2} \quad ,$$

where $\mathbf{F}$ is the resultant force acting on the body and m its mass according to one observer, $\mathbf{F}'$ and m' being the corresponding quantities according to the other, the quantities being defined in the same way by both, then the law is independent of the state of motion of an observer and may be considered a general law.

In practice it is difficult, nor is it necessary, to have two such observers making observations of the same phenomenon. It suffices to have an observer making the required observations and deducing the law concerning the phenomenon. Then, by means of a set of equations known as a *coordinate transformation*, agreed upon by the two observers for relating (x, y, z, t) and (x', y', z', t'), the law may be expressed in terms of quantities pertaining to a second observer. If the two expressions can be made identical by suitably defining the quantities involved in a consistent way, the law is a general law and the equation expressing it is said to be *covariant* under the transformation. The difficulty is that it is not at all obvious what should be the coordinate transformation. We have either to hypothesize certain principles which will specify a suitable transformation or to assume certain physical laws to be general and accordingly derive the transformation from their covariance property. Furthermore, we may have to restrict the mode of the relative motion of the observers, at least in the first instance.

In the Newtonian mechanics, the first law of motion or the law of inertia defines a set of coordinate systems to which Newton's laws of motion apply. These coordinate systems move with uniform translatory velocities relative to one another and are called *inertial systems* or *frames*. It is remarkable that a frame attached to the earth may to the first approximation be considered an inertial frame when the rotation of the earth may be neglected. A better approximation would be one attached to a distant star. If the Galilean transformation is used to relate the coordinates of the inertial frames, the laws of mechanics as formulated by Newton will retain the same form in every frame. This fact is known as the *Galilean relativity*. The *Galilean transformation* may be derived from the assumption of a universal time, i.e. a common time for all inertial observers, and the generality of Newton's second law on which Newtonian mechanics is based.

Consider two inertial frames $\Sigma(x, y, z)$ and $\Sigma'(x', y', z')$ where the respective coordinate axes are parallel and Σ' moves with a uniform velocity v relative to Σ along the x-direction. The origin of time is chosen such that at $t = 0$ the origins of Σ and Σ' coincide. As the simplest working assumption, we assume $\mathbf{F}$ and m to be invariant quantities, i.e., to have the same values in all inertial frames. Such an assumption would have to be modified if it should turn out that inconsistencies occur, which however has not been the case. The covariance of Newton's second law then yields

$$\frac{d^2x'}{dt^2} = \frac{d^2x}{dt^2} \quad ,$$

or, on integration,

$$\frac{dx'}{dt} = \frac{dx}{dt} + C$$

where C is a constant of integration. As a body at rest in Σ' will be observed to move with velocity v in Σ, $C = -v$. Further integration gives

$$x' = x - vt \quad ,$$

the constant of integration being zero as $x' = x$ at $t = 0$. The transverse axes are not affected by the relative motion so that $y' = y$, $z' = z$. The Galilean transformation for the simple geometry considered may therefore be represented by the equations

$$x' = x - vt \quad , \quad y' = y \quad , \quad z' = z \quad , \quad t' = t \quad . \tag{8.1}$$

We may now investigate the implication of the application of this transformation to electrodynamics. To do this we shall confine ourselves to vacuum or free space.

Maxwell's theory contains a constant $c = (\mu_0 \varepsilon_0)^{-\frac{1}{2}}$ which is interpreted as the velocity of propagation of electromagnetic waves or light in free space. The Galilean transformation for the components of the velocity of a point is obtained by differentiating Eq. (8.1):

$$\frac{dx'}{dt} = \frac{dx}{dt} - v \quad , \quad \frac{dy'}{dt} = \frac{dy}{dt} \quad , \quad \frac{dz'}{dt} = \frac{dz}{dt} \quad . \tag{8.2}$$

The application of this transformation would mean that the velocity of light in free space would depend on the velocity of the frame of the observer. It follows then that there is a unique frame in which the velocity of light is c, while in any other inertial frame the relative velocity v with respect to this unique frame must appear in the expression for the velocity of light, and hence in the electromagnetic laws. This unique frame referring to which Maxwell's equations are assumed to be valid is known as the *absolute frame*. It is presumably attached to a hypothetical medium called the *ether* which, prevading all space, vacuum or otherwise, is supposed to act as the carrier of electromagnetic waves.

Equations (8.2) imply that in a frame moving relative to the ether such as the earth, the velocity of light in free space has the values $c+v$ or $c-v$ according to whether the motion is parallel or antiparallel to v, and $\sqrt{c^2-v^2}$ in a transverse direction. Experiments can therefore be performed by which the velocity v of the apparatus with respect to the ether may be inferred. For instance we may observe the shift in the interference pattern, formed by coherent lights which have travelled different paths, when the apparatus is rotated.

Such an experiment was performed by Michelson and Morley using a Michelson interferometer of equal arm lengths, but negative results were obtained. To explain the failure in detecting v, Fitzgerald and Lorentz independently postulated that length was contracted in the direction of motion relative to the ether in such a way that it exactly compensated the change in the difference of the times of travel of the interfering rays which accompanied the rotation of the apparatus. This postulate however would still fail to explain the negative results of the Kennedy-Thorndike experiment performed later employing a Michelson interferometer of unequal arm lengths[a]. To preserve the concept of an absolute frame, one might still assume the ether to be dragged along completely with the apparatus in its motion relative to the absolute frame. The velocity of light would then be c relative to the apparatus in all directions. This hypothesis of ether-drag however contradicted the observations of star light abberation [Prob. (8.26)].

Both the hypotheses of length contraction and of complete ether-drag were the results of an attempt to explain the empirical fact that light appeared to travel with the same speed in all directions in a frame attached to the earth, which in the first approximation may be regarded as an inertial frame. Both failed to explain all the

---

[a] A discussion of these experiments is given, for example, in *Handbook of Physics*, 2nd ed., pp. 6-162 through 6-164, edited by E.U. Condon and H. Odishaw, McGraw-Hill, New York (1967).

observations. While other hypotheses of a similar nature might presumably be proposed, one asks if it would not be a more fruitful approach simply to accept the empirical fact that the speed of light in free space is the same in all directions and in all inertial frames by regarding it as a postulate in itself. Furthermore, since the motion of the earth relative to the absolute frame could not be demonstrated, would it still be meaningful to speculate on the existence of such a frame? It would certainly be more logical simply to assume that all inertial frames are equivalent in the description of the electromagnetic phenomena.

If we accept the postulates of the constancy of the velocity of light and the equivalence of all inertial frames, then there is no intrinsic reason to object to the generality of Maxwell's equations. Starting from this point of view the appropriate coordinate transformation may be sought. It turned out that the Lorentz transformation, and the Lorentz transformation alone, could satisfy these requirements.

Einstein extended the equivalence of all inertial frames to cover the entire field of physics. This postulate, together with that of the invariance of the velocity of light, forms the basis of his special theory of relativity. It then follows that the transformation appropriate to the theory is the Lorentz transformation.

8.2 Principle of Relativity and the Lorentz Transformation

In his special theory of relativity, Einstein proposed the *principle of relativity* as follows:

'The laws by which the states of physical systems alter are independent of the choice as to which of two coordinate frames in uniform translatory motion relative to each other these alterations of states are referred'.

A physical law that satisfies this principle is considered a *general law;* otherwise it must be reformulated to meet this requirement. Since the velocity of light c appears in Maxwell's theory of vacuum electrodynamics, the assumption that Maxwell's equations are general laws would require it to be the same in all inertial frames. The isotropy of space would require it to be the same in all directions. Without referring to electrodynamics, the invariance of the velocity of light in all inertial frames is often stated as a separate postulate. This and the principle of relativity form the basis of Einstein's theory of special relativity.

To derive the transformation between two intertial frames consistent with these postulates we first note that, as velocity is a differential dx_i/dt, for the velocity of light to be an invariant the transformation must involve t as well as x_i. Furthermore, as the frames under consideration are inertial, a body under no force must remain in the state of rest or of uniform motion. Hence its motion must be described by a linear equation of x_i and t in each frame. It follows then that the desired transformation must itself be linear in x_i and t.

Let Σ and Σ' be two inertial Cartesian coordinate frames with parallel axes, the x-axis being in the direction of the uniform relative velocity v of Σ' with respect to Σ. Choose as the zero of time measurements the instant when the two origins O and O' coincide. At that same instant, let a spherical light wave be emitted from the origin. The wavefront at a subsequent instant is described in the two frames by

$$x^2 + y^2 + z^2 - c^2t^2 = 0 \tag{8.3}$$

and

$$x'^2 + y'^2 + z'^2 - c^2t'^2 = 0 \tag{8.4}$$

respectively, in accordance with the principle of relativity and the invariance of the velocity of light.

Substitution of the desired transfromation equations in Eq. (8.4) must give rise to Eq. (8.3). This can be achieved only if

$$x'^2 + y'^2 + z'^2 - c^2t'^2 = k(x^2 + y^2 + z^2 - c^2t^2) \quad ,$$

where k is a constant depending on the scales chosen. It is most convenient to choose the scales so that $x_i' \to x_i$ and $t' \to t$ as $v \to 0$. We shall then have $k = 1$. Note that the condition $t' \to t$ as $v \to 0$ implies that the two clocks used for measuring time in Σ and Σ' must be synchronized so that both their readings and rates are the same by first placing them stationary relative to one another. Since the velocity of light is the same in all directions, such synchronization can be easily carried out using light signals even if the clocks are separated by a distance. Now, as only the x-axes have been given a preferred direction while the y- and z-axes remain largely arbitrary, the above must imply that $y' = y$, $z' = z$ and

$$x'^2 - c^2t'^2 = x^2 - c^2t^2 \quad . \tag{8.5}$$

It remains to find the linear relations $x' = x'(x, t)$ and $t' = t'(x, t)$ which will give rise to Eq. (8.5).

As the origin O' of Σ' has coordinates $x' = 0$ in Σ' and $x = vt$ in Σ, the linear equation for x' must have the form

$$x' = \gamma(x - vt) \quad ,$$

where γ is independent of the spatial and time coordinates but may depend on v. To satisfy Eq. (8.5) the transformation for t must also be homogeneous and linear. Let it be

$$t' = ax + bt \quad ,$$

where a and b are constants but may depend on v. Substituting these expressions in Eq. (8.5) and equating the coefficients of x^2, xt and t^2 on the two sides, three equations relating γ, a and b are obtained which have the solutions

$$\gamma = \pm \frac{1}{\sqrt{1 - \frac{v^2}{c^2}}} \quad , \quad b = \pm\gamma \quad , \quad a = \mp\gamma\frac{v}{c^2} \quad .$$

The requirement that as $v \to 0$, $t' \to t$ and $x' \to x$ means that as $v \to 0$, $\gamma \to +1$ and $b \to +1$. It is clear then that the upper signs are to be used.

Thus the transformation sought may be represented by the equations

$$x' = \gamma(x - \beta ct) \quad ,$$

$$y' = y \quad ,$$

$$z' = z \quad ,$$

$$ct' = \gamma(ct - \beta x) \quad , \tag{8.6}$$

where

$$\gamma = \frac{1}{\sqrt{1 - \beta^2}} \tag{8.7}$$

and

$$\beta = \frac{v}{c} \quad . \tag{8.8}$$

These equations relate the coordinates and time (x, y, z, t) in Σ to (x', y', z', t') in Σ' of the same event and have been uniquely determined for the geometry specified from the postulates of the special theory. It follows that a physical law that satisfies the principle of relativity must have the same form in all inertial frames if the above transformation is employed. The law is then said to be *Lorentz covariant* and may be considered a general law. Covariance under the Lorentz transformation therefore provides a criterion for the test of generality of a physical law.

The transformation represented by Eqs. (8.6) had been obtained earlier by Lorentz on the basis of an ether theory and is consequently known as the *Lorentz transformation*. In this theory the laws of electrodynamics are assumed to be valid in an absolute frame attached to the ether. In motions through the ether, longitudinal lengths are contracted and clocks are slowed down, in such manners as to give a negative result to the Michelson-Morley experiment. Taking into account such changes, the transformation to be used between the absolute frame and a uniformly moving frame is the Lorentz transformation [Prob. (8.1)]. The fact that Maxwell's equations turned out to be invariant was not attached much significance. In relativity, on the other hand, the generality of a physical law is tested by its covariant property under the Lorentz transformation.

The inverse of the Lorentz transformation must have exactly the same form as Eq. (8.6) by virtue of the principle of relativity. As the roles of Σ and Σ' are now interchanged, the relative velocity is $-v$. The equations for the inverse Lorentz transformation are therefore

$$x = \gamma(x' + \beta ct') \quad ,$$

$$y = y' \quad ,$$

$$z = z' \quad ,$$

$$ct = \gamma(ct' + \beta x') \quad . \tag{8.9}$$

It should be noted that as γ becomes infinite for v approaching c, the transformation applies only if $v < c$. This restriction will always be adhered to. In fact, velocities that are greater than the velocity of light have no place in special relativity. The factor γ, which plays an important role in relativity, is often referred to as the *Lorentz factor.*

8.3 The Lorentz-Fitzgerald Contraction and Time Dilation

The length of a rigid body[b] in a particular frame is defined as the distance between the locations of its extreme points which are determined at the same instant. In particular, the length in the frame in which the body is at rest i.e. the rest frame of the body, is called its *rest* or *proper length.* According to Eq. (8.6), a length $\ell_0 = x_2' - x_1'$ in the rest frame Σ' is observed in Σ to be $\ell = x_2 - x_1$,

---

[b] See the remark at the end of the section. Here the term rigid body is taken to mean a body which may be considered rigid in its rest frame.

where x_1 and x_2 are given by

$$x_2' = \gamma(x_2 - \beta ct_2)$$

and

$$x_1' = \gamma(x_1 - \beta ct_1)$$

with $t_2 = t_1$. It follows that

$$\ell_o = \gamma\ell \quad .$$

Since γ is always greater than unity, $\ell < \ell_o$. The length of a rigid body moving with a uniform velocity βc thus appears to contract by a factor $(1-\beta^2)^{-\frac{1}{2}}$ in the direction of motion. The contraction was first proposed by Lorentz and Fitzgerald independently for explaining the negative results of the Michelson-Morley experiment and is known as the *Lorentz-Fitzgerald contraction*. Note that as the contraction is a function of v^2 a length along the direction of relative motion and stationary in Σ will also be observed to contract by the same factor in Σ'. As $y_2' - y_1' = y_2 - y_1$, etc., no contraction will be observed for transverse lengths. We may summarize these results as follows:

$$\left(\frac{\delta x'}{\delta x}\right)_t = \gamma \quad , \quad \left(\frac{\delta y'}{\delta y}\right)_t = \left(\frac{\delta z'}{\delta z}\right)_t = 1 \quad . \tag{8.10}$$

Consider now the time interval $t_2' - t_1'$ between two events occuring at the same location x' in Σ'. When observed in Σ the interval $t_2 - t_1$ between the same two events is given by

$$ct_2 = \gamma(ct_2' + \beta x')$$

$$ct_1 = \gamma(ct_1' + \beta x') \quad .$$

Hence

$$t_2 - t_1 = \gamma(t_2' - t_1') \quad ,$$

or

$$\left(\frac{\delta t}{\delta t'}\right)_{x'} = \gamma \quad . \tag{8.11}$$

As $\gamma > 1$ always, $\delta t > \delta t'$, meaning that a moving clock will appear to go slow.

This effect is known as *time dilation* or *time dilatation*. Again note that a clock stationary in Σ will likewise appear to go slow to an observer in Σ'.

The concept of *simultaneity* is also relative. Consider two events occurring at the instant t' and the locations x_1' and x_2' according to Σ'. To an observer in Σ, the times of the events will be given by

$$ct_1 = \gamma(ct' + \beta x_1') \quad ,$$

and

$$ct_2 = \gamma(ct' + \beta x_2') \quad ,$$

so that

$$c(t_2 - t_1) = \gamma\beta(x_2' - x_1') \quad .$$

This means that $t_2 \neq t_1$ unless $x_2' = x_1'$. Thus two events simultaneous in one inertial system will not in general be simultaneous in another. Two events are generally simultaneous only if they occur at the same location as well as at the same instant.

The result that simultaneity at different places is only relative has important consequences in physics. For example, Newton's third law which asserts that the action and reaction between two interacting mass points are equal and opposite at any one instant admits no direct extension to special relativity unless the interaction takes place in contact. The definition of a rigid body encounters difficulty for the same reason.

8.4 Four-vectors

As time is involved in the Lorentz transformation, it is convenient to consider a four-dimensional space with ct as the fourth coordinate — the *space-time continuum* or the *'world'* — and write

$$x^1 = x \quad , \quad x^2 = y \quad , \quad x^3 = z \quad , \quad x^4 = ct \quad .$$

The set of quantities x^1, x^2, x^3, x^4 are said to form a *space-time four-vector*. Each such vector defines an event or a "world point" in the four-dimensional space-time "world". A Lorentz transformation may be considered as a coordinate rotation in the world since the transformation equations are linear and homogeneous in the four coordinates [Prob. (8.2)].

The differentials dx^1, dx^2, dx^3, dx^4 form the *space-time interval* and transform according to the rule of partial derivation[c]

$$dx'^{\alpha} = \frac{\partial x'^{\alpha}}{\partial x^{\beta}} dx^{\beta} = Q^{\alpha}_{\beta} dx^{\beta} \quad ,$$

where, as the Lorentz transformation, being linear and homogeneous, applies also to the differentiatials, Q^{α}_{β} are the elements of the symmetric 4×4 matrix

$$Q = \begin{bmatrix} \gamma & 0 & 0 & -\beta\gamma \\ 0 & 1 & 0 & 0 \\ 0 & 0 & 1 & 0 \\ -\beta\gamma & 0 & 0 & \gamma \end{bmatrix} \quad . \tag{8.12}$$

The inverse transformation is then

$$dx^{\gamma} = \frac{\partial x^{\gamma}}{\partial x'^{\alpha}} dx'^{\alpha} = \frac{\partial x^{\gamma}}{\partial x'^{\alpha}} Q^{\alpha}_{\beta} dx^{\beta} = \delta^{\gamma}_{\beta} dx^{\beta} \quad .$$

As the determinant $|Q| \neq 0$, the last relation shows that $\partial x^{\gamma}/\partial x'^{\alpha}$ is an element of the reciprocal matrix Q^{-1}. We may therefore write the inverse transformation as

$$dx^{\gamma} = (Q^{-1})^{\gamma}_{\alpha} dx'^{\alpha} \quad .$$

---

c We shall use Greek alphabets for indices which run from 1 to 4 and restrict Latin alphabets to indices which run from 1 to 3. Note that the alphabets i,j,k,... are also used as suffixes in the suffix notation, where no distinction is made between contravariant and covariant representations.

From the inverse Lorentz transformation we can immediately write

$$Q^{-1} = \begin{bmatrix} \gamma & 0 & 0 & \beta\gamma \\ 0 & 1 & 0 & 0 \\ 0 & 0 & 1 & 0 \\ \beta\gamma & 0 & 0 & \gamma \end{bmatrix} \quad . \tag{8.13}$$

It is easy to verify that $QQ^{-1} = I$, the unit matrix.

In general, we may define a *contravariant* four-vector as a set of four quantities T^α that transform like dx^α:

$$T'^\alpha = Q^\alpha_\beta \, T^\beta \quad ,$$

or

$$T^\alpha = (Q^{-1})^\alpha_\beta \, T'^\beta \quad . \tag{8.14}$$

A function $V(x,y,z,t)$ that is invariant under the Lorentz transformation, i.e. $V(x,y,z,t) = V(x',y',z',t')$, is called a *scalar*. The components of the gradient of a scalar V will transform according to

$$\frac{\partial V}{\partial x'^\alpha} = \frac{\partial x^\beta}{\partial x'^\alpha}\frac{\partial V}{\partial x^\beta} = (Q^{-1})^\beta_\alpha \frac{\partial V}{\partial x^\beta} \quad .$$

In general, we may define a *covariant* four-vector as any set of four quantities T_α that transform like $\frac{\partial V}{\partial x^\alpha}$, i.e.

$$T'_\alpha = (Q^{-1})^\beta_\alpha \, T_\beta \quad . \tag{8.15}$$

The inverse transformation for a covariant four-vector follows,

$$T_\alpha = Q^\beta_\alpha \, T'_\beta \quad . \tag{8.16}$$

8.5 Transformation of Electromagnetic Field Vectors

Maxwell's equations for vacuum are.

$$\nabla \times \mathbf{E} = -\frac{\partial \mathbf{B}}{\partial t} \quad , \tag{8.17}$$

$$\nabla \times \mathbf{B} = \frac{1}{c^2}\frac{\partial \mathbf{E}}{\partial t} + \mu_0 \mathbf{J} \quad , \tag{8.18}$$

$$\nabla \cdot \mathbf{E} = \frac{\rho}{\varepsilon_0} \quad , \tag{8.19}$$

$$\nabla \cdot \mathbf{B} = 0 \quad , \tag{8.20}$$

where as $c = (\mu_0\varepsilon_0)^{-\frac{1}{2}}$ is an invariant constant, μ_0 and ε_0 must also be treated as invariants. If these equations represent general laws and hence are Lorentz covariant, we must have in Σ'

$$\nabla' \times \mathbf{E}' = -\frac{\partial \mathbf{B}'}{\partial t'} \quad , \tag{8.21}$$

$$\nabla' \times \mathbf{B}' = \frac{1}{c^2}\frac{\partial \mathbf{E}'}{\partial t'} + \mu_0 \mathbf{J}' \quad , \tag{8.22}$$

$$\nabla' \cdot \mathbf{E}' = \frac{\rho'}{\varepsilon_0} \quad , \tag{8.23}$$

$$\nabla' \cdot \mathbf{B}' = 0 \quad . \tag{8.24}$$

The inverse transformation of $\frac{\partial}{\partial x^\alpha}$ is given by Eq. (8.16) to be

$$\frac{\partial}{\partial x} = \gamma\left(\frac{\partial}{\partial x'} - \frac{\beta}{c}\frac{\partial}{\partial t'}\right) \quad , \quad \frac{\partial}{\partial y} = \frac{\partial}{\partial y'} \quad , \quad \frac{\partial}{\partial z} = \frac{\partial}{\partial z'} \quad ,$$

$$\frac{1}{c}\frac{\partial}{\partial t} = \gamma\left(-\beta\frac{\partial}{\partial x'} + \frac{1}{c}\frac{\partial}{\partial t'}\right) \quad . \tag{8.25}$$

Applying these to Eq. (8.19) and the x-component of Eq. (8.18), we have

$$\gamma\frac{\partial E_x}{\partial x'} - \frac{\beta\gamma}{c}\frac{\partial E_x}{\partial t'} + \frac{\partial E_y}{\partial y'} + \frac{\partial E_z}{\partial z'} = \frac{\rho}{\varepsilon_0} \quad ,$$

$$-\frac{\beta\gamma}{c}\frac{\partial E_x}{\partial x'} + \frac{\gamma}{c^2}\frac{\partial E_x}{\partial t'} + \mu_o J_x = \frac{\partial B_z}{\partial y'} - \frac{\partial B_y}{\partial z'} \quad .$$

Eliminating $\partial E_x/\partial x'$ from the above we find

$$\frac{\partial}{\partial y'}\left\{\gamma(B_z - \frac{\beta}{c} E_y)\right\} - \frac{\partial}{\partial z'}\left\{\gamma(B_y + \frac{\beta}{c} E_z)\right\}$$

$$= \frac{1}{c^2}\frac{\partial E_x}{\partial t'} + \mu_o\gamma(J_x - \beta c\rho) \quad .$$

The remaining components of Eq. (8.18) similarly give

$$\frac{\partial B_x}{\partial z'} - \frac{\partial}{\partial x'}\left\{\gamma(B_z - \frac{\beta}{c} E_y)\right\} = \frac{1}{c^2}\frac{\partial}{\partial t'}\left\{\gamma(E_y - \beta c B_z)\right\} + \mu_o J_y \quad ,$$

$$\frac{\partial}{\partial x'}\left\{\gamma(B_y + \frac{\beta}{c} E_z)\right\} - \frac{\partial B_x}{\partial y'} = \frac{1}{c^2}\frac{\partial}{\partial t'}\left\{\gamma(E_z + \beta c B_y)\right\} + \mu_o J_z \quad .$$

A comparison of the last three equations with the component equations of (8.22) gives

$$E'_x = E_x \quad , \qquad B'_x = B_x \quad ,$$

$$E'_y = \gamma(E_y - \beta c B_z) \quad , \qquad B'_y = \gamma(B_y + \frac{\beta}{c} E_z) \quad ,$$

$$E'_z = \gamma(E_z + \beta c B_y) \quad , \qquad B'_z = \gamma(B_z - \frac{\beta}{c} E_y) \quad ; \qquad (8.26)$$

$$J'_x = \gamma(J_x - \beta c\rho) \quad ,$$

$$J'_y = J_y \quad ,$$

$$J'_z = J_z \quad . \qquad (8.27)$$

The same transformation equations would be obtained also if we had made use of Eqs. (8.17) and (8.20) and their counterpart in Σ'. This confirms the Lorentz covariance of Maxwell's equations.

The transformation equations (8.26) for the field vectors may be written concisely as

$$E'_{\parallel} = E_{\parallel} \quad , \qquad B'_{\parallel} = B_{\parallel} \quad ,$$

$$\mathbf{E}'_{\perp} = \gamma(\mathbf{E}_{\perp} + \mathbf{v} \times \mathbf{B}_{\perp}) \quad ,$$

$$\mathbf{B}'_{\perp} = \gamma(\mathbf{B}_{\perp} - \frac{\mathbf{v} \times \mathbf{E}_{\perp}}{c^2}) \quad ,$$

where the symbol $\parallel$ indicates components longitudinal to the relative velocity $\mathbf{v}$, and $\perp$ indicates transverse components. These equations relate the field vectors at a given point at a given instant as observed by two observers in uniform motion relative to each other.

It should be noted that as a purely electric or a purely magnetic field may be observed in a different frame as a combination of electric and magnetic fields, and vice versa, the division of the electromagnetic field into electric and magnetic fields is not fundamental but merely a reflection of the choice of the observer's frame.

The transformation equation for the charge density ρ is found by substituting the expressions (8.26) in Eq. (8.23) and making use of the inverse of Eq. (8.25) to be

$$\frac{\rho'}{\varepsilon_0} = \gamma(\nabla \cdot \mathbf{E}) - \gamma\beta c \left(\frac{\partial B_z}{\partial y} - \frac{\partial B_y}{\partial z} - \frac{1}{c^2}\frac{\partial E_x}{\partial t}\right)$$

or, using Eqs. (8.18) and (8.19),

$$\rho' = \gamma(\rho - \frac{\beta}{c} J_x) \quad . \tag{8.28}$$

This equation and Eqs. (8.27) show that J_x/c, J_y/c, J_z/c and ρ form a contravariant four-vector, the *current density four-vector*.

Since ρ is a component of a four-vector, it is not an invariant quantity. The total charge, on the other hand, is invariant. Let Σ' be the rest frame of a charge, then the inverse of Eq. (8.28) gives

$$\rho = \gamma(\rho' + \frac{\beta}{c} J_x') = \gamma\rho' \quad ,$$

as $\mathbf{J}' = 0$ in the rest frame. The total charge in Σ is by definition

$$\int \rho dxdydz = \int \gamma\rho' \frac{dx'}{\gamma} dy'dz' = \int \rho' dx'dy'dz' \quad ,$$

where use has been made of the Lorentz-Fitzgerald contraction relation (8.10).

Thus the total charge within some finite boundary in any inertial frame is equal to the total charge in the rest frame and is therefore an invariant. It follows directly that the electronic charge e is an invariant quantity. It may then be inferred that as the measure of a charge, which is simply the counting of the number of electronic charges present therein, is independent of its motion, charge conservation and hence the continuity equation are valid relativistically.

8.6 Transformation of Electromagnetic Potentials

Since Maxwell's equations and the continuity equation of charge are Lorentz covariant, the Lorentz condition

$$\nabla \cdot \mathbf{A} + \frac{1}{c^2}\frac{\partial \Phi}{\partial t} = 0 \quad , \tag{8.29}$$

which is equivalent to the continuity equation, and the differential equations for the vector and scalar potentials,

$$\nabla^2 \mathbf{A} - \frac{1}{c^2}\frac{\partial^2 \mathbf{A}}{\partial t^2} = -\mu_0 \mathbf{J} \tag{8.30}$$

and

$$\nabla^2 \Phi - \frac{1}{c^2}\frac{\partial^2 \Phi}{\partial t^2} = -\frac{\rho}{\varepsilon_0} \quad , \tag{8.31}$$

are also Lorentz covariant. The transformation relations for the potentials may then be obtained in the same manner as for the field vectors. Applying Eq. (8.25) to Eq. (8.29) we obtain

$$\frac{\partial}{\partial x'}\left\{\gamma(A_x - \frac{\beta}{c}\Phi)\right\} + \frac{\partial A_y}{\partial y'} + \frac{\partial A_z}{\partial z'} + \frac{1}{c^2}\frac{\partial}{\partial t'}\left\{\gamma(\Phi - \beta c A_x)\right\} = 0 \quad .$$

A comparison of this with the counterpart of Eq. (8.29) in the Σ' frame gives

$$A'_x = \gamma(A_x - \beta\frac{\Phi}{c}) \quad ,$$

$$A'_y = A_y \quad ,$$

$$A'_z = A_z \quad ,$$

$$\frac{\Phi'}{c} = \gamma(\frac{\Phi}{c} - \beta A_x) \quad . \tag{8.32}$$

Again we see that cA_x, cA_y, cA_z, and Φ form a contravariant four-vector, the *electromagnetic potential four-vector.*

These transformation equations may be used to find the potentials of a point charge q moving with a uniform velocity $\mathbf{v}$. Let $\mathbf{r} = (x,y,z)$ be the position vector of a field point fixed in Σ, and Σ' be the rest frame of the charge. As was pointed out in Sec. 4.4, electromagnetic effects propagate with the velocity of light c, which is the same in both frames. Without loss of generality we may assume the charge to be situated at the origin of Σ' and consider effects that leave the charge at the instant $t = t' = 0$ when the origins of the two frames coincide. They will arrive at the field point at the instant $t = r/c$ according to Σ, and $t' = r'/c$ according to Σ', where $\mathbf{r}'$ is the position vector of the field point in Σ' when the effects arrive. The Lorentz transformation gives

$$ct' = \gamma(ct - \beta x)$$

or

$$r' = \gamma(r - \beta x) = \gamma r(1 - \frac{v_r}{c}) \quad , \tag{8.33}$$

where $v_r = \frac{vx}{r}$ is the (retarded) radial velocity of the charge towards the field point.

Since the charge is stationary in Σ', $\mathbf{J}' = 0$ and the only non-vanishing potential is

$$\Phi' = \frac{1}{4\pi\varepsilon_0} \frac{q}{r'} \quad .$$

The inverse of the transformation equations (8.32) then gives

$$A_x = \frac{\gamma\beta}{c} \Phi' = \frac{\mu_0}{4\pi} \frac{qv}{r\left(1 - \frac{v_r}{c}\right)} \quad ,$$

$$A_y = A_y' = 0 \quad , \qquad A_z = A_z' = 0 \quad ,$$

$$\Phi = \gamma\Phi' = \frac{1}{4\pi\varepsilon_0} \frac{qv}{r(1 - \frac{v_r}{c})} \quad .$$

The vector potential may be written in the vector form

$$\mathbf{A} = \frac{\mu_0}{4\pi} \frac{q\mathbf{v}}{r(1 - \frac{v_r}{c})} \quad .$$

These expressions for the potentials of a moving point charge are identical with those for the Liénard-Wiechert potentials, Eqs. (4.29) and (4.30), which have been derived without the restriction of uniform motion for the charge. Since both are based, in the final analysis, on the validity of Maxwell's equations, it would appear that the assumption of uniform motion is unimportant in first order calculations. This demonstrates the general empirical rule that results obtained in special relativity can usually be applied to problems where the relative velocity v is not uniform, in spite of the fact that accelerated frames have no place in the theory. In the present case, if the charge has acceleration the rest frame Σ' is to be interpreted as the inertial frame in which the charge is momentarily at rest at the instant when the electromagnetic effects leave the source.

8.7 The Doppler Effect

The frequency of an electromagnetic wave measured by an observer depends on his motion relative to the source. This effect, called the *Doppler effect*, can usually be explained without using special relativity. The theory of relativity introduces an additional factor due to time dilation.

Consider a source of monochromatic electromagnetic waves moving with velocity **v** relative to an observer. Let the rest frames of the source and of the observer be Σ' and Σ respectively, chosen in such a way that the source is at the origin of Σ' and the x-axes are along the direction of **v**, and let the coordinates of the observer be (x, y, z) in Σ.

Suppose a wave is emitted at the instant t' from the source. This event is observed in Σ to have the coordinates and time

$$x_0 = \gamma(x' + \beta ct') = \gamma vt' \quad , \quad y_0 = 0 \quad , \quad z_0 = 0 \quad ,$$

$$t_0 = \gamma\left(t' + \frac{\beta x'}{c}\right) = \gamma t' \quad .$$

If the wave arrives at the observer at time t in his frame, then as the wave propagates with velocity c we have the relation for $\bar{r}$, the distance between the location of the source when the electromagnetic wave leaves and the observer:

$$\bar{r}^2 = (x - \gamma vt')^2 + y^2 + z^2 = c^2(t - \gamma t')^2 \quad .$$

Differentiating we obtain

$$dt = \gamma\left\{1 - \frac{v}{c}\frac{(x - x_0)}{\bar{r}}\right\} dt' \quad , \tag{8.34}$$

x, y, z being kept fixed since the observer is stationary in Σ.

We may interpret dt' as the interval between the instants of emission of two consecutive waves by the source and dt as the interval between the instants of arrival of the two waves. Then if ω_0

and ω are respectively the proper and observed angular frequencies of the waves, we have for a fixed number of waves

$$\omega_0 dt' = \omega dt \quad .$$

Using this relation in Eq. (8.34) we find

$$\omega_0 = \gamma(1 - \frac{v_r}{c})\omega \quad , \tag{8.35}$$

where v_r is the radial velocity of the source *towards* the observer. This equation describes the *relativistic Doppler effect* and differs from the non-relativistic expression by the factor γ, which is a correction second order in v/c. Note that Eq. (8.33) cannot be used to obtain the expression for the Doppler effect since here t' represents not the time of emission of a wave from the source, but the time of arrival of a wave at the moving observer according to the frame Σ'.

Two particular situations are of interest. If the source moves radially away from the observer, then $v_r = -v$ and Eq. (8.35) reduces to

$$\frac{\omega}{\omega_0} = \sqrt{\frac{1 - \frac{v}{c}}{1 + \frac{v}{c}}} \quad .$$

The change of frequency given by this equation is first order in v/c and is called the *longitudinal Doppler effect*. If the source has no radial motion, then $v_r = 0$ and

$$\frac{\omega}{\omega_0} = \sqrt{1 - \frac{v^2}{c^2}} \quad .$$

This effect is called the *transverse Doppler effect* and has no non-relativistic equivalent. It is second order in v/c and arises entirely from time dilation as is obvious from Eq. (8.34). The transverse Doppler effect has been used as the basis of an experimental verification of time dilation[d].

8.8 Covariant Formulation of Physical Laws

We have encountered many sets of quantities which constitute four-vectors. These are:

(a) the space-time four-vector : $x^\alpha = (x, y, z, ct)$,

(b) the space-time interval : $dx^\alpha = (dx, dy, dz, cdt)$,

(c) the current density four-vector: $J^\alpha = (\frac{J_x}{c}, \frac{J_y}{c}, \frac{J_z}{c}, \rho)$,

(d) the electromagnetic potential four-vector : $A^\alpha = (cA_x, cA_y, cA_z, \Phi)$.

As the components of all the four-vectors transform in the same way when a Lorentz transformation is applied to space and time co-ordinates, if we can express a physical law in the form of an equation of which each term transforms like the corresponding component of a four-vector then the equation will automatically retain the same form under the Lorentz transformation. For example, if A, B, C are four-vectors and a, b, c are scalars or invariant constants, a physical law of the form

$$aA'^\alpha + bB'^\alpha + cC'^\alpha = 0 \quad , \quad \alpha = 1,2,3,4 \tag{8.36}$$

will transform under a Lorentz transformation into

$$Q^\alpha_\beta(aA^\beta + bB^\beta + cC^\beta) = 0 \quad .$$

If we multiply both sides by $(Q^{-1})^\gamma_\alpha$ and sum over α, then as

---

[d] Hay, Schiffer, Cranshaw & Egelstaff, *Phys. Rev. Lett.* **4** (1960) 165.

$(Q^{-1})^{\gamma}_{\alpha}Q^{\alpha}_{\beta} = \delta^{\gamma}_{\beta}$ it becomes

$$aA^{\gamma} + bB^{\gamma} + cC^{\gamma} = 0 \quad , \qquad \gamma = 1,2,3,4 \quad .$$

Thus the law represented by the four-vector equation (8.36) is Lorentz covariant. A law expressible as a four-vector equation is therefore a general law.

Not all general laws, however, can be formulated in terms of four-vectors. More generally, tensors must be employed. A tensor is a set of quantities which transform like the products of the components of four-vectors. The rank of the tensor is the minimum number of the vectors involved. For example, there are three types of tensors of rank 2 each having 4^2 components:

Contravariant tensors which transform according to

$$T'^{\alpha\beta} = Q^{\alpha}_{\gamma}Q^{\beta}_{\delta}T^{\gamma\delta} \quad .$$

Covariant tensors which transform according to

$$T'_{\alpha\beta} = (Q^{-1})^{\gamma}_{\alpha}(Q^{-1})^{\delta}_{\beta}T_{\gamma\delta} \quad .$$

Mixed tensors which transform according to

$$T'^{\alpha}_{\beta} = Q^{\alpha}_{\gamma}(Q^{-1})^{\delta}_{\beta}T^{\gamma}_{\delta} \quad .$$

The extension of these definitions to a tensor of any rank is obvious. Furthermore, we may by definition consider a four-vector a tensor of rank 1 and a scalar a tensor of rank 0.

An important tensor operation is *contraction*. By equating a contravariant index and a covariant index in the product of two or more tensors or in a mixed tensor, the total rank of the product or the mixed tensor is reduced by 2. Consider for example the product of two four-vectors which has been contracted:

$$A'^{\alpha}B'_{\alpha} = Q^{\alpha}_{\beta}(Q^{-1})^{\gamma}_{\alpha}A^{\beta}B_{\gamma} = \delta^{\gamma}_{\beta}A^{\beta}B_{\gamma} = A^{\beta}B_{\beta} \quad .$$

The product is invariant and therefore a scalar.

A physical law formulated as an equation in which each term is a tensor of a given rank is clearly Lorentz covariant since all the terms involved transform in the same way. Conversely a law that cannot be so formulated may not be a general law and must be modified. The formulation of a physical law as a tensor equation is called a *covariant formulation.*

To every contravariant four-vector there corresponds a covariant four-vector. From their transformation properties it can be seen that they are related by

$$T_4 = T^4 \quad , \quad T_i = -T^i \quad , \quad i = 1,2,3 \quad .$$

These relations may be summarized as follows

$$T_\alpha = g_{\alpha\beta} T^\beta \quad , \quad T^\alpha = g^{\alpha\beta} T_\beta \quad , \tag{8.37}$$

where $g_{\alpha\beta}$ and $g^{\alpha\beta}$ may both be represented by the same diagonal matrix of constant elements

$$g = \begin{bmatrix} -1 & 0 & 0 & 0 \\ 0 & -1 & 0 & 0 \\ 0 & 0 & -1 & 0 \\ 0 & 0 & 0 & 1 \end{bmatrix} . \tag{8.38}$$

By applying the contraction operation it is possible to obtain a scalar for every four-vector

$$T^\alpha T_\alpha \equiv g_{\alpha\beta} T^\alpha T^\beta \quad . \tag{8.39}$$

The invariant function so obtained is called the *norm* of the four-vector T^α. It can usually be readily evaluated in some convenient inertial frame, often the rest frame. Equation (8.39) then provides a relation among the components of a four-vector in any inertial frame, which is often useful in the solution of a problem. Examples of the invariant functions obtained in this manner are

$$x^\alpha x_\alpha = g_{\alpha\beta}x^\alpha x^\beta = -x^2 - y^2 - z^2 + c^2t^2 \quad ,$$

and

$$(ds)^2 = dx^\alpha dx_\alpha = g_{\alpha\beta}dx^\alpha dx^\beta = -(dx)^2 - (dy)^2 - (dz)^2 + c^2(dt)^2 \ .$$

The latter is called the *line element* of the space-time continuum and is equal to $c^2(dt_0)^2$, where dt_0 is the proper time interval.

The sets of numbers $g_{\alpha\beta}$ and $g^{\alpha\beta}$ can be shown to be tensors. On the one hand

$$T'^\alpha T'_\alpha = g'_{\alpha\beta}T'^\alpha T'^\beta \quad ,$$

while on the other, as $T'^\alpha T'_\alpha$ is an invariant,

$$T'^\alpha T'_\alpha = T^\gamma T_\gamma = g_{\gamma\delta}T^\gamma T^\delta = g_{\gamma\delta}(Q^{-1})^\gamma_\alpha(Q^{-1})^\delta_\beta T'^\alpha T'^\beta \quad .$$

As the above is true for any four-vector T^α we must have

$$g'_{\alpha\beta} = (Q^{-1})^\gamma_\alpha(Q^{-1})^\delta_\beta g_{\gamma\delta} \quad , \tag{8.40}$$

showing that $g_{\alpha\beta}$ is a covariant tensor of rank 2. Similarly $g^{\alpha\beta}$ can be shown to be a contravariant tensor. These are called the *metric tensors* because of their appearance in the expression for the line element of the space-time continuum. The metric tensors are used to change a contravariant index into a covariant one, and vice versa.

8.9 Covariant Formulation of Mechanics

Since physical laws are to be written as four-vector or tensor equations we shall first attempt to represent physical quantities as four-vectors or tensors.

Consider first the laws of mechanics. In the reformulation of these laws we are constantly guided by the requirement that as

$v/c \to 0$ they should tend to the Newtonian form, since in the limit of small velocities the Lorentz transformation becomes identical with the Galilean transformation as is obvious from a comparison of Eqs. (8.6) with Eqs. (8.1) when terms of second order in v/c are neglected.

Since a scalar is invariant, the multiplication or division of a four-vector by a scalar will reproduce a four-vector. The velocity $\mathbf{u} = d\mathbf{r}/dt$ of a point P does not constitute the first three components of a four-vector, for while $d\mathbf{r}$ represents the first three components of the space-time interval dx^α, dt is not a scalar. On the other hand, the line element $(ds)^2 = -(dx)^2 - (dy)^2 - (dz)^2 + (cdt)^2$ is a scalar and is equal to $(cdt_0)^2$, where dt_0 is the time interval in the rest frame Σ_0 of the point. It follows that

$$ds = cdt_0 = \frac{cdt}{\gamma_0} \tag{8.41}$$

is an invariant, where $\gamma_0 = (1 - \beta_0^2)^{-\frac{1}{2}}$, $\beta_0 c$ being the velocity of the rest frame relative to Σ. Equation (8.41) also shows ds to be a measure of dt. The four-vector formed by dividing dx^α with the scalar ds,

$$u^\alpha \equiv \frac{dx^\alpha}{ds} = \left(\frac{\gamma_0 \mathbf{u}}{c}, \gamma_0\right) \quad , \tag{8.42}$$

is a measure of the velocity $\mathbf{u}$ and is commonly called the *Minkowski velocity four-vector* or simply the *four-velocity*. It should be noted that the four-velocity, unlike the Newtonian velocity, is a dimensionless quantity.

The transformation of a four-vector gives the following relations for the components of $\mathbf{u}$ in Σ and Σ':

$$\gamma_0' u_x' = \gamma\gamma_0 (u_x - v) \quad ,$$

$$\gamma_0' u_y' = \gamma_0 u_y \quad , \qquad \gamma_0' u_z' = \gamma_0 u_z \quad ,$$

$$\gamma_0' = \gamma\gamma_0 \left(1 - \frac{\beta}{c} u_x\right) \quad .$$

Eliminating from the above γ_0 and γ'_0, the Lorentz factors of P in Σ and Σ' respectively, we have

$$u'_x = \frac{u_x - v}{1 - \frac{vu_x}{c^2}} \quad ,$$

$$u'_y = \frac{u_y}{\gamma\left(1 - \frac{vu_x}{c^2}\right)} \quad ,$$

$$u'_z = \frac{u_z}{\gamma\left(1 - \frac{vu_x}{c^2}\right)} \quad . \tag{8.43}$$

As $\mathbf{u}'$ is the velocity of P in Σ' and $\mathbf{v}$ is the velocity of Σ' relative to Σ we may interpret the velocity $\mathbf{u}$ of P in Σ as the sum of $\mathbf{u}'$ and $\mathbf{v}$. The inverse transformation

$$u_x = \frac{u'_x + v}{1 + \frac{vu'_x}{c^2}} \quad , \quad u_y = \frac{u'_y}{\gamma\left(1 + \frac{vu'_x}{c^2}\right)} \quad , \quad u_z = \frac{u'_z}{\gamma\left(1 + \frac{vu'_x}{c^2}\right)}$$

then shows that it will not be possible to add up velocities which are smaller than or equal to c to achieve a velocity greater than c. It should be noted however that two points moving away from each other may appear to have a relative velocity up to 2c to a stationary observer in an inertial frame.

By analogy to the ordinary momentum vector, the *momentum four-vector* p^α of a particle is defined as the product of the velocity four-vector with an invariant constant m_0c^2 characteristic of the particle:

$$p^\alpha \equiv m_0c^2u^\alpha = (m_0\gamma_0\mathbf{u}c, \; m_0\gamma_0c^2) \quad . \tag{8.44}$$

The principle of conservation of linear momentum for a system of two free particles in a given frame may then be formulated as

$$p_1^{\alpha} + p_2^{\alpha} = P^{\alpha} \quad , \tag{8.45}$$

where P^{α}, the total momentum of the system is a constant four-vector. The first three components of Eq. (8.45) give

$$m_{o1}\gamma_{o1}\mathbf{u}_1 c + m_{o2}\gamma_{o2}\mathbf{u}_2 c = \text{constant} \quad ,$$

or in the Newtonian form

$$m_1\mathbf{u}_1 + m_2\mathbf{u}_2 = \text{constant} \quad ,$$

if we identify $m_o\gamma_o$ with the mass m of a particle. It follows that the mass m of a particle is not a constant but depends on its velocity $\beta_o c$ relative to the observer. In the rest frame Σ_o, $\gamma_o = 1$ and $m = m_o$. m_o is called the *rest mass* or *proper mass* of the particle, while m is often called the relative mass.

It should be noted that, while in the Newtonian mechanics the conservation of linear momentum is a consequence of the law of action and reaction, the same reasoning does not apply in special relativity unless the particles involved interact only in contact (see Sec. 8.3). The conservation of momentum therefore has the status of a hypothesis in special relativity. Rather, the principle of conservation of momentum is used to define the relative mass. For a system of n particles the principle states that $\sum_{s=1}^{n} p_s^{\alpha}$ is a constant.

The fourth component of Eq. (8.45) is

$$m_{o1}\gamma_{o1}c^2 + m_{o2}\gamma_{o2}c^2 = \text{constant} \quad . \tag{8.46}$$

To find its physical significance consider the small velocity limit of $m_o\gamma_o c^2$:

$$m_o\gamma_o c^2 = m_o c^2(1-\beta_o^2)^{-\frac{1}{2}} = m_o c^2 + \frac{1}{2}m_o\beta_o^2 c^2 + \ldots$$

As the lowest order velocity-dependent term in the expansion represents the kinetic energy in the Newtonian mechanics, we must interpret $m_o\gamma_o c^2$ as a measure of energy and Eq. (8.46) as expressing the conservation of energy of a system of free particles. The expression $m_o c^2$ must then represent the energy possessed by a particle when it is at rest. It is called the *rest energy* of a particle and the quantity $m_o\gamma_o c^2$ or mc^2 its *total energy*.

In terms of the momentum $\mathbf{p} = m\mathbf{u}$ and the total energy $E = mc^2$, the momentum four-vector of a particle is

$$p^\alpha = (\mathbf{p}c,\ E) \quad . \tag{8.47}$$

Its first three components give a measure of the momentum of the particle and the fourth component its total energy, all components having the dimensions of energy.

Contracting the product of the contravariant and covariant four-momenta of a particle, we have

$$p^\alpha p_\alpha = g_{\alpha\beta}p^\alpha p^\beta = -\ p^2c^2 + E^2 = m_o^2c^4 \quad ,$$

evaluating it in the rest frame. This equation is Lorentz covariant and is usually written as

$$E^2 = p^2c^2 + m_o^2c^4 \quad . \tag{8.48}$$

It gives the total energy of a particle in terms of its momentum and rest mass.

Newton's second equation $\mathbf{F} = \frac{d\mathbf{p}}{dt}$ is used to define the *Minkowski force four-vector:*

$$F^\alpha \equiv \frac{dp^\alpha}{ds} = (\gamma_o \frac{d\mathbf{p}}{dt},\ c\gamma_o \frac{dm}{dt}) \quad , \tag{8.49}$$

or in terms of the Newtonian force $\mathbf{F}$,

$$F^\alpha = (\gamma_o\mathbf{F},\ \frac{\gamma_o}{c}\frac{d(mc^2)}{dt}) \quad . \tag{8.50}$$

Contracting F^{α} with u_{α} we obtain the Lorentz invariant function

$$F^{\alpha}u_{\alpha} = -\frac{\gamma_0^2}{c}\,\mathbf{F}\cdot\mathbf{u} + \frac{\gamma_0^2}{c}\frac{d}{dt}(mc^2) \quad .$$

In the rest frame, $\mathbf{u} = 0$, $mc^2 = m_0c^2$, and the function vanishes. Hence

$$\frac{d}{dt}(mc^2) = \mathbf{F}\cdot\mathbf{u} \quad , \tag{8.51}$$

which equates the rate of doing work by a force $\mathbf{F}$ to the rate of increase of the quantity mc^2 of the particle acted upon, consistent with the interpretation of the latter as energy. It also follows from Eq. (8.51) that the fourth component of F^{α} is $\gamma_0\,\mathbf{u}\cdot\mathbf{F}/c$ and is a measure of the rate at which $\mathbf{F}$ does work.

8.10 Covariant Formulation of Vacuum Electrodynamics

As Maxwell's electrodynamics is inherently Lorentz covariant, what is required here is simply to rewrite the relevant equations in covariant form, i.e. as vector and tensor equations.

The definition of the current density $\mathbf{J} = \rho\mathbf{u}$ can be carried over to special relativity by defining the *current density four-vector* as

$$j^{\alpha} \equiv \rho_0 \frac{dx^{\alpha}}{ds} = \left(\rho_0\gamma_0\frac{\mathbf{u}}{c}, \rho_0\gamma_0\right) \quad , \tag{8.52}$$

where ρ_0 is an invariant constant. The first three components of j^{α} will be proportional to those of $\mathbf{J}$ if $\rho_0\gamma_0$ is interpreted as charge density. We may then define the current density four-vector as

$$j^{\alpha} = \left(\rho\frac{\mathbf{u}}{c}, \rho\right) = \left(\frac{\mathbf{J}}{c}, \rho\right) \quad . \tag{8.53}$$

Note that the definition (8.52) is analogous to the definition (8.44) of the momentum four-vector. Similarly we may interpret ρ_0 as the proper charge density or the charge density in the rest frame Σ_0 of the charge element under consideration, whose Lorentz factor is γ_0.

The continuity equation of charge,

$$\nabla \cdot \mathbf{J} + \frac{\partial \rho}{\partial t} = 0 \quad ,$$

may be written as

$$\frac{1}{c} \nabla \cdot \mathbf{J} + \frac{\partial \rho}{\partial x^4} = \frac{\partial j^\alpha}{\partial x^\alpha} = 0 \quad . \tag{8.54}$$

Similarly, the Lorentz condition,

$$\nabla \cdot \mathbf{A} + \frac{1}{c^2} \frac{\partial \Phi}{\partial t} = 0$$

may be written in the covariant form

$$\frac{\partial \Phi^\alpha}{\partial x^\alpha} = 0 \quad , \tag{8.55}$$

where the *electromagnetic potential four-vector* is defined as

$$\Phi^\alpha \equiv (c\mathbf{A}, \Phi) \quad . \tag{8.56}$$

To reformulate the differential equations for the potentials we first rewrite the D'Alembert operator

$$\Box \equiv \nabla^2 - \frac{1}{c^2} \frac{\partial^2}{\partial t^2} = - \frac{\partial^2}{\partial x_\alpha \partial x^\alpha} \quad , \tag{8.57}$$

where $dx_\alpha = g_{\alpha\beta} dx^\beta$, having components $dx_i = -dx^i$, $dx_4 = dx^4$. We have seen in Sec. 8.4 that the operator $\frac{\partial}{\partial x^\alpha}$ transforms like a covariant four-vector. The D'Alembert operator which may be considered as the contraction of a covariant and a contravariant four-vector is therefore an invariant.

The inhomogeneous wave equations (8.30) and (8.31) for the potentials may be written as

$$\Box(c\mathbf{A}) = - \frac{\mathbf{J}}{\varepsilon_0 c}$$

and

$$\Box\Phi = -\frac{\rho}{\varepsilon_0} \quad .$$

Combining these we obtain the covariant equation

$$\Box\Phi^{\alpha} = -\frac{j^{\alpha}}{\varepsilon_0} \quad . \tag{8.58}$$

This equation emphasizes the interrelationship between the vector and scalar potentials; they describe different aspects of the same field, the electromagnetic field.

The components of **E** and **B** do not transform like the components of a four-vector, while the equations

$$B_i = \frac{\partial A_k}{\partial x_j} - \frac{\partial A_j}{\partial x_k} \equiv B_{jk} = -B_{kj} \tag{8.59}$$

suggest that B_i may be represented by an antisymmetric tensor of rank 2, which changes sign when the indices are interchanged. Such a tensor would have $4!/(2!)^2$ or 6 non-zero independent components. However, as **E** and **B** are merely different manifestations of the same field, we might expect them to be representable by the same antisymmetric tensor so that the numbers of components would correspond. The expression for E_i in terms of the potentials,

$$E_i = -\frac{\partial\Phi}{\partial x_i} - \frac{\partial A_i}{\partial t} \quad , \tag{8.60}$$

is similar to Eq. (8.59). This suggests that we define an antisymmetric contravariant *electromagnetic field tensor* by

$$F^{\alpha\beta} \equiv \frac{\partial\Phi^{\beta}}{\partial x_{\alpha}} - \frac{\partial\Phi^{\alpha}}{\partial x_{\beta}} \quad . \tag{8.61}$$

Comparing this with Eqs. (8.59) and (8.60) we see that the components of $F^{\alpha\beta}$ are related to the components of the field vectors by

$$F^{ij} = - c\varepsilon_{ijk}B_k \quad ,$$

$$F^{i4} = E_i \quad , \qquad i, j = 1,2,3 \quad . \tag{8.62}$$

We may then represent $F^{\alpha\beta}$ by a matrix

$$(F^{\alpha\beta}) = \begin{bmatrix} 0 & -cB_3 & cB_2 & E_1 \\ cB_3 & 0 & -cB_1 & E_2 \\ -cB_2 & cB_1 & 0 & E_3 \\ -E_1 & -E_2 & -E_3 & 0 \end{bmatrix} \quad . \tag{8.63}$$

An antisymmetric covariant field tensor $F_{\alpha\beta}$ may also be define:

$$F_{\alpha\beta} \equiv \frac{\partial\Phi_\beta}{\partial x^\alpha} - \frac{\partial\Phi_\alpha}{\partial x^\beta} = g_{\alpha\gamma}g_{\beta\delta}F^{\gamma\delta} \quad , \tag{8.64}$$

where $g_{\alpha\beta}$ is the metric tensor defined by Eq. (8.38). The components of the covariant field tensor are related to the components of the field vectors by

$$F_{ij} = - c\varepsilon_{ijk}B_k \quad ,$$

$$F_{i4} = - E_i \quad , \qquad i, j = 1,2,3 \quad . \tag{8.65}$$

Having defined the electromagnetic field tensor we can now rewrite Maxwell's equations in the covariant form. The ith component of Eq. (8.18),

$$\varepsilon_{ijk} \frac{\partial B_k}{\partial x_j} = \frac{1}{c^2} \left(\frac{J_i}{\varepsilon_o} + \frac{\partial E_i}{\partial t}\right) \quad ,$$

may be written as

$$\frac{\partial F^{ji}}{\partial x^j} + \frac{\partial F^{4i}}{\partial x^4} = \frac{j^i}{\varepsilon_o} \quad , \qquad i = 1,2,3 \quad . \tag{8.66}$$

Equation (8.19) may be written as

$$\frac{\partial F^{j4}}{\partial x^j} = \frac{\rho}{\varepsilon_o} \quad . \tag{8.67}$$

The four equations (8.66) and (8.67) may be combined into a covariant equation

$$\frac{\partial F^{\beta\alpha}}{\partial x^\beta} = \frac{j^\alpha}{\varepsilon_o} \quad . \tag{8.68}$$

Of the remaining Maxwell's equations, Eq. (8.17) has the component form

$$\varepsilon_{ijk} \frac{\partial E_k}{\partial x_j} = - \frac{\partial B_i}{\partial t} \quad .$$

Multiplying both sides by $\varepsilon_{i\ell m}$ and summing over the index i we obtain

$$\varepsilon_{i\ell m}\varepsilon_{ijk} \frac{\partial E_k}{\partial x_j} = (\delta_{\ell j}\delta_{mk} - \delta_{\ell k}\delta_{mj}) \frac{\partial E_k}{\partial x} = - \varepsilon_{i\ell m} \frac{\partial B_i}{\partial t} \quad ,$$

or

$$\frac{\partial E_m}{\partial x_\ell} - \frac{\partial E_\ell}{\partial x_m} = - \frac{1}{c} \frac{\partial}{\partial t} (c\varepsilon_{i\ell m} B_i) \quad , \qquad \ell, m = 1,2,3 \quad \text{and}$$

$$\ell \neq m \quad . \tag{8.69}$$

Equation (8.20) may be written as

$$\frac{\partial B_1}{\partial x_1} + \frac{\partial B_2}{\partial x_2} + \frac{\partial B_3}{\partial x_3} = 0 \quad . \tag{8.70}$$

To combine the above four equations, we note that if we represent the vectors by the contravariant electromagnetic field tensor, and the coordinates by the contravariant space-time four-vector, Eq. (8.69) would become

$$\frac{\partial F^{m4}}{\partial x^{\ell}} + \frac{\partial F^{4\ell}}{\partial x^{m}} - \frac{\partial F^{\ell m}}{\partial x^{4}} = 0 \quad , \quad \ell, m = 1,2,3 \quad ,$$

and Eq. (8.70) would become

$$\frac{\partial F^{23}}{\partial x^{1}} + \frac{\partial F^{31}}{\partial x^{2}} + \frac{\partial F^{12}}{\partial x^{3}} = 0 \quad .$$

While in these equations the terms may be generated by cyclic permutation of three indices, symmetry is violated by the negative sign of the last term in the first set of equations. This may be remedied by using the components of the covariant space-time four-vector to represent the coordinates. The equations may then be combined as

$$\frac{\partial F^{\alpha\beta}}{\partial x_{\gamma}} + \frac{\partial F^{\beta\gamma}}{\partial x_{\alpha}} + \frac{\partial F^{\gamma\alpha}}{\partial x_{\beta}} = 0 \quad . \tag{8.71}$$

Note that the left-hand side of this equation is an anti-symmetric contravariant tensor of rank 3. It changes sign when any two indices are interchanged and vanishes identically when any two indices are equal. Thus the number of non-trivial equations given by (8.71) is only $\frac{4!}{3!} = 4$, in agreement with that of the original set of Eqs. (8.69) and (8.70).

Equations (8.68) and (8.71) are Maxwell's equations in the covariant form.

8.11 The Lorentz Force

Consider a point charge q moving with velocity $\mathbf{u}$ in an electromagnetic field of electric vector $\mathbf{E}$ and magnetic vector $\mathbf{B}$ in Σ and let Σ' be the instantaneous rest frame of the charge, the x-axes being chosen parallel to $\mathbf{u}$. In Σ', the charge is stationary at the instant under consideration and the only force acting on it is the electrostatic force $\mathbf{F}' = q\mathbf{E}'$ so that the Minkowski force on the charge is

$$F'^{\alpha} = (q\mathbf{E}', 0) \quad .$$

If we transform F'^{α} by the inverse of Eq. (8.14) and then $\mathbf{E}'$ by Eq. (8.26), we find the first three components of F^{α} as

$$F^1 = \gamma q E'_x = \gamma q E_x = \gamma F_x \quad ,$$

$$F^2 = q E'_y = \gamma q (E_y - uB_z) = \gamma F_y \quad ,$$

$$F^3 = q E'_z = \gamma q (E_z + uB_y) = \gamma F_z \quad ,$$

use having been made of the expression for the Minkowski force, Eq. (8.50). We then obtain the force $\mathbf{F}$ acting on the charge in Σ as

$$\mathbf{F} = q(\mathbf{E} + \mathbf{u} \times \mathbf{B}) \quad . \tag{8.72}$$

This force is known as the *Lorentz force*. It may be recalled that the electrostatic force on a stationary charge q is by definition $q\mathbf{E}$ and the magnetic force on a charge q of velocity $\mathbf{u}$ is by definition $q\mathbf{u} \times \mathbf{B}$. The electric force on a moving charge q was postulated to be $q\mathbf{E}$ by assuming the force to be unaffected by the state of motion of the charge. If a magnetic field is also present the moving charge will of course be subjected to both the electric and magnetic forces, with the resultant given by Eq. (8.72). This expression therefore has only the status of a postulate in Maxwell's theory. In special relativity it is readily derived from the definition of the electrostatic force using only the transformation properties of the physical quantities involved. However, as the velocity of a charge in an electromagnetic field is constantly changing, the derivation requires the assumption that the application of the Lorentz transformation is not affected by the presence of acceleration in the motion of a frame relative to an inertial frame, while $\mathbf{u}$ is to be interpreted as the instantaneous velocity.

8.12 Motion of a Charge in Electromagnetic Field

The equation of motion of a particle of charge q, rest mass m_0 and velocity $\mathbf{u}$ in an electromagnetic field of potentials $\mathbf{A}(\mathbf{r}, t)$ and $\Phi(\mathbf{r},t)$ is given by the covariant Newton's equation (8.49)

$$\frac{dp^\alpha}{ds} = F^\alpha \quad . \tag{8.73}$$

Using the expression (8.72) for the Lorentz force, the i-th component of the above may be written as

$$\frac{d}{dt}(m_o\gamma u_i) = q\left(-\frac{\partial\Phi}{\partial x_i} - \frac{\partial A_i}{\partial t} + \varepsilon_{ijk}u_j\varepsilon_{k\ell m}\frac{\partial A_m}{\partial x_\ell}\right) \quad , \tag{8.74}$$

where $\gamma = (1 - \frac{u^2}{c^2})^{-\frac{1}{2}}$. As

$$\varepsilon_{ijk}\varepsilon_{k\ell m}u_j\frac{\partial A_m}{\partial x_\ell} = (\delta_{i\ell}\delta_{jm} - \delta_{im}\delta_{j\ell})u_j\frac{\partial A_m}{\partial x_\ell}$$

$$= u_j\frac{\partial A_j}{\partial x_i} - \frac{\partial A_i}{\partial x_j}\frac{dx_j}{dt}$$

$$= u_j\frac{\partial A_j}{\partial x_i} - \frac{dA_i}{dt} + \frac{\partial A_i}{\partial t} \quad ,$$

Equation (8.74) may be written as

$$\frac{d}{dt}(m_o\gamma u_i) = q\left(-\frac{\partial\Phi}{\partial x_i} + u_j\frac{\partial A_j}{\partial x_i} - \frac{dA_i}{dt}\right) \quad .$$

If we regard $u_i \equiv \frac{dx_i}{dt}$ and x_i to be independent variables for the purpose of partial differentiation, the above may further be rewritten as

$$\frac{d}{dt}(m_o\gamma u_i + qA_i) - \frac{\partial}{\partial x_i}(-q\Phi + qu_jA_j) = 0 \quad .$$

The last equation will have the form of Lagrange's equation

$$\frac{d}{dt}\left(\frac{\partial L}{\partial u_i}\right) - \frac{\partial L}{\partial x_i} = 0 \quad , \tag{8.75}$$

provided we define the Lagrangian L as

$$L = - q\Phi + qu_jA_j + K(\mathbf{u}) \quad ,$$

with the condition that $\frac{\partial L}{\partial u_i} = m_o\gamma u_i + qA_i$.

This condition implies that

$$\frac{\partial K}{\partial u_i} = m_o\gamma u_i$$

or

$$K = -\frac{m_oc^2}{\gamma} \quad .$$

The constant of integration, which is immaterial as the derivative of L only appears in Lagrange's equation, has been omitted. The Lagrangian to be used is therefore

$$L = -\frac{m_oc^2}{\gamma} - q\Phi + q\mathbf{u}\cdot\mathbf{A} \quad . \tag{8.76}$$

The motion of the particle is now completely described by Eq. (8.75) with L given by Eq. (8.76).

Two conservation laws corresponding to the conservation of energy and the conservation of momentum may now be obtained. If we multiply both sides of Eq. (8.75) by u_i and sum over i, then as $L = L(x_j, u_j, t)$ we shall find

$$\frac{d}{dt}\left(u_j\frac{\partial L}{\partial u_j} - L\right) = -\frac{\partial L}{\partial t} \quad .$$

This equation shows that if $\mathbf{A}$, and hence L, does not depend on t

explicitly there is a constant of the motion H given by

$$H = u_j \frac{\partial L}{\partial u_j} - L \quad . \tag{8.77}$$

Substitution of L from Eq. (8.76) gives

$$H = m_0 c^2 \gamma + q\Phi \quad . \tag{8.78}$$

H is therefore the energy of the charge which includes the rest, kinetic and potential energies. In general, H as given by Eq. (8.77) is called the *Hamiltonian*. The above shows that the Hamiltonian of a charge moving in an electromagnetic field of potentials $\mathbf{A}(\mathbf{r})$ and $\Phi(\mathbf{r})$ is conserved.

From Eq. (8.75) we see that if L does not depend on x_i explicitly the quantity

$$\frac{\partial L}{\partial u_i} = m_0 \gamma u_i + qA_i \equiv p_i \tag{8.79}$$

is conserved. Since $m_0 \gamma u_i$ is the i-th component of the momentum of the particle, we must interpret p_i as the component of a momentum. It is known as the momentum conjugate to x_i and the set of quantities x_i, p_i is said to be *canonical variables*. Moving in an electromagnetic field, the mechanical momentum $m_0 \gamma \mathbf{u}$ of a charge is not conserved. It is the mechanical momentum plus a momentum $q\mathbf{A}$ contributed by the field that may be conserved. In terms of the canonical variables, the Hamiltonian has the expression

$$H = c\sqrt{m_0^2 c^2 + (p_i - qA_i)^2} + q\Phi \quad ,$$

which is the classical basis of Dirac's equation of the electron.

The analysis of the orbit of a charged particle in an arbitrary electromagnetic field is a complicated problem, especially if radiation reaction is taken into account. If the field is weak and the acceleration is small, the radiation reaction may be treated as a small perturbation to be taken care of separately by successive approximation. Even if the radiation reaction is neglected, the motion is still generally very complex except when **E** and **B** are both static and uniform. In particular, when one of the electric and magnetic fields is absent or when one of these fields may be transformed away the motion can be easily visualized.

Consider first the simple case of a pure magnetic field **B**. The fourth component of the Lorentz force equation (8.73) gives

$$\frac{d}{dt}(\gamma m_0 c^2) = q\mathbf{u}\cdot\mathbf{u}\times\mathbf{B} = 0 \quad ,$$

showing that γ and hence $|\mathbf{u}|$ are constants of the motion. Thus kinetic energy is conserved in the motion of a charge in a pure magnetic field. This confirms the general result on the conservation of the Hamiltonian (8.78), for as the vector potential **A** does not contain t explicitly and the scalar potential Φ is a constant in a pure magnetic field the kinetic energy is conserved.

The first three components of Eq. (8.73) may then be written as

$$\frac{d\mathbf{u}}{dt} = \mathbf{u}\times\boldsymbol{\omega} \quad , \qquad (8.80)$$

where

$$\boldsymbol{\omega} = \frac{q\mathbf{B}}{m_0\gamma} \quad .$$

Equation (8.80) can be readily solved if **B** is static and uniform. Taking the z-axis parallel to **B**, the component equations are

$$\ddot{x} = \dot{y}\omega \quad , \quad \ddot{y} = -\dot{x}\omega \quad , \quad \ddot{z} = 0 \quad ,$$

where the dot signifies time derivation $\frac{d}{dt}$. The z-equation shows that the component $u_{||}$ of the velocity in the direction of the field is a constant. *The other two equations may be combined into a complex* equation

$$\ddot{\xi} + i\omega\dot{\xi} = 0$$

by putting $x+iy=\xi$. This equation has the general solution $\xi=\rho e^{-i(\omega t+\phi)}+\varepsilon_0$, where ρ and ϕ are real constants and ε_0 is a complex constant. The solution is equivalent to $x-x_0=\rho\cos(\omega t+\phi)$ and $y-y_0=-\rho\sin(\omega t+\phi)$, showing that the motion in a plane perpendicular to the field is circular with a radius ρ given by $u_\perp=\rho\omega$, where $u_\perp=\sqrt{\dot{x}^2+\dot{y}^2}$ is the velocity component in this plane, or, in terms of the transverse momentum $p_\perp$, given by

$$p_\perp = qB\rho \quad .$$

Thus if $u_{||}$ is initially zero, the motion is confined to a plane transverse to the field and the orbit is a circle of radius ρ. In general, the orbit is a helix of radius ρ and pitch angle arctan $(u_{||}/\rho\omega)$. ω is the angular velocity of revolution of the charge about the axis of the circle or the helix and is the relativistic version of the gyromagnetic or Larmor frequency introduced in Sec. 6.10.

Another simple case is the motion of a charge in a combination of an electric field **E** and a magnetic field **B**, which are static, uniform and mutually perpendicular, for then if $E\neq Bc$ it will be possible to find an inertial frame in which one field or the other vanishes. Let **v** be the velocity of the new frame Σ' relative to Σ. If **v** is chosen to be transverse to both **E** and **B**, then $E'_{||}=E_{||}=0$, $B'_{||}=B_{||}=0$, only the transverse components remaining in Σ'.

Suppose $E<Bc$, then by selecting $\mathbf{v}=(\mathbf{E}\times\mathbf{B})/B^2$ we can make $E'_\perp$ vanish also. The field in Σ' is now purely magnetic with an induction $\mathbf{B}'=\frac{1}{\gamma}\mathbf{B}$, which maintains the same direction as **B** but is weaker by a factor γ. The motion in this frame is therefore a helix. Returning to the original frame, the charge is seen to spiral around the lines of force of **B**, being at the same time subject to a uniform drift velocity **v** transverse to both **B** and **E**. If the motion in Σ' is circular, i.e. confined to a plane transverse to **B'**, the orbit will appear in Σ to be approximately a cycloid.

Since **v** is physically meaningful only if its magnitude is less than c, the above transformation cannot be carried out if $E>Bc$.

In this case, we may instead transform to a frame Σ'' where the field is purely electric. The relative velocity required is now $\mathbf{v} = (\mathbf{E} \times \mathbf{B}c^2)/E^2$ and the electric intensity in this frame is $\mathbf{E}'' = \frac{1}{\gamma}\mathbf{E}$. The orbit of the particle as observed in this frame is a caternary with the axis of symmetry along $\mathbf{E}''$ [Prob. (8.17)]. In the Σ frame, a uniform drift transverse to both $\mathbf{B}$ and $\mathbf{E}$ is also observed. In this case, however, the magnetic field is not sufficient to prevent the charge from moving in the direction of $\mathbf{E}$ with increasing speed.

If $\mathbf{E}$ and $\mathbf{B}$ have parallel components, it will not be possible to get rid of one of these by a Lorentz transformation and both fields must be considered together. The problem can still be readily solved if both are static and uniform.

8.13 Electromagnetic Momentum — Energy Tensor

In electrodynamics we are dealing with charges and currents, and electromagnetic fields with which they interact and to which they themselves also contribute. As currents are simply moving charges, ultimately we have to consider charged particles moving under the action of the fields. The dynamical quantities to be considered are forces, energy and momentum.

Consider a continuous distribution of charges in an electromagnetic field and let $\mathbf{u}$ be the velocity of the charges in a volume element dV where the density is ρ. The Lorentz force per unit volume acting on the charges is

$$\mathbf{f} = \rho(\mathbf{E} + \mathbf{u} \times \mathbf{B}) \quad . \tag{8.81}$$

To see if $\mathbf{f}$ may be represented by a four-vector consider its i-th component

$$f_i = \rho E_i + \rho\varepsilon_{ijk}u_jB_k = \rho F^{i4} - \frac{\rho}{c}u_jF^{ij} \quad .$$

The right-hand side of this equation may be written as the contraction of the contravariant field tensor $F^{\alpha\beta}$ with the covariant current density four-vector $j_\beta = (-\rho \frac{\mathbf{u}}{c}, \rho)$:

$$f^\alpha = F^{\alpha\beta} j_\beta \quad . \tag{8.82}$$

The *Lorentz force density* therefore forms the first three components of a four-vector. Its fourth component is

$$f^4 = F^{4\beta} j_\beta = \frac{\rho}{c} \mathbf{E} \cdot \mathbf{u} = \frac{1}{c} \mathbf{f} \cdot \mathbf{u} \quad ,$$

which is equal to $1/c$ times the rate of mechanical and thermal work done by $\mathbf{f}$, or $1/c$ times the energy expended by the electric field per unit volume per unit time in moving the charges. The magnetic force however does no work since it always acts transverse to the velocity $\mathbf{u}$. Thus the Lorentz force density four-vector has components

$$f^\alpha = (\mathbf{f}, \frac{1}{c} \mathbf{f} \cdot \mathbf{u}) \quad . \tag{8.83}$$

Using the covariant counterpart of Maxwell's equation (8.68) we have

$$f^\alpha = \varepsilon_0 F^{\alpha\beta} \frac{\partial F_{\gamma\beta}}{\partial x_\gamma} = \varepsilon_0 \left\{ \frac{\partial}{\partial x_\gamma} (F^{\alpha\beta} F_{\gamma\beta}) - F_{\gamma\beta} \frac{\partial F^{\alpha\beta}}{\partial x_\gamma} \right\} \quad . \tag{8.84}$$

As the indices β and γ in the terms $F_{\gamma\beta} \frac{\partial F^{\alpha\beta}}{\partial x_\gamma}$ are dummies, the latter may be written as

$$-F_{\beta\gamma} \frac{\partial F^{\alpha\beta}}{\partial x_\gamma} = -\frac{1}{2} (F_{\beta\gamma} \frac{\partial F^{\alpha\beta}}{\partial x_\gamma} + F_{\gamma\beta} \frac{\partial F^{\alpha\gamma}}{\partial x_\beta})$$

$$= -\frac{1}{2} F_{\beta\gamma} (\frac{\partial F^{\alpha\beta}}{\partial x_\gamma} + \frac{\partial F^{\gamma\alpha}}{\partial x_\beta})$$

$$= \frac{1}{2} F_{\beta\gamma} \frac{\partial F^{\beta\gamma}}{\partial x_\alpha} \quad ,$$

making use also of the antisymmetric property of the field tensor and Maxwell's equation (8.71). Furthermore, we can write the covariant field tensor in terms of its contravariant counterpart so that

$$F_{\beta\gamma}\frac{\partial F^{\beta\gamma}}{\partial x_\alpha} = g_{\beta\delta}g_{\gamma\zeta}F^{\delta\zeta}\frac{\partial F^{\beta\gamma}}{\partial x_\alpha}$$

$$= F^{\delta\zeta}\frac{\partial}{\partial x_\alpha}(g_{\beta\delta}g_{\gamma\zeta}F^{\beta\gamma})$$

$$= F^{\delta\zeta}\frac{\partial F_{\delta\zeta}}{\partial x_\alpha} = F^{\beta\gamma}\frac{\partial F_{\beta\gamma}}{\partial x_\alpha} \quad .$$

Thus

$$F_{\gamma\beta}\frac{\partial F^{\alpha\beta}}{\partial x_\gamma} = \frac{1}{4}\left(F_{\beta\gamma}\frac{\partial F^{\beta\gamma}}{\partial x_\alpha} + F^{\beta\gamma}\frac{\partial F_{\beta\gamma}}{\partial x_\alpha}\right) = \frac{1}{4}\frac{\partial}{\partial x_\alpha}(F_{\beta\gamma}F^{\beta\gamma})$$

$$= \frac{1}{4}\delta^\alpha_\gamma\frac{\partial}{\partial x_\gamma}(F_{\beta\delta}F^{\beta\delta}) \quad .$$

Equation (8.84) may now be written as

$$f^\alpha + \frac{\partial}{\partial x_\gamma}T^\alpha_\gamma = 0 \quad , \tag{8.85}$$

where

$$T^\alpha_\gamma = -\varepsilon_0 F^{\alpha\beta}F_{\gamma\beta} + \frac{\varepsilon_0}{4}\delta^\alpha_\gamma F^{\beta\delta}F_{\beta\delta} \quad , \tag{8.86}$$

δ^α_γ being the Kronecker delta of suffixes α and γ. It can be easily shown that δ^α_γ is a mixed tensor of rank 2, so that T^α_γ is also a mixed tensor of rank 2.

To evaluate the components of the tensor T^α_γ we first calculate the following subsums:

$$\varepsilon_0 F^{i\delta}F_{j\delta} = \varepsilon_0(c^2\varepsilon_{ik\ell}B_\ell\varepsilon_{jkm}B_m - E_iE_j)$$

$$= \delta_{ij}\mu_0 H^2 - \mu_0 H_i H_j - \varepsilon_0 E_i E_j \quad ,$$

$$\varepsilon_o F^{i\delta} F_{4\delta} = \varepsilon_o F^{ij} F_{4j} = -\varepsilon_o c \varepsilon_{ijk} E_j B_k = -\frac{1}{c} (\mathbf{E} \times \mathbf{H})_i \quad ,$$

$$\varepsilon_o F^{4\delta} F_{i\delta} = \frac{1}{c} (\mathbf{E} \times \mathbf{H})_i \quad ,$$

$$\varepsilon_o F^{4\delta} F_{4\delta} = \varepsilon_o F^{4j} F_{4j} = -\varepsilon_o E^2 \quad .$$

Note that Latin letters have been used to denote indices which take the values 1, 2 and 3 only. From the above we find that

$$\varepsilon_o F^{i\delta} F_{i\delta} = 3\mu_o H^2 - \mu_o H^2 - \varepsilon_o E^2 = 2\mu_o H^2 - \varepsilon_o E^2 \quad ,$$

$$\varepsilon_o F^{\beta\delta} F_{\beta\delta} = \varepsilon_o F^{i\delta} F_{i\delta} + \varepsilon_o F^{4\delta} F_{4\delta} = 2\mu_o H^2 - 2\varepsilon_o E^2 \quad .$$

The components of T^α_γ are therefore

$$T^i_j = \varepsilon_o E_i E_j + \mu_o H_i H_j - \frac{1}{2} \delta_{ij} (\mu_o H^2 + \varepsilon_o E^2) = T^j_i \quad ,$$

$$T^i_4 = \frac{N_i}{c} = -T^4_i \quad ,$$

$$T^4_4 = U \quad , \tag{8.87}$$

where

$$U = \frac{1}{2} (\varepsilon_o E^2 + \mu_o H^2) \quad \text{and} \quad \mathbf{N} = \mathbf{E} \times \mathbf{H} \quad .$$

On account of the asymmetry of T^i_4 the mixed tensor T^α_γ is not symmetric. A symmetric tensor may however be obtained from the identity

$$T^{\alpha\beta} = g^{\alpha\gamma} T^\beta_\gamma \quad .$$

As the sum on the right-hand side may be written as $g^{\alpha i} T^\beta_i + g^{\alpha 4} T^\beta_4$, we see from the elements of $g^{\alpha\gamma}$ and Eq. (8.87) that

$$T^{jk} = -T^k_j = -T^j_k = T^{kj} \quad ,$$

$$T^{4j} = T^j_4 = - T^4_j = T^{j4} \qquad ,$$

showing that $T^{\alpha\beta}$ is symmetric.

The symmetric tensor $T^{\alpha\beta}$, which is known as the *electromagnetic momentum-energy tensor*, may be represented by the matrix

$$(T^{\alpha\beta}) = \begin{bmatrix} -T_{11} & -T_{12} & -T_{13} & \frac{N_1}{c} \\ -T_{21} & -T_{22} & -T_{23} & \frac{N_2}{c} \\ -T_{31} & -T_{32} & -T_{33} & \frac{N_3}{c} \\ \frac{N_1}{c} & \frac{N_2}{c} & \frac{N_3}{c} & U \end{bmatrix} \quad , \tag{8.88}$$

where T_{ij} are the components of a symmetric 3×3 tensor

$$(T_{ij}) = \begin{bmatrix} \varepsilon_o E_1^2 + \mu_o H_1^2 - U & \varepsilon_o E_1 E_2 + \mu_o H_1 H_2 & \varepsilon_o E_1 E_3 + \mu_o H_1 H_3 \\ \varepsilon_o E_1 E_2 + \mu_o H_1 H_2 & \varepsilon_o E_2^2 + \mu_o H_2^2 - U & \varepsilon_o E_2 E_3 + \mu_o H_2 H_3 \\ \varepsilon_o E_1 E_3 + \mu_o H_1 H_3 & \varepsilon_o E_2 E_3 + \mu_o H_2 H_3 & \varepsilon_o E_3^2 + \mu_o H_3^2 - U \end{bmatrix} \quad ,$$

known as *Maxwell's electromagnetic stress tensor*.

Historically, to account for the action-at-a-distance nature of electromagnetic forces, Faraday and later Maxwell considered the region of space occupied by an electromagnetic field as being in a state of stress, where electric and magnetic forces are transmitted as tension or compression by elastic-band like lines of force. Then the forces acting on the charges in a volume V by external agencies must be transmitted across its boundary surface S with the result that we may express the i-th component of the total force $\mathbf{F}$ by the surface integral

$$F_i = \oint_S T_{ij} dS_j \quad ,$$

where dS_j is the component of a surface element $\mathbf{dS}$ parallel to the x_j-axis. By virtue of the corollary of the divergence theorem (A11) and the fact that $\mathbf{F}$ is a volume force we may write the above in terms of volume integrals:

$$\int_V f_i dV = \int_V \frac{\partial T_{ij}}{\partial x_j} dV \quad ,$$

where f_i is the i-th component of the force per unit volume acting on the volume element dV. As V is arbitrary, it follows that

$$f_i = \frac{\partial T_{ij}}{\partial x_j} \quad . \tag{8.89}$$

The quantities T_{ij} are the components of a symmetric tensor called the *stress tensor*.

The tensor (8.88) obtained in special relativity has, in addition to the components of Maxwell's stress tensor, components involving the Poynting vector $\mathbf{N}$ and the electromagnetic field energy density U. For the interpretation of these additional components, we rewrite Eq. (8.85) as

$$f^\alpha = -\frac{\partial}{\partial x_\gamma}(g_{\gamma\beta} T^{\alpha\beta}) = -g_{\gamma\beta}\frac{\partial T^{\alpha\beta}}{\partial x_\gamma} = -\frac{\partial T^{\alpha\beta}}{\partial x^\beta} \quad . \tag{8.90}$$

The first three components of this equation may be written as

$$f_i = \frac{\partial T_{ij}}{\partial x_j} - \frac{\partial}{\partial t}\left(\frac{N_i}{c^2}\right) \quad , \tag{8.91}$$

while the fourth component is

$$\mathbf{f} \cdot \mathbf{u} = -\frac{\partial N_i}{\partial x_i} - \frac{\partial U}{\partial t} \quad . \tag{8.92}$$

As $\mathbf{f} \cdot \mathbf{u} = \mathbf{E} \cdot \mathbf{J}$, Eq. (8.92) may be written as

$$\nabla \cdot \mathbf{N} + \mathbf{E} \cdot \mathbf{J} + \frac{\partial U}{\partial t} = 0 \quad ,$$

which is identical with the Poynting equation (2.33). It balances for a unit volume the rate of mechanical and thermal work, plus the rate of outward radiation flow, with the rate of field energy change, thus expressing the principle of energy conservation.

To interpret Eq. (8.91) we note that as $\mathbf{f}$ is the total force acting on the charges contained in a unit volume it must be identified with the rate of increase of the mechanical momentum of the charges $d\mathbf{p}/dt$. Rewriting the equation and integrating over a volume V of surface S, we obtain

$$\frac{d}{dt}\left(\int_V p_i dV + \int_V \frac{N_i}{c^2} dV\right) = \oint_S T_{ij} dS_j \quad . \tag{8.93}$$

If we consider an isolated system of charges and fields the surface integral may be made to vanish by taking the volume sufficiently large. The above then gives

$$\int \mathbf{p}\, dV + \int \frac{\mathbf{N}}{c^2} dV = \text{constant} \quad . \tag{8.94}$$

This clearly must represent the law of momentum conservation: the total momentum of the charges plus a quantity $\int \frac{\mathbf{N}}{c^2} dV$ is a constant of the motion. As the latter quantity is related to the field vectors it must be interpreted as the momentum of the electromagnetic field. The conservation law here is similar to the conservation of the canonical momentum of a charge particle which also includes a contribution from the field in which it moves. Then in view of Eq. (8.94) we must interpret the quantity

$$\mathbf{G} \equiv \frac{\mathbf{N}}{c^2} = \frac{\mathbf{E} \times \mathbf{H}}{c^2} \tag{8.95}$$

as the *momentum density* of the electromagnetic field, assuming that momentum, like energy, is localized in the field. Equation (8.91) then expresses the balance between the volume force $\frac{\partial T_{ij}}{\partial x_j}$ exerted across the boundaries of the unit volume and the rate of increase of the density of the i-th component of the total momentum of the charges and field.

The above analysis holds even if no charge and current are present, in which case the mechanical momentum of course vanishes. Consider now an arbitrary volume V in the region of the electromagnetic field so that the surface integral in Eq. (8.93) does not generally vanish. The equation may then be written as

$$-\frac{d}{dt}\int_V G_i dV = \oint_S (-T_{ij}n_j)dS \quad ,$$

where n_j is the j-th component of the outward normal to the surface element dS. As the left-hand side represents the rate of decrease of the i-th component of the electromagnetic field momentum contained in V, the right-hand side must represent the outward flow through the surface S of the momentum component per unit time. It follows then that $-T_{ij}n_j$ must be interpreted as the outward force acting parallel to the x_i-axis on the surface per unit area, or the i-th component of the *radiation pressure* transmitted across S. If we consider a unit cube with surfaces normal to the coordinate axes, then $-T_{ij}$ gives the force parallel to the x_i-axis acting on the surface normal to the x_j-axis.

8.14 Lorentz's Electromagnetic Model of the Electron

The inadequacy of the classical theory of electrodynamics in dealing with microscopic phenomena has already been pointed out in Chap. VI. As a further example we shall present in this section some aspects of Lorentz's electromagnetic model of the electron, according to which the mass of the electron is determined entirely by its charge distribution. Apart from this *electromagnetic mass* there is assumed no other true or material mass. For its calculation the electron is assumed, in its rest frame Σ', to be a symmetrical sphere of radius a having a charge density $\rho(r')$ which depends only on the radial distance r' from its centre.

Suppose that the electron moves with a uniform velocity $\mathbf{v}$ parallel to the x-axis in the inertial frame Σ. In the rest frame Σ' of the electron, whose centre is taken to be the origin, it sets up an electrostatic field such that at any point $\mathbf{r}'$ the electromagnetic potentials are $\mathbf{A}' = 0$ and $\Phi' = \Phi'(r')$. Define the polar angle θ' and azimuth angle ϕ' as shown in Fig. 8.1. Then the field vectors at a point $\mathbf{r}' = r'(\cos\theta', \sin\theta'\cos\phi', \sin\theta'\sin\phi')$ are

$$\mathbf{E}' = E'_r(\cos\theta', \sin\theta'\cos\phi', \sin\theta'\sin\phi') \quad ,$$

and

$$\mathbf{B}' = 0 \quad , \tag{8.96}$$

where E'_r depends on the model assumed for the electron.

The fields in the Σ frame due to the electron may be obtained from the expressions (8.96) and the inverse of the transformation equations (8.26). These are:

$$\mathbf{E} = E'_r(\cos\theta', \gamma\sin\theta'\cos\phi', \gamma\sin\theta'\sin\phi') \quad ,$$

and

$$\mathbf{B} = \frac{1}{c}\gamma\beta E'_r(0, -\sin\theta'\sin\phi', \sin\theta'\cos\phi') \quad , \tag{8.97}$$

where

$$\gamma = (1-\beta^2)^{-\frac{1}{2}} \quad \text{and} \quad \beta = \frac{v}{c} \quad .$$

The total momentum of the electromagnetic field of the moving electron may be evaluated by integrating the momentum density $(\mathbf{E}\times\mathbf{H})/c^2$

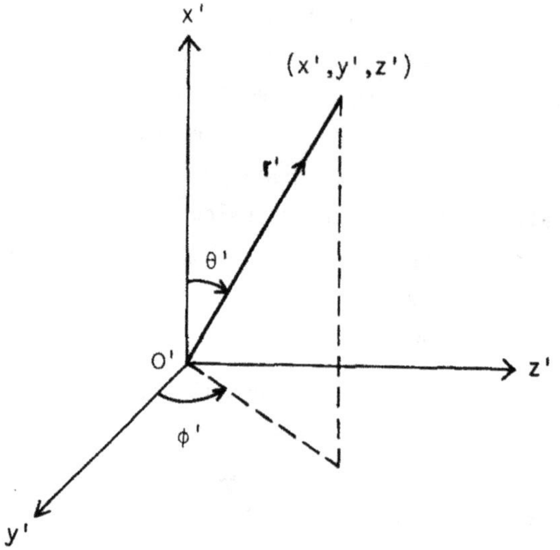

Fig. 8.1 Polar and azimuth angles in the rest frame Σ' of an electron, whose centre coincides with the origin O'.

over the entire space. Its x-component is

$$P_x = \frac{1}{c^2}\int (E_yH_z - E_zH_y)dxdydz$$

$$= \frac{\varepsilon_0}{c}\gamma\beta \int E_r'^2\sin^2\theta' dx'dy'dz' \quad ,$$

where the length contraction formula (8.10) has been used in transforming the volume element. Since ϕ' is not involved in the integral we may write

$$dx'dy'dz' = 2\pi r'^2 \sin\theta' d\theta' dr'$$

so that

$$P_x = \frac{\varepsilon_0}{c}\gamma\beta \int_0^{\pi} \sin^3\theta' d\theta' \int_0^{\infty} E_r'^2 2\pi r'^2 dr' = \frac{4W'}{3c}\gamma\beta \quad , \qquad (8.98)$$

where

$$W' = \frac{\varepsilon_0}{2}\int_0^\infty E_r'^2 4\pi r'^2 dr' \tag{8.99}$$

is the electrostatic field energy of the electron in the rest frame. By similar calculations or from a physical consideration we find $p_y = p_z = 0$. Hence, the total momentum of the field of the electron is

$$\mathbf{p} = \frac{4W'}{3c^2}\gamma\mathbf{v} \quad . \tag{8.100}$$

If it is assumed that the inertial mass m of the electron arises solely from its electromagnetic field, then the momentum of the field must be identified with the mechanical momentum $m\mathbf{v}$ of the electron. m is then called the *electromagnetic mass*. Equation (8.100) gives

$$m = \frac{4W'}{3c^2}\gamma \quad .$$

This equation shows that the mass of the electron depends on its velocity through γ. For a stationary electron, $\gamma = 1$ and the rest mass m_0 is

$$m_0 = \frac{4W'}{3c^2} \quad . \tag{8.101}$$

In general, for a moving electron the mass is

$$m = m_0\gamma \quad . \tag{8.102}$$

This is just the mass-velocity relationship required by special relativity.

The value of the rest mass m_0 depends on the charge distribution assumed for the electron. Consider the following simple models:

(a) The electron is assumed to be a thin uniform spherical shell of radius a and total charge e. As

$$E'_r = \frac{e}{4\pi\varepsilon_0 r'^2} \qquad \text{for} \qquad r' \geq a \quad ,$$

and

$$E'_r = 0 \qquad \text{for} \qquad r' < a \quad ,$$

it follows that

$$W' = \frac{e^2}{8\pi\varepsilon_0 a} \quad ,$$

giving

$$m_0 = \frac{e^2}{6\pi\varepsilon_0 c^2 a} \quad . \tag{8.103}$$

(b) The electron is assumed to have a total charge e uniformly distributed in a sphere of radius a. As

$$E'_r = \frac{e}{4\pi\varepsilon_0 r'^2} \qquad \text{for} \qquad r' \geq a \quad ,$$

and

$$E'_r = \frac{er'}{4\pi\varepsilon_0 a^3} \qquad \text{for} \qquad r' < a \quad ,$$

we find that

$$W' = \frac{3e^2}{20\pi\varepsilon_0 a} \quad ,$$

which gives

$$m_0 = \frac{e^2}{5\pi\varepsilon_0 c^2 a} \quad . \tag{8.104}$$

Special relativity requires that if the electromagnetic field carries a momentum density $\mathbf{E}\times\mathbf{H}/c^2$, it must also carry an energy density $\frac{1}{2}(\varepsilon_0 E^2 + \mu_0 H^2)$, giving a total field energy of

$$W = \frac{\varepsilon_0}{2}\int (E^2 + c^2B^2)dxdydz$$

$$= \frac{1}{2\gamma}\int_0^{\pi}\{\cos^2\theta' + \gamma^2(1+\beta^2)\sin^2\theta'\}\ \sin\theta' d\theta' \int_0^{\infty}\frac{\varepsilon_0}{2}E_r'^2 4\pi r'^2 dr'$$

$$= (1 + \frac{\beta^2}{3})\gamma W' \quad . \tag{8.105}$$

If this energy is identified with the total energy mc^2 of the electron, then

$$m = (1 + \frac{\beta^2}{3})\gamma\,\frac{W'}{c^2} \quad . \tag{8.106}$$

The velocity dependence of mass obtained from energy consideration is thus different from that obtained from momentum consideration. Furthermore, Eq. (8.105) becomes, for small velocities,

$$W \approx W' + \frac{5W'}{6c^2}v^2 \quad .$$

Then, if we identify W' with the rest energy m_oc^2 of the electron and $\frac{5W'v^2}{6c^2}$ with the kinetic energy $\frac{1}{2}m_ov^2$ for small velocities, a discrepancy occurs in the mass values obtained. This is of course to be expected since Eq. (8.106) differs from the correct relativistic mass-velocity relationship.

From the above discussion it is clear that quite independent of any detailed model that may be assumed for the electron, the classical electromagnetic model does not lead to a consistent mass-velocity relationship. Furthermore, such a simple model cannot explain the spin and magnetic moment of the electron. This example again demonstrates the failure of the classical theory when applied to atomic phenomena.

The expression $\frac{5W'}{3c^2}$ for the electromagnetic mass combined with a uniform spherical charge distribution leads to the so-called *"classical radius"* of the electron

$$r_o = \frac{e^2}{4\pi\varepsilon_o m_o c^2} \quad .$$

This, as the other rest mass formulas, gives the correct order of magnitude for the dimensions of an electron if such a classical concept as a definite radius could still be used in the description of atomic particles.

PROBLEMS

1. The Lorentz transformation may be derived on the assumptions of time dilation and length contraction for moving clocks and lengths. Derive the transformation equations for inertial frames Σ and Σ', where Σ' is the rest frame of the clock and length, from the following assumptions:

$$\left(\frac{\partial x}{\partial t}\right)_{x'} = v \quad , \quad \left(\frac{\partial x'}{\partial t'}\right)_{x} = -v \quad ,$$

$$\left(\frac{\partial t}{\partial t'}\right)_{x'} = \gamma \quad , \quad \left(\frac{\partial x'}{\partial x}\right)_{t} = \gamma \quad ,$$

$$y = y' \quad , \quad z = z' \quad ,$$

where $\gamma = (1 - \frac{v^2}{c^2})^{-\frac{1}{2}}$. Explain the physical meaning of these assumptions.

2. Show that the Lorentz transformation between inertial frames Σ and Σ' may be represented by

$$x' = x\cosh\phi - ct\sinh\phi \quad ,$$

$$ct' = -x\sinh\phi + ct\cosh\phi \quad ,$$

$$y' = y \quad , \quad z' = z \quad ,$$

where $\tanh\phi = \beta$, βc being the velocity of Σ' relative to Σ.

Hence show that (i) the transformation corresponds to a rotation through an angle $i\phi$ in the four-dimensional (x,y,z,ict)-space, (ii) two successive Lorentz transformations with relative velocities $\beta_1 c$ and $\beta_2 c$ are equivalent to a single Lorentz transformation of relative velocity βc, where β is given by

$$\beta = \frac{\beta_1 + \beta_2}{1 + \beta_1\beta_2} \quad .$$

Show also that for three successive Lorentz transformations L_1, L_2, L_3, the associative law holds: $(L_1L_2)L_3 = L_1(L_2L_3)$.

As for every Lorentz transformation there exists an inverse and the identity transformation may also be considered a Lorentz transformation, the set of Lorentz transformations forms a group. Furthermore $L_1L_2 = L_2L_1$, showing that the group is Abelian.

3. If $\mathbf{r}$ and $\mathbf{r}'$ are the positions of a particle in inertial frames Σ and Σ' respectively at an instant t according to a clock in Σ and t' according to a clock in Σ', the Lorentz transformation may be written as

$$\mathbf{r}' = \mathbf{r} + \left\{ \frac{\mathbf{r}\cdot\mathbf{v}}{v^2}(\gamma - 1) - \gamma t \right\} \mathbf{v} \quad ,$$

$$t' = \gamma\left(t - \frac{\mathbf{r}\cdot\mathbf{v}}{c^2}\right) \quad ,$$

where $\gamma = (1 - \beta^2)^{-\frac{1}{2}}$ and $\beta = \frac{v}{c}$. These equations obviously apply to the case where the velocity $\mathbf{v}$ of Σ' relative to Σ is in an arbitrary direction.

If the particle has velocity $\mathbf{u} = \frac{d\mathbf{r}}{dt}$ in Σ and $\mathbf{u}' = \frac{d\mathbf{r}'}{dt'}$ in Σ', show that

$$\mathbf{u}' = \left\{ \mathbf{u} + \left[\frac{\mathbf{u}\cdot\mathbf{v}}{v^2}(\gamma - 1) - \gamma \right] \mathbf{v} \right\} \left\{ \gamma\left(1 - \frac{\mathbf{u}\cdot\mathbf{v}}{c^2}\right) \right\}^{-1} \quad .$$

4. An event P is observed to be the cause of another event Q. If the time and spatial separations of P and Q are Δt and Δx respectively, the speed at which the physical effect causing Q propagates is defined as $u = \frac{\Delta x}{\Delta t}$. Show that the principle of relativity requires u not to exceed the velocity of light c.

Hint: Show that if u were greater than c, then there would exist inertial frames Σ' in which $\Delta t'$ becomes negative.

5. A relativistic particle of rest mass m moves in a field of potential $V(\mathbf{r})$. Show that $m\gamma c^2 + V = \text{constant}$, where $\gamma = (1-\beta^2)^{-\frac{1}{2}}$, βc being the speed of the particle.

6. Show that in the direction of an observer the Doppler effect is given approximately by $\frac{v}{c} \approx \frac{\omega - \omega_0}{\omega_0}$.

7. The distribution of the velocity component v along any given direction of the molecules of a dilute gas of absolute temperature T is known to be

$$\frac{d\mathcal{N}(v)}{\mathcal{N}} = \frac{\sqrt{m}}{\sqrt{2\pi kT}} \exp\left(-\frac{v^2 m}{2kT}\right) dv \quad ,$$

where k is Boltzmann's constant and m the mass of a gas molecule. Assuming the radiation of the gas molecules to be monochromatic with angular frequency ω_0 in the rest frame and the spectral line widening to be due entirely to the Doppler effect, find the frequency distribution of the radiation.

If the maximum intensity occurs at the wavelength λ_0 and the intensity drops to half the maximum at $\lambda_0 \pm \Delta\lambda$, $\Delta\lambda$ is called the half-width of the line. Show that the half width of the above line is given approximately by

$$\Delta\lambda \simeq \frac{\lambda_0}{c} (2kT \ln 2)^{\frac{1}{2}} \quad .$$

What is the Doppler half-width of the H_α-line (wavelength 6563A) radiated by hydrogen atoms at 50°C?

8. The *acceleration four-vector* of a particle is defined as $a^\alpha = \frac{du^\alpha}{ds}$, where u^α is the four-velocity and $(ds)^2$ the line element. At any instant the particle may be assigned a rest frame Σ' in which it is momentarily stationary. If Σ' moves with velocity $\mathbf{v}$ relative to Σ along the x-direction, use the transformation formulas for a four-vector to show that the acceleration vector $\mathbf{a}$ is transformed according to

$$\mathbf{a}' = \gamma^2 \mathbf{a} + \gamma^2(\gamma - 1)\,\frac{\mathbf{v}\cdot\mathbf{a}}{v^2}\,\mathbf{v} \quad ,$$

where $\gamma = (1 - \frac{v^2}{c^2})^{-\frac{1}{2}}$. Hence show that the rate of radiation loss of a charge of velocity $\mathbf{v}$ and acceleration $\mathbf{a}$ is

$$-\frac{dW}{dt'} = \frac{q^2}{6\pi\varepsilon_0 c^3}\,\gamma^6 \left\{ a^2 - \left(\frac{\mathbf{u}\times\mathbf{a}}{c}\right)^2 \right\} \quad .$$

Note that this expression is identical with Eq. (4.58).

9. Show that a particle having a finite total energy but moving with the velocity of light c must have zero rest mass. It follows that for such a particle the energy and momentum are related by $E = pc$.

Planck first proposed the theory that the electromagnetic radiation of angular frequency ω may be consisting of photons, moving with the velocity of light, each having the energy $E = \hbar\omega$, where $\hbar$ is a constant. Show that if $\hbar$ may be considered an invariant constant, the Planck relation is Lorentz covariant.

10. Show from the principles of the conservation of energy and the conservation of momentum that if a photon of energy $\hbar\omega$ is scattered by a free electron, initially at rest, through an angle θ the scattered photon will have an energy $\hbar\omega'$ given by

$$\hbar\omega' = \hbar\omega\left\{1 + \frac{\hbar\omega}{mc^2}(1 - \cos\theta)\right\}^{-1} ,$$

where m is the rest mass of the electron. Show also that the electron acquires in the process a kinetic energy

$$T = \frac{(\hbar\omega)^2}{mc^2} \frac{1 - \cos\theta}{\left\{1 + \frac{\hbar\omega}{mc^2}(1 - \cos\theta)\right\}} .$$

11. Since a photon of energy $\hbar\omega$ carries a momentum $\hbar\mathbf{k}$, where $\mathbf{k}$, having the magnitude ω/c and the direction of propagation, is the propagation vector, its incidence on a surface must be accompanied by the action of an impulse on the latter. Considering an electromagnetic wave as a beam of photons, show that the pressure exerted by waves of energy density U incident at an angle θ to the normal on a plane surface, whose coefficient of reflection is R, is $U(1+R)\cos^2\theta$. Show also that the tangential force exerted on a unit area of the surface is $\frac{1}{2}U(1-R)\sin 2\theta$.

12. A point charge q moving with a uniform velocity $\mathbf{v}$ has coordinates $(vt, 0, 0)$ at an instant t. Show that the potentials of its electromagnetic field at a point (x,y,z) at that instant are

$$\mathbf{A} = \frac{\mu_0}{4\pi}\frac{q\mathbf{v}}{r^*} , \qquad \Phi = \frac{1}{4\pi\varepsilon_0}\frac{q}{r^*} ,$$

where

$$r^* = \{(x - vt)^2 + (1 - \beta^2)(y^2 + z^2)\}^{\frac{1}{2}}, \qquad \beta = \frac{v}{c} .$$

If electromagnetic effects leaving the charge at the instant

$t = 0$ reach the field point at time t, then $r = ct$, where r is the distance between the charge and the field point at $t = 0$. Show that r^* is now equal to $r(1 - \frac{v_r}{c})$, where v_r is the radial velocity of the charge towards the field point. The expressions for **A** and Φ are then identical with those for the Liénard-Wiechert potentials, Eqs. (4.29) and (4.30).

Show also that the electric intensity and magnetic induction due to the charge are

$$\mathbf{E} = \frac{1}{4\pi\varepsilon_0} \frac{q\mathbf{r}}{\gamma^2 r^{*3}} \quad , \qquad \mathbf{B} = \frac{\mathbf{v} \times \mathbf{E}}{c^2} \quad ,$$

where $\mathbf{r} = (x - vt, y, z)$.

13. Show that the electric field due to a point charge of velocity $\mathbf{v}$ in the direction of motion is reduced by a factor γ^2, and that in the transverse direction it is increased by a factor γ when compared with the Coulomb field, where $\gamma = (1 - \frac{v^2}{c^2})^{-\frac{1}{2}}$.

14. A particle of charge q and rest mass m moves in a uniform electromagnetic field of vectors $\mathbf{E} = (E, 0, 0)$ and $\mathbf{B} = (0, 0, E/c)$. Write down Eq. (8.73) in the component form, and show that if the particle starts from rest from the plane $x = 0$ its velocity is $\mathbf{v} = (c/\sqrt{2}, -c/2, 0)$ at $x = mc^2/eE$.

15. An unstable particle has a rest mass m_0 and a lifetime t_0 when at rest. It is observed to be in uniform motion with total energy E_0 in the laboratory until it disintegrates into two equal particles, each of rest mass m. (i) Show that the parent particle has travelled a distance $(E_0^2 - m_0^2c^4)^{\frac{1}{2}} t_0/m_0c$ in the laboratory before it disintegrates. (ii) Show that the speed of each secondary particle in the rest frame of the parent particle is $c(1 - 4m^2/m_0^2)^{\frac{1}{2}}$. (iii) If in the laboratory frame one of the secondary particles has total energy E, show that its path makes

an angle θ with that of the parent particle where

$$\cos\theta = \frac{2EE_0 - m_0^2c^4}{2E_0(E^2 - m^2c^4)^{\frac{1}{2}}} \quad .$$

16. A particle of positive charge q and rest mass m is projected with an initial velocity u on a positive charge Q, which has a much greater rest mass, at the origin in such a way that it would pass Q at a distance b if the repulsive Coulomb force was absent. Set up the Lagrangian of the particle. Show that the motion of q is two dimensional and derive the following equations of the motion in polar coordinates:

$$\gamma_0 - \gamma = \frac{Qq}{4\pi\varepsilon_0 mc^2 r} \quad , \qquad \gamma r^2\dot{\theta} = \gamma_0 bu \quad ,$$

$$\frac{d}{dt}(\gamma\dot{r}) - \gamma r\dot{\theta}^2 = \frac{Qq}{4\pi\varepsilon_0 mr^2} \quad ,$$

where $\gamma = (1 - \frac{\dot{r}^2 + r^2\dot{\theta}^2}{c^2})^{-\frac{1}{2}}$ and $\gamma_0 = (1 - \frac{u^2}{c^2})^{-\frac{1}{2}}$.

To obtain the orbit of the projectile, $r = r(\theta)$, put $w = r^{-1}$ and deduce

$$\frac{d^2w}{d\theta^2} + \lambda^2 w + \mu = 0 \quad ,$$

where $\mu = \frac{Qq}{4\pi\varepsilon_0 m\gamma_0 b^2u^2}$ and $\lambda^2 = 1 - \frac{\mu^2b^2u^2}{c^2}$.

Hence show that the orbit is given by

$$\frac{1}{r} = \frac{\sin(\lambda\theta)}{b\lambda} + \frac{\mu}{\lambda^2}\{\cos(\lambda\theta) - 1\} \quad .$$

Show further that the distance of closest approach is $\mu b^2 + b(\lambda^2 + b^2\mu^2)^{\frac{1}{2}}$ and that q is deflected through an angle

$$\pi - \frac{2}{\lambda} \operatorname{arccot} \left(\frac{\lambda}{b\mu}\right).$$

17. A particle of charge q and rest mass m moves in a uniform electric field of intensity $\mathbf{E}$ whose direction is taken to be the x-direction. Initially the particle is at the origin and has a momentum $(0, p_0, 0)$. Show that the x-coordinate of the particle at time t is given by

$$\left(x + \frac{U_0}{qE}\right)^2 = c^2t^2 + \left(\frac{U_0}{qE}\right)^2$$

and that the orbit is given by

$$x = \frac{U_0}{qE} \left\{\cosh\left(\frac{eEy}{p_0 c}\right) - 1\right\} \quad ,$$

where $U_0 = \sqrt{p_0^2c^2 + m^2c^4}$.

18. A particle of charge q and rest mass m starts from the origin with a velocity $\mathbf{v} = (0, v_0, 0)$ in a static and uniform electromagnetic field with $\mathbf{E} = (0, 0, E)$ and $\mathbf{B} = (0, 0, B)$. Show that its coordinates at time t are related by

$$x^2 + y^2 = \rho^2 \quad , \quad (z + z_0)^2 = c^2t^2 + z_0^2 \quad ,$$

where $\rho = \dfrac{m\gamma_0 v_0}{qB}$, $z_0 = \dfrac{m\gamma_0 c^2}{qE}$ and $\gamma_0 = \left(1 - \dfrac{v^2}{c^2}\right)^{-\frac{1}{2}}$.

Describe the particle trajectory.

19. A particle of charge q and rest mass m starts from rest at the origin in a static and uniform electromagnetic field with $\mathbf{E} = (0, E_2, E_3)$ and $\mathbf{B} = (0, 0, B)$. Show that the coordinates at time t are related by

$$x^2 + y^2 + (z + z_0)^2 = c^2t^2 + z_0^2 \quad ,$$

where $z_0 = \frac{mc^2}{qE_3}$.

20. Lagrange's equations may be written in the covariant form

$$\frac{d}{ds}\left(\frac{\partial L}{\partial u^\alpha}\right) - \frac{\partial L}{\partial x^\alpha} = 0 \quad ,$$

where $u^\alpha \equiv \frac{dx^\alpha}{ds}$, if the Lagrangian $L(x^\alpha, u^\alpha)$ is formulated as a scalar. Show that if the Lagrangian for a particle of charge q and rest mass m_0 moving in an electromagnetic field of four-potential Φ^α is chosen to be

$$L = \frac{1}{2} m_0 c^2 u^\alpha u_\alpha + q u^\alpha \Phi_\alpha$$

Lagrange's equations reduce to

$$\frac{d}{ds}(m_0 c^2 u_\alpha) = q F_{\alpha\beta} u^\beta \quad ,$$

which is identical with Eq. (8.73).

The Hamiltonian is by definition $H = p_\alpha u^\alpha - L$, where $p_\alpha = \frac{\partial L}{\partial u^\alpha}$. Show that Hamilton's equations

$$\frac{\partial H}{\partial p_\alpha} = \frac{dx^\alpha}{ds} \quad , \quad \frac{\partial H}{\partial x^\alpha} = -\frac{dp_\alpha}{ds}$$

also lead to the correct equations of motion.

Hint: Note that as $u^\alpha u_\alpha = g_{\alpha\beta} u^\alpha u^\beta$, $\frac{\partial}{\partial u^\alpha}(u^\beta u_\beta) = 2u_\alpha$.

21. Show that the equations (2.20) and (2.24) expressing the gauge transformation may be written in the covariant form

$$\Phi^\alpha = \Phi_0^\alpha + \frac{\partial \chi}{\partial x_\alpha} \quad ,$$

where χ is a scalar.

Hence show that (i) the electromagnetic field tensor $F^{\alpha\beta}$ is invariant with respect to a gauge transformation, and (ii) Lagrange's equations of motion for a charge particle, Eqs. (8.75), are unchanged when the four-potential Φ_α in L undergoes a gauge transformation.

22. From the definition of the electromagnetic field tensor, $F^{\alpha\beta}$, and the transformation relation for a tensor of rank 2, obtain the transformation equations for the components of **E** and **B**.

23. The symbol $\varepsilon_{\alpha\beta\gamma\delta}$ is defined to have the properties

$$\begin{aligned}\varepsilon_{\alpha\beta\gamma\delta} &= 0 \quad \text{if any two indices are equal,}\\ &= +1 \text{ (or } -1) \quad \text{if the indices are in an even (or odd) permutation of 1234.}\end{aligned}$$

In another inertial frame Σ', define

$$\varepsilon'_{\alpha\beta\gamma\delta} = (Q^{-1})^\lambda_\alpha (Q^{-1})^\mu_\beta (Q^{-1})^\nu_\gamma (Q^{-1})^\sigma_\delta \, \varepsilon_{\lambda\mu\nu\sigma}$$

and show that $\varepsilon'_{\alpha\beta\gamma\delta}$ has all the properties required for the symbol. This will mean that $\varepsilon_{\alpha\beta\gamma\delta}$ transforms like an antisymmetric tensor of the fourth rank. It follows that $\varepsilon_{\alpha\beta\gamma\delta}F^{\alpha\beta}F^{\gamma\delta}$ is a scalar.

Using the above result show that $\mathbf{B}\cdot\mathbf{E}$ is invariant under the Lorentz transformation. Hence show that if **E** and **B** are not mutually perpendicular it will not be possible to find a frame in which the field is purely electric or purely magnetic.

24. From the fact that $F^{\alpha\beta}F_{\alpha\beta}$ is a scalar, show that $B^2c^2 - E^2$ is invariant under a Lorentz transformation.

25. Show that an electromagnetic wave that is observed to be plane, sinusoidal and linearly polarized in an inertial frame will be observed to be so in any other inertial frame. Show also that

the *propagation four-vector* may be defined as

$$k^{\alpha} = (\mathbf{k}, \frac{\omega}{c}) \quad ,$$

where $\mathbf{k}$ and ω are the propagation vector and angular frequency of the wave respectively.

Hint: Show that the exponent $\mathbf{k} \cdot r - \omega t$ in the wave function is invariant and use the results of Problems (8.23) and (8.24).

26. A monochromatic electromagnetic wave from a source fixed at the origin O' of Σ' is observed at a point which has coordinates (x', y', z') in Σ' and (x, y, z) in Σ. Show that (i) the frequency ω' and ω observed in the two frames are related by (the Doppler effect)

$$\omega' = \gamma\omega(1 - \beta\cos\theta) \quad ,$$

where βc is the velocity of Σ' relative to Σ, and (ii) the angles θ' and θ made by the directions of propagation in the two frames with the x-axes are related by (aberration of starlight)

$$\tan\theta = \frac{\sin\theta}{\gamma(\cos\theta - \beta)} \quad .$$

Hence show that if the average velocity of the earth in its orbit is 30 km s^{-1}, a star at the zenith will appear to describe a circle of radius 20.5 seconds of arc. This phenomenon was first observed by Bradley in 1727. In the classical theory this could be explained without assuming any ether drag [Prob. (8.27)]. Note that as $\frac{v}{c} \approx 10^{-4}$, Galilean transformation may be applied.

27. Electromagnetic waves propagate with a phase velocity $\frac{c}{n}$ in a stationary medium of refractive index n. Show that in a frame Σ in which the medium has velocity $\mathbf{v}$, the waves propagating in the direction of $\mathbf{v}$ has a phase velocity $c(\frac{1 + \beta n}{n + \beta})$, where $\beta = \frac{v}{c}$. Furthermore, show that the phase velocity in Σ is $\frac{c}{n} + (1 - \frac{1}{n^2})v$

for $v \ll c$.

According to Fresnel's theory of ether-drag, the velocity of light in the ether permeating a medium of refractive index n is c/n. If the medium moves with a velocity v with respect to the ether, the velocity of light relative to the medium will be $\frac{c}{n} - v$ in the direction of v. This however would be increased to $\frac{c}{n} - \alpha v$, where $0 < \alpha < 1$, if the ether is partially dragged along by the medium. The velocity of light in the absolute frame would then be $\frac{c}{n} - \alpha v + v = \frac{c}{n} + (1-\alpha)v$. Fresnel suggested that $\alpha = n^{-2}$, which was confirmed by a later experiment of Fizeau. This is in agreement with the relativistic result to the first order in β. The coefficient $1 - n^{-2}$ is called the *Fresnel-Fizeau drag coefficient.*

28. An electromagnetic wave has a propagation vector $\mathbf{k}$ in the frame Σ which makes an angle θ with the x-axis. The plane containing $\mathbf{k}$ and the x-axis makes an angle ϕ with the y-axis. Show that in the frame Σ' the angles θ' and ϕ' are given by

$$\phi' = \phi \quad , \quad \cos\theta' = \frac{\cos\theta - \beta}{1 - \beta\cos\theta} \quad ,$$

where βc is the velocity of Σ' relative to Σ.

Hence show that a beam of light which subtends a solid angle $d\Omega$ in Σ will be observed in Σ' to subtend a solid angle

$$d\Omega' = \frac{1 - \beta^2}{(1 - \beta\cos\theta)^2}\, d\Omega \quad .$$

29. A plane electromagnetic wave polarized with $\mathbf{E}$ in the x-direction propagates parallel to the z-axis. Show that the average radiation pressure on an xy-plane is in the z-direction and is equal to the energy density of the wave, using (i) Maxwell's stress tensor, (ii) the expression for the field momentum density.

30. A plane electromagnetic wave of intensity I in free space is reflected at normal incidence from a medium of refractive index n. Show that the pressure exerted by the free space radiation on the interface is $\frac{2(1+n^2)}{(1+n)^2}\frac{I}{c}$.

APPENDIX

NOTES ON SUFFIX NOTATION AND VECTOR ANALYSIS

A.1 Suffix Notation and Summation Convention

Vectors and vector differential operators are defined independent of any particular coordinate system. Their manipulation, however, may be facilitated by the use of Cartesian coordinates and the suffix notation[a]. The results obtained, expressed in vectors and the differential operators, are again independent of the coordinate system employed.

In the Cartesian coordinate system, the components x, y, z of the position vector $\mathbf{r}$ of a point along the coordinate axes may be conveniently designated x_1, x_2, x_3 respectively. When the symbols employed for these components are to be emphasized, we may denote $\mathbf{r}$ by (x_i), where the index i is understood to run from 1 to 3. Thus we may use any of the following notations to denote the location of a point:

$$\mathbf{r}, \quad (x_i) \quad , \quad (x, y, z) \quad \text{or} \quad (x_1, x_2, x_3) \quad .$$

Similarly, a vector $\mathbf{A}$ may be represented by

$$\mathbf{A}, \quad (A_i) \quad , \quad (A_x, A_y, A_z) \quad \text{or} \quad (A_1, A_2, A_3) \quad .$$

It is frequently required to sum up terms arising from permutation of the values of the indices denoting the components of vectors or tensors, or elements of matrices, in expressions such as

$$\mathbf{A} \cdot \mathbf{B} = \sum_{k=1}^{3} A_k B_k \quad ,$$

---

[a] First proposed by A. Einstein and is known as Einstein's summation rule. See A. Einstein, *Annalen der Physik* **49** (1916) 760.

$$A^2 = \sum_{k=1}^{3} A_k^2 = \sum_{k=1}^{3} A_k A_k \quad ,$$

$$F_{ij} = \sum_{k=1}^{3} C_{ik} D_{kj} \quad , \quad \text{etc.}$$

On all such occasions, it is seen that the index k over which the summation is to be made appears repeatedly, while the non-summation indices occur only singly in a term. This suggests that we may dispense with the summation symbol altogether and let the repetition of an index imply summation over all its possible values. This is the summation convention used in tensor analysis and the one which we shall adopt here. Accordingly the above summations will be written simply as

$$\mathbf{A} \cdot \mathbf{B} = A_k B_k \quad ,$$

$$A^2 = A_k^2 \quad ,$$

$$F_{ij} = C_{ik} D_{kj} \quad , \quad \text{etc.}$$

The repeated index over whose values the summation is to be carried out is called a *dummy index*, since the value of the summation does not depend on the symbol employed for this index.

The summation convention will be used only for indices employed to identify components along coordinate axes or elements of matrices or tensors, but not otherwise. For instance, when summing over n charges of a system the summation sign is still retained:

$$\mathbf{E} = \frac{1}{4\pi\varepsilon_0} \sum_{s=1}^{n} \frac{q_s \mathbf{r}_s}{r_s^3} \quad ,$$

$$Q = \sum_{s=1}^{n} q_s \quad , \quad \text{etc.}$$

A.2 Kronecker Delta and the Epsilon ϵ_{ijk}

The *Kronecker delta* δ_{ij} is a function of two indices i and j defined by

$$\delta_{ij} = 1 \quad , \quad \text{if} \quad i = j$$

ar.:

$$\delta_{ij} = 0 \quad , \quad \text{if} \quad i \neq j \quad .$$

By means of the Kronecker delta we may write

$$A_i = \delta_{ij}A_j \quad ,$$

$$A^2 = \delta_{ij}A_iA_j \quad ,$$

$$\frac{\partial x_i}{\partial x_j} = \delta_{ij} \quad , \quad \text{etc.}$$

whenever it is found convenient to do so.

The components of the cross-product of two vectors **A** and **B**,

$$\mathbf{A} \times \mathbf{B} = (A_2B_3 - A_3B_2,\ A_3B_1 - A_1B_3,\ A_1B_2 - A_2B_1) \quad ,$$

can be represented analytically if we introduce a function epsilon[b] ε_{ijk} of the indices such that

$\varepsilon_{ijk} = 0$ if any two of the indices i, j, k have the same value,

$\varepsilon_{ijk} = 1$ if i, j, k are all different and occur in an even permutation of 123,

$\varepsilon_{ijk} = -1$ if i, j, k are all different and occur in an odd permutation of 123.

---

[b] Also known as the Levi-Civita density.

Thus ε_{ijk} has non-zero elements:

$$\varepsilon_{123} = \varepsilon_{231} = \varepsilon_{312} = 1 \quad ,$$

$$\varepsilon_{213} = \varepsilon_{132} = \varepsilon_{321} = -1 \quad ,$$

and

$$(\mathbf{A}\times\mathbf{B})_1 = \varepsilon_{123}A_2B_3 + \varepsilon_{132}A_3B_2 = \varepsilon_{1jk}A_jB_k \qquad \text{etc} \quad .$$

Generally, we may write

$$(\mathbf{A}\times\mathbf{B})_i = \varepsilon_{ijk}A_jB_k \quad .$$

In the derivation of vector relations we often come across $\varepsilon_{ijk}\varepsilon_{i\ell m}$ as coefficients of terms in a summation such as in

$$\{\mathbf{A}\times(\mathbf{B}\times\mathbf{C})\}_j = \varepsilon_{jki}\varepsilon_{i\ell m}A_kB_\ell C_m = \varepsilon_{ijk}\varepsilon_{i\ell m}A_kB_\ell C_m \quad .$$

For assigned values of j and k, consider the terms with all possible assignments of i, ℓ and m. For the coefficients to be non-zero we require that $j \neq k$, which is assumed to be the case, and $i \neq j \neq k$, $\ell \neq m \neq i$. Then the index i can have only one value which is different from those of j and k, while ℓ and m must take on the same values as j and k, if the coefficients are not to vanish. The non-vanishing coefficients are therefore

$$\varepsilon_{ijk}\varepsilon_{i\ell m} = +1 \quad \text{for} \quad \ell = j \quad \text{and} \quad m = k \quad ,$$

and

$$\varepsilon_{ijk}\varepsilon_{i\ell m} = -1 \quad \text{for} \quad \ell = k \quad \text{and} \quad m = j \quad .$$

Thus, when summation is taken over the indices i, ℓ and m we have the following result

$$\varepsilon_{ijk}\varepsilon_{i\ell m} = \delta_{\ell j}\delta_{mk} - \delta_{\ell k}\delta_{mj} \quad . \tag{A.1}$$

Using this identity, we have

$$\{\mathbf{A}\times(\mathbf{B}\times\mathbf{C})\}_j = A_k B_j C_k - A_k B_k C_j \quad ,$$

or

$$\mathbf{A}\times(\mathbf{B}\times\mathbf{C}) = (\mathbf{A}\cdot\mathbf{C})\mathbf{B} - (\mathbf{A}\cdot\mathbf{B})\mathbf{C} \quad .$$

A.3 Vector Differential Operators

Consider a region of space R where there exist a scalar field and a vector field such that a scalar function $u(\mathbf{r})$ and a vector function $\mathbf{v}(\mathbf{r})$ are defined at every point $\mathbf{r}$ of R. It is assumed that the region R is regular and that $u(\mathbf{r})$ and $\mathbf{v}(\mathbf{r})$ and their derivatives are continuous, i.e. $u(\mathbf{r})$ and $\mathbf{v}(\mathbf{r})$ are continuously differentiable, in R. Three differential operations may be applied to these functions.

(a) *Gradient of u*

The equation $u(\mathbf{r}) = C$, where C is a constant, represents a surface. If c is a constant very much smaller than C, $u(\mathbf{r}+d\mathbf{r}) = C + c$ represents a neighbouring surface. Let $|d\mathbf{r}| = ds$, the spatial rate of change of u along $d\mathbf{r}$ is $\frac{du}{ds}$. As $du = c$ is a constant for the surfaces, it is obvious that the maximum rate of change of u at $\mathbf{r}$ occurs in the direction of the unit normal $\mathbf{n}$ to the surfaces. The quantity

$$\text{grad } u \equiv \mathbf{n}\,\frac{du}{dn} \quad ,$$

where dn is the path element along the normal, is called the *gradient* of u at $\mathbf{r}$. Along any other direction given by a unit vector $\mathbf{t}$ the spatial rate of change of u is

$$\frac{du}{ds} = \frac{du}{dn}\,\mathbf{n}\cdot\mathbf{t} = \mathbf{t}\cdot\text{grad } u \quad . \qquad \text{(A.2)}$$

In the Cartesian coordinate system, the component of grad u along the x_i-axis is therefore

$$\frac{\partial u}{\partial x_i} = \mathbf{i} \cdot \text{grad } u = (\text{grad } u)_i \quad . \tag{A.3}$$

(b) *Divergence of* $\mathbf{v}$

The *divergence* of a vector function $\mathbf{v}(\mathbf{r})$ at $\mathbf{r}$ is defined as

$$\text{div } \mathbf{v}(\mathbf{r}) = \lim_{V \to 0} \frac{1}{V} \oint_S \mathbf{v} \cdot d\mathbf{S} \tag{A.4}$$

where V is the volume bounded by a closed surface S and the limit is to be taken such that the point $\mathbf{r}$ is always contained in S. If $\mathbf{v}$ is interpreted as a flux density, div $\mathbf{v}$ is the total outward flux per unit volume.

To express div $\mathbf{v}$ in the Cartesian coordinate system, we consider an elementary volume dV which has edges dx, dy and dx with the point $\mathbf{r}$ at its centre. The net outward flux crossing the parallel surfaces $dydz$ is

$$(v_x + \frac{\partial v_x}{\partial x}\frac{dx}{2})dydz - (v_x - \frac{\partial v_x}{\partial x}\frac{dx}{2})dydz = \frac{\partial v_x}{\partial x} dxdydz = \frac{\partial v_x}{\partial x} dV \quad .$$

Similar expressions are obtained for the net fluxes crossing the other pairs of parallel surfaces. Hence

$$\text{div } \mathbf{v} = \frac{1}{dV}(\frac{\partial v_x}{\partial x} + \frac{\partial v_y}{\partial y} + \frac{\partial v_z}{\partial z})dV = \frac{\partial v_x}{\partial y} + \frac{\partial v_y}{\partial y} + \frac{\partial v_z}{\partial z} = \frac{\partial v_i}{\partial x_i} \quad . \tag{A.5}$$

(c) *Curl of* $\mathbf{v}$

The *curl* of $\mathbf{v}(\mathbf{r})$ is defined in a similar way:

$$\text{curl } \mathbf{v} = \lim_{V \to 0} \frac{1}{V} \oint_S d\mathbf{S} \times \mathbf{v} \tag{A.6}$$

where V, S and the limit sign have the same meanings as in the definition of div **v**.

Curl **v** is a vector. Its component along the x_j-axis is

$$\mathbf{i}_j \cdot \text{curl } \mathbf{v} = \lim_{V \to 0} \frac{1}{V} \oint_S \mathbf{i}_j \cdot d\mathbf{S} \times \mathbf{v}$$

$$= \lim_{V \to 0} \frac{1}{V} \oint_S \mathbf{v} \times \mathbf{i}_j \cdot d\mathbf{S}$$

$$= \text{div}(\mathbf{v} \times \mathbf{i}_j) \quad .$$

The base vector $\mathbf{i}_j$ has a component of magnitude unity along the x_j-axis and no component along the other axes. In the suffix notation its m-th component is therefore δ_{jm}. Thus, according to the above,

$$(\text{curl } \mathbf{v})_j = \frac{\partial}{\partial x_k} (\varepsilon_{k\ell m} v_\ell \delta_{jm})$$

$$= \varepsilon_{k\ell j} \frac{\partial v_\ell}{\partial x_k} = \varepsilon_{j\ell k} \frac{\partial v_k}{\partial x_\ell} \quad , \qquad \text{(A.7)}$$

where we have first rotated the indices j, ℓ, k without changing the permutation and then interchanged the dummy indices ℓ and k. Writing out in full we have

$$\text{curl } \mathbf{v} = \left(\frac{\partial v_z}{\partial y} - \frac{\partial v_y}{\partial z}\right) \mathbf{i} + \left(\frac{\partial v_x}{\partial z} - \frac{\partial v_z}{\partial x}\right) \mathbf{j} + \left(\frac{\partial v_y}{\partial x} - \frac{\partial v_x}{\partial y}\right) \mathbf{k}$$

$$= \begin{vmatrix} \mathbf{i} & \mathbf{j} & \mathbf{k} \\ \frac{\partial}{\partial x} & \frac{\partial}{\partial y} & \frac{\partial}{\partial z} \\ v_x & v_y & v_z \end{vmatrix} \quad .$$

If **v** represents a velocity field, then curl **v** is a measure of the angular velocity. Consider for instance a rigid body rotating with angular velocity $\boldsymbol{\omega}$ about some axis through a point O in the body.

The velocity of a point P with radius vector $\mathbf{r}$ from O is $\mathbf{v} = \mathbf{v}_0 + \boldsymbol{\omega} \times \mathbf{r}$, where $\mathbf{v}_0$ is the velocity of O relative to some origin fixed in space. As $\mathbf{v}_0$ and $\boldsymbol{\omega}$ are both independent of $\mathbf{r}$,

$$(\text{curl } \mathbf{v})_i = \varepsilon_{ijk} \frac{\partial}{\partial x_j} (\varepsilon_{k\ell m} \omega_\ell x_m) = \varepsilon_{kij} \varepsilon_{k\ell m} \omega_\ell \frac{\partial x_m}{\partial x_j}$$

$$= (\delta_{i\ell}\delta_{jm} - \delta_{im}\delta_{j\ell})\omega_\ell \delta_{mj}$$

$$= \omega_i \delta_{jj} - \omega_j \delta_{ij} = 3\omega_i - \omega_i = 2\omega_i \quad ,$$

i.e., $\text{curl } \mathbf{v} = 2\boldsymbol{\omega}$.

In fluid dynamics, the curl of the velocity is related to the circulation round a closed path moving with the fluid. The component of curl $\mathbf{v}$ at $\mathbf{r}$ in the direction of a unit vector $\boldsymbol{\upsilon}$ is

$$\boldsymbol{\upsilon} \cdot \text{curl } \mathbf{v} = \lim_{V \to 0} \frac{1}{V} \oint_S \boldsymbol{\upsilon} \cdot \mathbf{n} \times \mathbf{v} \, dS \quad ,$$

$\mathbf{n}$ being the unit normal to the surface element $d\mathbf{S}$. Take for the closed surface S a right cylinder of a small height h parallel to $\boldsymbol{\upsilon}$ constructed on a base which is formed by a simple planar closed curve C enclosing the point $\mathbf{r}$ as shown in Fig. A.1.

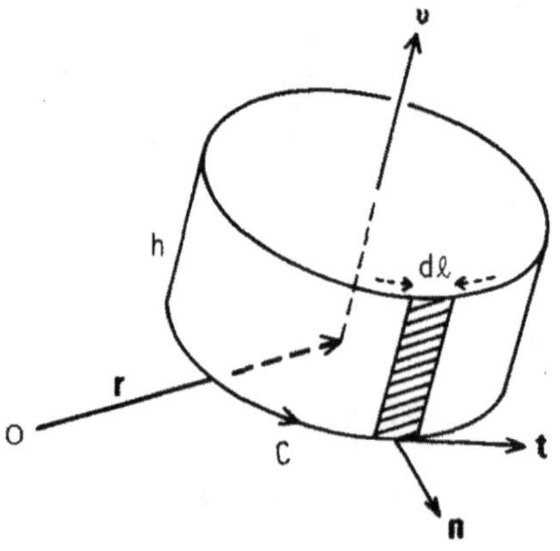

Fig. A.1 A short cylinder of a small height h parallel to $\boldsymbol{\upsilon}$ is constructed on a base formed by a simple closed curve C.

On the end surfaces, $\mathbf{n} \parallel \upsilon$ and the integrand vanishes. To find the contribution of the curved surface to the integral, take for the surface element dS a strip parallel to υ of length h and width $d\ell$. The normal $\mathbf{n}$ to this element is perpendicular to both υ and the tangent $\mathbf{t}$ to the curve C. Thus

$$\begin{aligned}\upsilon \cdot \mathbf{n} \times \mathbf{v}\, h d\ell &= h\mathbf{v} \cdot \upsilon \times \mathbf{n}\, d\ell \\ &= h\mathbf{v} \cdot \mathbf{t}\, d\ell \\ &= h\mathbf{v} \cdot d\mathbf{r} \quad .\end{aligned}$$

Writing $\mathbf{t}d\ell = d\mathbf{r}$ and $V = Ah$, we have

$$\upsilon \cdot \text{curl } \mathbf{v} = \lim_{A \to 0} \frac{1}{A} \oint_C \mathbf{v} \cdot d\mathbf{r} \quad , \tag{A.8}$$

A being the area bounded by C. The line integral is known as the circulation of $\mathbf{v}$ round C. For any fluid motion, the circulation round a closed curve moving with the fluid is constant for all time provided the external forces are conservative.

The form of the differential operators in the suffix notation, (A.3), (A.5) and (A.7), suggests that $\frac{\partial}{\partial x_i}$ may be considered as components of a vector operator ∇ and that we may write

$$\text{grad } u \equiv \nabla u \quad ,$$

$$\text{div } \mathbf{v} \equiv \nabla \cdot \mathbf{v} \quad ,$$

and

$$\text{curl } \mathbf{v} \equiv \nabla \times \mathbf{v} \quad .$$

∇ is known as the *del* or *nabla*.

Grad u and curl $\mathbf{v}$ are vectors and we can form the divergence and curl of these. The resultant can be readily obtained using the suffix notation method:

(a) $$\nabla \cdot \nabla u \equiv \nabla^2 u = \frac{\partial^2 u}{\partial x_i^2} \quad .$$

(b) $$\nabla \cdot \nabla \times \mathbf{v} = \frac{\partial}{\partial x_i}\left(\varepsilon_{ijk}\frac{\partial v_k}{\partial x_j}\right) = \varepsilon_{ijk}\frac{\partial^2 v_k}{\partial x_i \partial x_j} = \varepsilon_{jik}\frac{\partial^2 v_k}{\partial x_j \partial x_i}$$

$$= -\varepsilon_{ijk}\frac{\partial^2 v_k}{\partial x_i \partial x_j} = 0 \quad ,$$

where interchange has been made of the dummy indices i and j.

(c) $$(\nabla \times \nabla u)_i = \varepsilon_{ijk}\frac{\partial^2 u}{\partial x_j \partial x_k} = \varepsilon_{ikj}\frac{\partial^2 u}{\partial x_k \partial x_j} = -\varepsilon_{ijk}\frac{\partial^2 u}{\partial x_j \partial x_k} = 0 \quad ,$$

i.e. $\nabla \times \nabla u = 0$.

(d) $$\{\nabla \times (\nabla \times \mathbf{v})\}_i = \varepsilon_{ijk}\frac{\partial}{\partial x_j}\left(\varepsilon_{k\ell m}\frac{\partial v_m}{\partial x_\ell}\right)$$

$$= \varepsilon_{kij}\varepsilon_{k\ell m}\frac{\partial^2 v_m}{\partial x_j \partial x_\ell}$$

$$= (\delta_{i\ell}\delta_{jm} - \delta_{im}\delta_{j\ell})\frac{\partial^2 v_m}{\partial x_\ell \partial x_j}$$

$$= \frac{\partial}{\partial x_i}\left(\frac{\partial v_j}{\partial x_j}\right) - \frac{\partial^2 v_i}{\partial x_j^2} \quad ,$$

i.e. $\nabla \times (\nabla \times \mathbf{v}) = \nabla(\nabla \cdot \mathbf{v}) - \nabla^2 \mathbf{v}$ with $\nabla^2 \mathbf{v}$ interpreted as $\frac{\partial^2 \mathbf{v}}{\partial x_j^2}$.

Other combinations such as $\nabla \times (\mathbf{A} \times \mathbf{B})$ or $\nabla \cdot (\mathbf{A} \times \mathbf{B})$ can be similarly expanded.

A vector field $\mathbf{v}$ of which the curl, $\nabla \times \mathbf{v}$, is zero everywhere is said to be *irrotational*. A vector field $\mathbf{v}$ whose divergence, div $\mathbf{v}$, is zero everywhere is said to be *solenoidal*. (b) and (c) above suggest that a solenoidal vector $\mathbf{v}$ can always be expressed as the curl of a vector function ω, i.e. $\mathbf{v} = \nabla \times \omega$, and that an irrotational vector $\mathbf{v}$ can always be expressed as the gradient of a scalar function u, i.e. $\mathbf{v} = \nabla u$. These functions ω and u are not unique and can be obtained by integration.

A.4 Integral Theorems

Results of partial integration of the differential operators are known as *integral theorems*. It will be assumed that the vector and scalar functions involved are continuously differentiable and that the region of space which forms the domain of integration is regular.

The Divergence Theorem[c]

$$\int_V \nabla \cdot \mathbf{v} \, dV = \oint_S \mathbf{v} \cdot d\mathbf{S} \quad , \tag{A.9}$$

where V is the volume bounded by a closed surface S, the area element $d\mathbf{S}$ being taken in the direction of the outward normal.

To prove this theorem, we subdivide V into N cells of elementary volumes ΔV_r which may be considered rectangular boxes or boxes bounded by plane surfaces and portions of S. Let the centre of ΔV_r be P_r. By definition we have

$$\nabla \cdot \mathbf{v}(P_r) = \lim_{\Delta V_r \to 0} \frac{1}{\Delta V_r} \oint_{S_r} \mathbf{v} \cdot d\mathbf{S} \quad ,$$

where S_r is the boundary surface of ΔV_r. The above equality holds without the limit sign provided a small correction ε_r is included, i.e.

$$\oint_{S_r} \mathbf{v} \cdot d\mathbf{S} = \nabla \cdot \mathbf{v} \; \Delta V_r + \varepsilon_r \Delta V_r \quad .$$

Note that no summation is implied in the last term of the right-hand side as the index r does not specify the component of a vector. We

---

[c] Also known as Gauss', Green's or Ostrogradsky's theorem.

then sum up all the N cells:

$$\sum_{r=1}^{N} \oint_{S_r} \mathbf{v} \cdot d\mathbf{S} = \sum_{r=1}^{N} \nabla \cdot \mathbf{v}\, \Delta V_r + \sum_{r=1}^{N} \varepsilon_r \Delta V_r \quad , \tag{A.10}$$

and then let $N \to \infty$. The surface integrals of $\mathbf{v}$ over the common boundary between neighbouring cells cancel out since the outward normals to S_r and S_{r+1} are opposite in direction. The surviving terms are those due to the integrals over the outer surface S, and these sum up to $\oint_S \mathbf{v} \cdot d\mathbf{S}$. The first summation on the right-hand side becomes the volume integral $\int_V \nabla \cdot \mathbf{v} dV$ as $N \to \infty$. Let ε be the greatest of the corrections ε_r's. As the subdivision becomes finer, i.e. $N \to \infty$, $\varepsilon \to 0$. Thus

$$\sum_r \varepsilon_r \Delta V_r \leq \varepsilon \sum_r \Delta V_r = \varepsilon V \to 0$$

as $N \to \infty$. Equation (A.10) therefore takes the form (A.9) in the limit of $N \to \infty$.

Note that this theorem still applies if the volume V is bounded both externally and internally by closed surfaces, in which case surface integration is to be taken over all the boundary surfaces with $d\mathbf{S}$ in all cases pointing outwards from the volume V.

Corollary to the Divergence Theorem

Suppose $\mathbf{v}$ has only one component v_1. The divergence theorem takes the form

$$\int_V \frac{\partial v_1}{\partial x_1} dV = \oint_S v_1 dS_1 \quad . \tag{A.11}$$

This holds for any scalar function v_1. We may call it u and write the above as

$$\iiint_V \frac{\partial u}{\partial x_1} dx_1 dx_2 dx_3 = \iint_S u dx_2 dx_3 \quad .$$

Note that dS_1 is the component of $d\mathbf{S}$ in the direction of x_1 and may be written as $dx_2 dx_3$ in Cartesian coordinates. In general, we have

$$\int_V \frac{\partial u}{\partial x_i} dV = \oint_S u dS_i \tag{A.12}$$

as a corollary to the divergence theorem.

Green's Formula

Letting $\mathbf{v} = \Phi\nabla\psi - \psi\nabla\Phi$, where Φ and ψ are arbitrary scalar functions which are continuously differentiable, and applying the divergence theorem, we have

$$\int_V (\Phi\nabla^2\psi - \psi\nabla^2\Phi) dV = \oint_S (\Phi\nabla\psi - \psi\nabla\Phi) \cdot d\mathbf{S} \quad . \tag{A.13}$$

This relation is known as *Green's formula.*

Stokes' Theorem

$$\int_S \nabla \times \mathbf{v} \cdot d\mathbf{S} = \oint_C \mathbf{v} \cdot d\mathbf{r} \quad . \tag{A.14}$$

Here S is an arbitrary open two-sided surface with a single (non-intersecting) closed curve C as boundary. The normal to the surface S which gives the direction of $d\mathbf{S}$ is considered positive if it is in the direction of advance of a right-handed screw which is turned in the sense C is traversed.

To prove the theorem, we subdivide S into N approximately planar elements, say, triangular areas, ΔS_r, bounded respectively by simple closed contours C_r as shown in Fig. A.2. Let P_r be a point

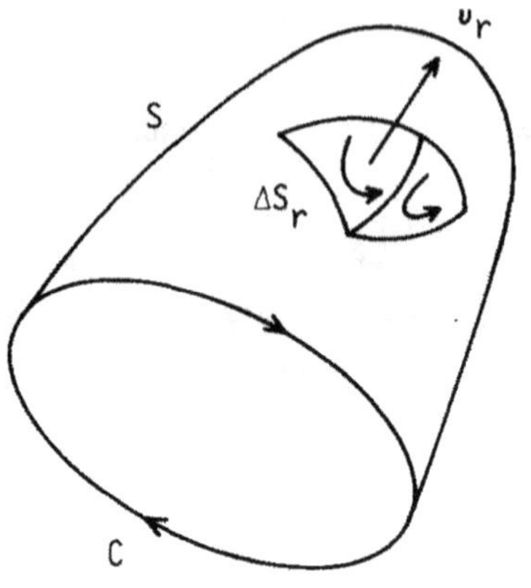

Fig. A.2 S is an open two-sided surface bounded by C. It may be subdivided into approximately planar elements ΔS_r.

on ΔS_r and υ_r be the unit normal at P_r to the area element, its direction conforming to the right-handed screw convention with respect to C_r. Equation (A.8) takes the form

$$\upsilon_r \cdot \nabla \times \mathbf{v}(P_r) = \lim_{\Delta S_r \to 0} \frac{1}{\Delta S_r} \oint_{C_r} \mathbf{v} \cdot d\mathbf{r}$$

and may be written as a finite equation

$$\nabla \times \mathbf{v}(P_r) \cdot \upsilon_r \Delta S_r = \oint_{C_r} \mathbf{v} \cdot d\mathbf{r} + \varepsilon_r \Delta S_r$$

with the inclusion of a small correction ε_r. As $N \to \infty$, $\varepsilon \to 0$, ε being the greatest of the ε_r's. Summing over all such contours, we have

$$\sum_{r=1}^{N} \nabla \times \mathbf{v}(P_r) \cdot \Delta \mathbf{S}_r = \sum_{r=1}^{N} \oint_{C_r} \mathbf{v} \cdot d\mathbf{r} + \sum_{r=1}^{N} \varepsilon_r \Delta S_r \quad . \qquad \text{(A.15)}$$

The line integrals of $\mathbf{v}$ along the common boundary between two neighbouring cells cancel as the $d\mathbf{r}$'s for the two cells run in opposite directions. Only the integral along the contour C survives. If we let $N \to \infty$, the summation on the left-hand side becomes a surface integral, and

$$\sum_{r=1}^{N} \varepsilon_r \Delta S_r \le \varepsilon \sum_{r=1}^{N} \Delta S_r = \varepsilon S \to 0 \quad .$$

Equation (A.15) thus reduces to Stokes' theorem as $N \to \infty$.

Stokes' theorem implies that an irrotational field is always conservative.

A.5 Curvilinear Coordinates

Consider a set of equations

$$\xi_i = \xi_i(x_1, x_2, x_3) \quad , \quad i = 1, 2, 3 \quad . \tag{A.16}$$

where ξ_1, ξ_2 and ξ_3 are independent, single-valued and continuous, which transforms a set of three numbers (x_1, x_2, x_3) into (ξ_1, ξ_2, ξ_3). If (x_1, x_2, x_3) are the coordinates of a point P in the Cartesian system, the numbers (ξ_1, ξ_2, ξ_3), which may also be used to represent the same point, are said to be the coordinates in a more general system, a *curvilinear system*.

Geometrically, with ξ_1, ξ_2 and ξ_3 given, (A.16) defines three surfaces S_1, S_2 and S_3 which intersect at the point $P(\xi_1, \xi_2, \xi_3)$. ξ_1 is constant on S_1 and ξ_2 is constant on S_2. If ξ_1 and ξ_2 are kept constant while ξ_3 is allowed to change, P will move along the intersection curve of S_1 and S_2. This curve is called the coordinate line of ξ_3. Similarly the intersections of S_2 with S_3 and of S_3 with S_1 give the coordinate lines of ξ_1 and ξ_2 respectively. Three unit vectors $\mathbf{e}_1$, $\mathbf{e}_2$ and $\mathbf{e}_3$ at P along the tangents to the coordinate lines ξ_1, ξ_2 and ξ_3 respectively are used as the base vectors in the curvilinear system, in terms of which a vector may be expressed. These base vectors are local in nature in the sense that their directions vary from point to point, unlike the base vectors $\mathbf{i}_1$, $\mathbf{i}_2$, $\mathbf{i}_3$ in the Cartesian system which are fixed in direction.

Let $\mathbf{r} = (x_1, x_2, x_3)$ be the radius vector of P. Its components may be transformed using Eq. (A.16):

$$\mathbf{r} = \mathbf{r}(x_i(\varepsilon_j)) \quad .$$

Differentiating we have

$$d\mathbf{r} = \frac{\partial \mathbf{r}}{\partial \xi_j} d\xi_j \quad . \tag{A.17}$$

The partial derivative $\frac{\partial \mathbf{r}}{\partial \xi_1}$ which has been obtained keeping ξ_2 and ξ_3 fixed, is a vector tangential to coordinate line ξ_1 and may be written as $h_1\mathbf{e}_1$. In general, we may write

$$d\mathbf{r} = h_1 d\xi_1 \mathbf{e}_1 + h_2 d\xi_2 \mathbf{e}_2 + h_3 d\xi_3 \mathbf{e}_3 \quad . \tag{A.18}$$

Note that the components of the length element in the curvilinear system along $\mathbf{e}_1$, $\mathbf{e}_2$ and $\mathbf{e}_3$ are $h_1 d\xi_1$, $h_2 d\xi_2$ and $h_3 d\xi_3$ respectively. As $hd\xi$ has the dimensions of length, ξ_i need not have the dimensions of length.

The length element $ds = |d\mathbf{r}|$ is given by

$$(ds)^2 = d\mathbf{r} \cdot d\mathbf{r} = dx_j dx_j = \frac{\partial x_j}{\partial \xi_\ell} \frac{\partial x_j}{\partial \xi_m} d\xi_\ell d\xi_m = g_{\ell m} d\xi_\ell d\xi_m \quad . \tag{A.19}$$

The coefficients

$$g_{\ell m} = \frac{\partial x_i}{\partial \xi_\ell} \frac{\partial x_j}{\partial \xi_m} \tag{A.20}$$

are elements of a tensor known as the *metric tensor*.

A curvilinear system is orthogonal if $\mathbf{e}_1$, $\mathbf{e}_2$ and $\mathbf{e}_3$ are mutually perpendicular. Equation (A.18) gives for an orthogonal system

$$(ds)^2 = h_1^2(d\xi_1)^2 + h_2^2(d\xi_2)^2 + h_3^2(d\xi_3)^2 \quad . \qquad \text{(A.21)}$$

Comparing this with Eq. (A.19) we have

$$h_1^2 = g_{11} \quad , \quad h_2^2 = g_{22} \quad , \quad h_3^2 = g_{33} \quad . \qquad \text{(A.22)}$$

In the following we shall confine ourselves to orthogonal curvilinear systems only.

Gradient

The length element along coordinate line ξ_1 is $h_1 d\xi_1$ according to Eq. (A.18). Equation (A.2) then gives

$$\mathbf{e}_1 \cdot \text{grad } u = \frac{1}{h_1}\frac{\partial u}{\partial \xi_1} \quad .$$

Similar expressions are obtained for the other components. Thus

$$\text{grad } u = \frac{1}{h_1}\frac{\partial u}{\partial \xi_1}\mathbf{e}_1 + \frac{1}{h_2}\frac{\partial u}{\partial \xi_2}\mathbf{e}_2 + \frac{1}{h_3}\frac{\partial u}{\partial \xi_3}\mathbf{e}_3 \quad . \qquad \text{(A.23)}$$

Divergence

As for the Cartesian coordinates, we make use of the definition (A.4) and choose for the volume V an elementary rectangular box dV of sides $h_1 d\xi_1$, $h_2 d\xi_2$ and $h_3 d\xi_3$. The net outward flux in the direction of the coordinate line ξ_1 is

$$\left\{ v_1 h_2 h_3 d\xi_2 d\xi_3 + \frac{\partial}{\partial \xi_1}(v_1 h_2 h_3 d\xi_2 d\xi_3)\frac{d\xi_1}{2} \right\}$$

$$- \left\{ v_1 h_2 h_3 d\xi_2 d\xi_3 - \frac{\partial}{\partial \xi_1}(v_1 h_2 h_3 d\xi_2 d\xi_3)\frac{d\xi_1}{2} \right\}$$

$$= \frac{\partial}{\partial \xi_1}(v_1 h_2 h_3) d\xi_1 d\xi_2 d\xi_3 = \frac{1}{h_1 h_2 h_3}\frac{\partial}{\partial \xi_1}(v_1 h_2 h_3) dV \quad .$$

Similar expressions are obtained for the net fluxes in the directions of the other coordinate lines. Equation (A.4) then gives

$$\operatorname{div} \mathbf{v} = \frac{1}{h_1 h_2 h_3}\left\{\frac{\partial}{\partial \xi_1}(v_1 h_2 h_3) + \frac{\partial}{\partial \xi_2}(v_2 h_3 h_1) + \frac{\partial}{\partial \xi_3}(v_3 h_1 h_2)\right\} \quad . \tag{A.24}$$

It follows that

$$\nabla^2 u = \nabla \cdot \nabla u = \frac{1}{h_1 h_2 h_3}\left\{\frac{\partial}{\partial \xi_1}\left(\frac{h_2 h_3}{h_1}\frac{\partial u}{\partial \xi_1}\right) + \frac{\partial}{\partial \xi_2}\left(\frac{h_3 h_1}{h_2}\frac{\partial u}{\partial \xi_2}\right) + \frac{\partial}{\partial \xi_3}\left(\frac{h_1 h_2}{h_3}\frac{\partial u}{\partial \xi_3}\right)\right\} \quad . \tag{A.25}$$

Curl

Curl $\mathbf{v}$ can be obtained in a manner similar to that for $\operatorname{div} \mathbf{v}$ using Eq. (A.8). Consider $\mathbf{e}_1 \cdot \operatorname{curl} \mathbf{v}$, taking for C a rectangle OABD of sides $h_2 d\xi_2$ and $h_3 d\xi_3$ as shown in Fig. A.3.

The line integral along OA and BD is

$$-\left\{v_2 h_2 d\xi_2 + \frac{\partial}{\partial \xi_3}(v_2 h_2 d\xi_2)\frac{d\xi_3}{2}\right\}$$

$$+\left\{v_2 h_2 d\xi_2 - \frac{\partial}{\partial \xi_3}(v_2 h_2 d\xi_2)\frac{d\xi_3}{2}\right\}$$

$$= -\frac{\partial}{\partial \xi_3}(v_2 h_2) d\xi_2 d\xi_3 \quad .$$

The line integral along AB and DO is similarly $\frac{\partial}{\partial \xi_2}(v_3 h_3) d\xi_2 d\xi_3$. Equation (A.8) then gives

$$(\operatorname{curl} \mathbf{v})_1 = \frac{1}{h_2 h_3 d\xi_2 d\xi_3}\left\{\frac{\partial}{\partial \xi_2}(v_3 h_3) - \frac{\partial}{\partial \xi_3}(v_2 h_2)\right\} d\xi_2 d\xi_3$$

$$= \frac{1}{h_2 h_3}\left\{\frac{\partial}{\partial \xi_2}(v_3 h_3) - \frac{\partial}{\partial \xi_3}(v_2 h_2)\right\} \quad .$$

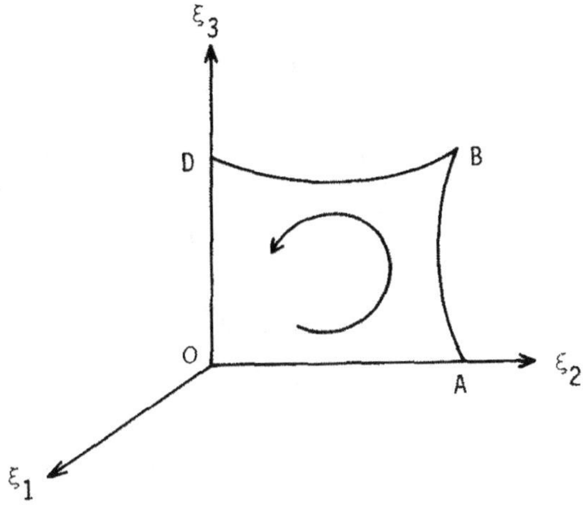

Fig. A.3 The closed loop C is a rectangle of sides $h_2 d\xi_2$ and $h_3 d\xi_3$.

In general we have

$$\text{curl } \mathbf{v} = \frac{1}{h_1 h_2 h_3} \begin{vmatrix} h_1\mathbf{e}_1 & h_2\mathbf{e}_2 & h_3\mathbf{e}_3 \\ \dfrac{\partial}{\partial\xi_1} & \dfrac{\partial}{\partial\xi_2} & \dfrac{\partial}{\partial\xi_3} \\ h_1 v_1 & h_2 v_2 & h_3 v_3 \end{vmatrix} . \tag{A.26}$$

A.6 Cylindrical and Spherical Coordinates

As shown in Fig. A.4, PP' is perpendicular to the xy-plane. Let angle $zOP = \theta$, angle $xOP' = \phi$, $OP = r$, $OP' = \rho$ and $P'P = z$. The coordinates in the cylindrical system are (ρ, ϕ, z) and those in the spherical system are (r, θ, ϕ).

Cylindrical Coordinates

Transformation equations are

$$x = \rho\cos\phi \quad , \quad y = \rho\sin\phi \quad , \quad z = z \quad .$$

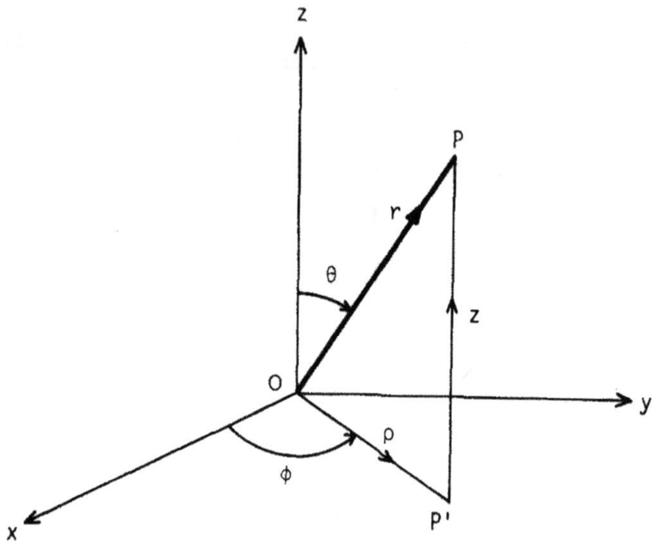

Fig. A.4 Relationship among the Cartesian coordinates (x, y, z), cylindrical coordinates (ρ, φ, z) and spherical coordinates (r, θ, φ), PP' being perpendicular to the xy plane.

Equations (A.20) and (A.22) give

$$h_\rho^2 = \frac{\partial x_i}{\partial \rho}\frac{\partial x_i}{\partial \rho} = \cos^2\phi + \sin^2\phi = 1 \quad ,$$

$$h_\phi^2 = \frac{\partial x_i}{\partial \phi}\frac{\partial x_i}{\partial \phi} = \rho^2\sin^2\theta + \rho^2\cos^2\phi = \rho^2 \quad .$$

Hence

$$h_\rho = 1 \quad , \quad h_\phi = \rho \quad , \quad \text{and} \quad h_z = 1 \quad .$$

Spherical coordinates

Transformation equations are

$$x = r\sin\theta\cos\phi \quad , \quad y = r\sin\theta\sin\phi \quad , \quad z = r\cos\theta \quad .$$

Equations (A.20) and (A.22) give

$$h_r^2 = \sin^2\theta\cos^2\phi + \sin^2\theta\sin^2\phi + \cos^2\theta = 1 \quad ,$$

$$h_\theta^2 = r^2\cos^2\theta\cos^2\phi + r^2\cos^2\theta\sin^2\phi + r^2\sin^2\theta = r^2 \quad ,$$

$$h_\phi^2 = r^2\sin^2\theta\sin^2\phi + r^2\sin^2\theta\cos^2\phi = r^2\sin^2\theta \quad .$$

Hence

$$h_r = 1 \quad , \quad h_\theta = r \quad \text{and} \quad h_\phi = r\sin\theta \quad .$$

Note that h_1, h_2 and h_3 can also be obtained from a geometrical consideration using Eq. (A.21).

INDEX